DATE DUE

ADVANCES IN GENETICS

VOLUME 31

Contributors to This Volume

Luke Alphey

Sanford I. Bernstein

Joseph B. Duffy

Beatriz Ferreiro

J. Peter Gergen

David Glover

Cayetano Gonzalez

Jeffrey C. Hall

William A. Harris

Dianne Hodges

C. P. Kyriacou

George A. Marzluf

ADVANCES IN GENETICS

Edited by

JEFFREY C. HALL

Department of Biology
Brandeis University
Waltham, Massachusetts

JAY C. DUNLAP

Department of Biochemistry
Dartmouth Medical School
Hanover, New Hampshire

VOLUME 31

ACADEMIC PRESS

San Diego   New York   Boston
London   Sydney   Tokyo   Toronto

Academic Press, Inc.
A Division of Harcourt Brace & Company
525 B Street, Suite 1900, San Diego, California 92101-4495

United Kingdom Edition published by
Academic Press Limited
24–28 Oval Road, London NW1 7DX

International Standard Serial Number: 0065-2660

International Standard Book Number: 0-12-017631-9

PRINTED IN THE UNITED STATES OF AMERICA
94 95 96 97 98 99   BC   9 8 7 6 5 4 3 2 1

CONTENTS

Sex, Segments, and the Central Nervous System: Common Genetic Mechanisms of Cell Fate Determination

JOSEPH B. DUFFY AND J. PETER GERGEN

Neurogenesis in *Xenopus*: A Molecular Genetic Perspective

BEATRIZ FERREIRO AND WILLIAM A. HARRIS

Cell Cycle Genes of *Drosophila*

CAYETANO GONZALEZ, LUKE ALPHEY, AND DAVID GLOVER

v

283118

Genetic and Molecular Analysis of *Drosophila* Behavior

C. P. KYRIACOU AND JEFFREY C. HALL

Genetics and Molecular Genetics of Sulfur Assimilation in the Fungi

GEORGE A. MARZLUF

Genetic and Biochemical Analysis of Alternative RNA Splicing

DIANNE HODGES AND SANFORD I. BERNSTEIN

CONTENTS vii

CONTENTS

CONTRIBUTORS TO VOLUME 31

Numbers in parentheses indicate the pages on which the authors' contributions begin.

LUKE ALPHEY (79), *Department of Anatomy and Physiology, Medical Sciences Institute, University of Dundee, Dundee DD1 4HN, Scotland*

SANFORD I. BERNSTEIN (207), *Biology Department and Molecular Biology Institute, San Diego State University, San Diego, California 92182*

JOSEPH B. DUFFY (1), *Department of Genetics, Harvard Medical School, Boston, Massachusetts 02115*

BEATRIZ FERREIRO (29), *Department of Biology and Center for Molecular Genetics, University of California at San Diego, La Jolla, California 92093, and La Jolla Cancer Research Foundation, La Jolla, California 92037*

J. PETER GERGEN (1), *Department of Biochemistry and Cell Biology, State University of New York at Stony Brook, Stony Brook, New York 11794*

DAVID GLOVER (79), *Department of Anatomy and Physiology, Medical Sciences Institute, University of Dundee, Dundee DD1 4HN, Scotland*

CAYETANO GONZALEZ (79), *Department of Anatomy and Physiology, Medical Sciences Institute, University of Dundee, Dundee DD1 4HN, Scotland*

JEFFREY C. HALL (139), *Department of Biology, Brandeis University, Waltham, Massachusetts 02254*

WILLIAM A. HARRIS (29), *Department of Biology and Center for Molecular Genetics, University of California at San Diego, La Jolla, California 92093*

DIANNE HODGES (207), *Biology Department and Molecular Biology Institute, San Diego State University, San Diego, California 92182, and Isis Pharmaceuticals, Carlsbad, California 92008*

C. P. KYRIACOU (139), *Department of Genetics, University of Leicester, Leicester, LE1 7RH, United Kingdom*

GEORGE A. MARZLUF (187), *Department of Biochemistry, The Ohio State University, Columbus, Ohio 43210*

PREFACE

In recent decades, especially this and the last one, the explosion of molecular genetics has capitalized on its genetic antecedents. Genetic and molecular analyses have thus proved to be a powerful combination for studying phenotypic variation. This synergism is apparent to all who follow investigations in these fields, and we hope to keep both the genetic and molecular approaches in focus as we engage authors to contribute to *Advances in Genetics*.

Our goals for this series are to identify emerging problems in genetics as they coalesce and to solicit contributions that delve into purely genetic phenomena and issues (such as chromosome structure, behavior, and recombination), as well as those involving "other" areas of biology in which genetic and molecular analyses are making a great difference (cell biology, development, neurobiology, and human biology).

Advances in Genetics will continue to publish comprehensive up-to-date articles about topics of contemporary interest. We will also encourage, when warranted, shorter and more discursive (even provocative) treatments of certain phenomena and issues that have come to the fore.

Advances in Genetics was first published in 1946 under the editorship of Milislav Demerec, who was then acting as the director of the Genetics Department of the Carnegie Institute of Washington in Cold Spring Harbor, New York. His goal was to produce a series such that "critical summaries of outstanding genetic problems, written by prominent geneticists in such a form that they will be useful as reference material for geneticists and also a source of information to nongeneticists, may appear in a single publication." Although times have changed and the face of genetics research is remarkably different than it was 48 years ago, Demerec's goals still provide a fine guide.

We close by thanking the two geneticists—John Scandalios and Theodore Wright—who carried the editorship of *Advances in Genetics* into the 1990s, and we are grateful to them for soliciting a review (co-authored by one of us) that appears in the current volume.

JEFFREY C. HALL
JAY C. DUNLAP

SEX, SEGMENTS, AND THE CENTRAL NERVOUS SYSTEM: COMMON GENETIC MECHANISMS OF CELL FATE DETERMINATION

Joseph B. Duffy* and J. Peter Gergen†

*Department of Genetics, Harvard Medical School, Boston, Massachusetts 02115;
and †Department of Biochemistry and Cell Biology, State University of New York
at Stony Brook, Stony Brook, New York 11794

For the real amazement, if you wish to be amazed, is this process. You start out as
a single cell derived from the coupling of a sperm and an egg; this divides in two,
then four, then eight, and so on, and at a certain stage there emerges a single cell
which has as all its progeny the human brain. The mere existence of such a cell
should be one of the great astonishments of the earth. People ought to be walking
around all day, all through their waking hours calling to each other in endless
wonderment, talking of nothing except that cell.

Lewis Thomas (1979)

I. Introduction

As mature organisms we are composed of an astonishing array of
diverse cell types—all derived from a single-celled zygote. When faced

1

with the task of generating such cellular diversity in a reproducible fashion, how has the embryo chosen to respond? Recent work in a number of developmental systems has suggested that the embryo has employed two approaches. First, given finite resources, the embryo has efficiently chosen to reutilize a limited set of proteins in different temporal and spatial contexts to create cellular diversity. Second, the embryo has also chosen to install molecular redundancies to ensure the reproducibility of these patterns from individual to individual.

Reutilization predicts the existence of molecules, which we will call promiscuous developmental regulators, that function in numerous developmental pathways. Such molecules provide direct molecular links between diverse developmental pathways. Why not use these same molecules for all pathways? Such an extreme view of reutilization would present a number of problems, including one of reproducibility. Alteration of any one gene would affect all pathways, making developmental patterns difficult to maintain from generation to generation. One approach to increasing reproducibility, which also maintains some level of reutilization, is redundancy through gene duplication. Gene duplication helps ensure reproducibility by providing redundancy and simultaneously produces substrates for use in other pathways. Although initially redundant, overlapping or unique roles for homologues can be generated through divergence. This provides direct and indirect links between distinct developmental pathways because it can lead to the reutilization of a single gene or a related gene in these distinct pathways.

Because of their ability to regulate gene expression, transcription factors are excellent candidates for promiscuous developmental regulators. Different combinations and concentrations of these molecules can control distinct classes of subordinate genes, leading to the elaboration of specific developmental programs. One striking example of a promiscuous transcriptional regulator that links three developmental pathways in the fruitfly, *Drosophila melanogaster,* is the *runt* gene. Initially identified on the basis of its role in segmentation, this gene encodes a novel nuclear regulatory protein that also has key roles in sex determination and neurogenesis (Nüsslein-Volhard and Wieschaus, 1980; Duffy *et al.,* 1991; Duffy and Gergen, 1991; Torres and Sanchez, 1992). The analysis of these three pathways in *Drosophila* has revealed other examples of developmental promiscuity, involving unique and related genes with overlapping or homologous roles in these pathways (Table 1).

In this chapter we examine the molecular mechanisms governing cell fate in sex determination, segmentation, and neurogenesis, and

TABLE 1
Sex, Segments, and the CNS: Promiscuous Transcriptional Regulators

| Gene | Developmental roles | | | Motifs |
	Sex determination	Segmentation	Neurogenesis	
sisterless-a	Numerator	No	No	bzip
sisterless-b (T4)	Numerator	No	Proneural	bHLH
runt	Numerator	Pair-rule	NB identity	Runt domain, WRPY
daughterless	Maternal transducer	No	Proneural	bHLH
lethal of scute (T3)	Redundant + element?	No	Proneural	bHLH
achaete (T5)	Redundant + element?	No	Proneural	bHLH
deadpan	Denominator	No	Inhibitor (PNS)	bHLH, WRPW
extramacrochaete	Maternal inhibitor	No	Inhibitor (PNS)	HLH
hairy	Redundant − element?	Pair-rule	Inhibitor (PNS)	bHLH, WRPW
even-skipped	No	Pair-rule	GMC/neuronal identity	Homeodomain
fushi tarazu	No	Pair-rule	GMC/neuronal identity	Homeodomain

the significance that the reutilization and redundancy of transcription factors has for these decisions. The theme emerging from the characterization of these pathways supports the tenet that diverse developmental pathways often employ similar molecules and may therefore employ similar mechanisms to determine cell fate. Furthermore, the evolutionary conservation of many of these molecules across a wide variety of species suggests that these principles will apply to cell fate determination in other organisms as well.

II. Determination of Somatic Sexual Fate

A. The Signal (X/A Ratio) and the Target (Sex-lethal)

The establishment of somatic sexual fate is one of the first pathways to generate cellular diversity in *Drosophila*. This simple binary decision (female vs. male) is reflected in the activity of the *Sex-lethal* (*Sxl*) gene, which is ON in females and OFF in males (Cline, 1978). Once activated in females, *Sxl* controls sexual dimorphism and dosage compensation (Cline, 1983). Loss-of-function mutations in the *Sxl* gene result in female-specific lethality, while gain-of-function mutations result in male-specific lethality; both are due to altered dosage compensation (Cline, 1978; Lucchesi and Skripsky, 1981; Gergen, 1987). *Sxl* regulation undergoes two distinct phases: initiation and maintenance (Sanchez and Nothiger, 1983; Cline, 1984; Maine *et al.*, 1985; Keyes *et al.*, 1992). Initiation of *Sxl* activity is controlled through the transcriptional regulation of an embryonic-specific promoter (p_e) by the X/A ratio (Keyes *et al.*, 1992). A ratio of 1 (2X:2A) results in *Sxl* expression from p_e leading to the ON or female mode, while a ratio of 0.5 (X:2A) represses *Sxl* expression from p_e, leading to the OFF or male mode (Keyes *et al.*, 1992). After initiation, maintenance of either mode involves the post-transcriptional regulation of *Sxl* through the differential splicing of *Sxl* transcripts from a constitutive promoter (Bell *et al.*, 1988; Salz *et al.*, 1989). Since the p_e provides the initial presence of *Sxl* protein in females, the ON mode of *Sxl* activity is maintained in females through an autoregulatory loop involving the ability of the *Sxl* protein to positively regulate its own splicing (Bell *et al.*, 1991).

B. Molecular Description of the X/A Ratio

Simply stated, the difference between females and males in *Drosophila* is one X chromosome; females have two (XX) and males have one (XY). The Y chromosome of males is not required for sex determination (Bridges, 1921). Analysis of triploid intersexes (2X:3A) demonstrated that this difference is measured through the balance between X chromosomal and autosomal elements (Bridges, 1921). This balance was termed the X/A ratio and is directly reflected in the dosage of specific X-linked zygotic genes, termed numerator elements, relative to the dosage of autosomal genes, termed denominator elements (Bridges, 1921, 1925; Cline, 1988). The twofold difference in numerator element gene dose in females versus males results in the activation of *Sxl* transcription in all somatic cells and the subsequent initiation of female development. Thus, numerator elements are positive and denominator elements negative regulators of *Sxl*.

Currently only three genes, *sisterless-a* (*sis-a*), *sisterless-b* (*sis-b*), and *runt*, have been identified as numerator elements (Cline, 1986, 1988; Duffy and Gergen, 1991; Torres and Sanchez, 1992). The best characterized of these numerator elements is *sis-b*. Molecularly, *sis-b* corresponds to the T4 transcription unit of the *achaete-scute* complex (*AS-C*) (Cline, 1988; Torres and Sanchez, 1989; Parkhurst *et al.*, 1990; Erickson and Cline, 1991). This complex of four related genes is a striking example of gene duplication, providing both redundancy and substrates for divergence to unique roles. All the members of the *AS-C* function during neural development, in which they have partially overlapping roles (for reviews see Ghysen and Dambly-Chaudiere, Campos-Ortega, 1993; Ghysen *et al.*, 1993; Jan and Jan, 1993). However, only T4 is absolutely required for sex determination (Erickson and Cline, 1991).

The *AS-C* consists of four genes encoding highly related proteins of the basic helix–loop–helix (bHLH) family of transcription factors (Campuzano *et al.*, 1985; Villares and Cabrera, 1987; Alonso and Cabrera, 1988; Murre *et al.*, 1989a). This motif consists of a basic DNA-binding domain (b) and an adjacent dimerization domain (HLH) capable of mediating protein–protein interactions (Murre *et al.*, 1989a,b). In a manner analogous to mammalian bHLH interactions in which a ubiquitously expressed partner is usually required for the function of bHLHs expressed in more specific patterns, the maternally expressed bHLH factor DA is required for *sis-b* to activate *Sxl* (Cline, 1986, 1988; Murre *et al.*, 1989a; Parkhurst *et al.*, 1990; Bopp *et al.*, 1991). Consistent with this role as a partner protein, DA also functions in neural development (Caudy *et al.*, 1988a, b; Cronmiller *et al.*, 1988).

The role of *runt* in sex determination was discovered because of its lethal interaction with T4. Like T4 and *da*, *runt* is also expressed and functions in neural development. Thus, the basis of this lethality could have been the result of defects in sex determination or neural development (Duffy *et al.*, 1991; Duffy, 1992). However, characterization of this interaction led to the demonstration that *runt* was acting as a numerator element during sex determination (Duffy *et al.*, 1991). Interestingly, unlike either *sis-b* or *da*, *runt* is not a member of the bHLH family (Kania *et al.*, 1990). Instead, recent work has identified murine and human genes containing a domain with significant sequence similarity to *runt*, termed the Runt domain (Daga *et al.*, 1992; Erickson *et al.*, 1992; Kagoshima *et al.*, 1993). *In vitro* experiments have indicated that this domain is capable of interacting with an unrelated protein partner and binding DNA (Kagoshima *et al.*, 1993). These results indicate that *runt* encodes a member of a novel class of proteins capable of interacting with other proteins and binding DNA (Kagoshima *et al.*, 1993). It is provocative to note that as with *sis-b*, the activity of *runt* may also depend on a partner protein (Kagoshima *et al.*, 1993). As we discuss later for *sis-b*, this dependence also provides a method of entry for denominator elements.

More recently, the molecular identity of the numerator element *sis-a* has revealed a similar scenario. As with *sis-b* and *runt*, *sis-a* appears to encode a transcription factor capable of DNA binding and protein–protein interactions. Sequence analysis of *sis-a* has indicated that it encodes a protein of the basic leucine zipper (bzip) family of transcription factors (Erickson and Cline, 1993).

On the denominator or inhibitory side, two genes have been identified (Younger-Shepherd *et al.*, 1992). The zygotic gene *deadpan* (*dpn*) and the maternal gene *extramacrochaete* (*emc*) are required in males to prevent activation of *Sxl* (Younger-Shepherd *et al.*, 1992). Supporting a model involving protein–protein interactions, *dpn* and *emc* encode proteins of the bHLH family (Ellis *et al.*, 1990; Garrell and Modolell, 1990; Bier *et al.*, 1992). *dpn* contains both the basic (b) and HLH domains, while *emc* contains only an HLH domain. Furthermore, as with *sis-b*, *da*, and *runt*, *emc* and *dpn* are also expressed during neural development (Ellis *et al.*, 1990; Garrell and Modolell, 1990; Bier *et al.*, 1992). In neurogenesis, like their roles in sex determination, *emc* and *dpn* inhibit the activity of members of the *AS-C* (Botas *et al.*, 1982; Moscoso del Prado and Garcia-Bellido, 1984a,b; Garcia-Alonso and Garcia-Bellido, 1988; Rushlow *et al.*, 1989; Ellis *et al.*, 1990; Garrell and Modolell, 1990; Skeath and Carroll, 1991; Cubas *et al.*, 1991; Bier *et al.*, 1992). While the nature of *dpn*'s role in neural development has not been elucidated by loss-of-function mutations, overexpression studies have demonstrated that *dpn* has effects consistent with inhibition of the *AS-C* (Bier *et al.*, 1992).

These types of inhibitory interactions, coupled with DNA binding, dimerization, and transcriptional studies, as well as analogies with mammalian HLH interactions, favor the following scenario for *sis-b*, *da*, *dpn*, and *emc* in the regulation of *Sxl*. In females, the increased dose of the *sis-b* gene (2X) leads to increased protein concentrations, thereby causing *sis-b/da* heterodimers, which are capable of stimulating *Sxl* transcription, to predominate. Conversely, in males where there is a reduced dose of *sis-b* (X), heterodimers containing *dpn* and *emc* predominate. In this case *Sxl* is not activated because *dpn* and *emc* inhibit the DNA-binding and transcriptional activity of *sis-b* and/or *da*.

Although there is considerable suggestive evidence for this model, alternative models in which the proteins with a basic (b) domain (SIS-B, SIS-A, DA, and DPN) do not function as heterodimers have not been ruled out. Since the zip domain of *sis-a* and the Runt domain are also capable of mediating protein–protein interactions, a similar set of heterodimeric interactions may also exist for *sis-a* and *runt* with the above or as-yet-unidentified partners. Furthermore, while there is no direct evidence of interactions between *sis-a* or *runt* with bHLH proteins, recent work has demonstrated that the HLH domain is capable of interacting with other protein motifs (Bengal *et al.*, 1992). Given the potential for or lack of interactions among these factors, it is important to determine how they cooperate to generate a binary switch of ON or OFF with respect to *Sxl* transcription, all based on a twofold difference in gene dose.

C. A Second Level of Complexity: Spatial Regulation of *Sxl*

To this point we have assumed that the regulation of *Sxl* is identical in all somatic cells. However, when the requirement for these genes in the regulation of *Sxl* was examined, a somewhat surprising discovery was made. While *sis-a*, *sis-b*, and *da* are required in all somatic cells for the proper regulation of *Sxl*, *runt* and *dpn* appear to be required only in the presegmental (central) body region (Bopp *et al.*, 1991; Duffy and Gergen, 1991; Younger-Shepherd *et al.*, 1992). *runt* is required for activation, while *dpn* is required for inhibition of *Sxl* expression in this region (Duffy and Gergen, 1991; Duffy, 1992; Younger-Shepherd *et al.*, 1992; Erickson and Cline, 1993).

This antagonism between *runt* and *dpn* raises a number of evolutionary and mechanistic questions. Evolutionarily, it is intriguing to note that *dpn* is most closely related at the amino acid level to the segmentation gene *hairy* (*h*) (Bier *et al.*, 1992). In segmentation, *h* and *runt* are expressed in complementary and mutually exclusive patterns (Ingham and Gergen, 1988; Kania *et al.*, 1990). Furthermore, they have opposite effects on the regulation of downstream genes (Carroll and Scott, 1986; Frasch and Levine, 1987; Ingham and Gergen, 1988). Thus, the sequence identity between *dpn* and *h* appears to represent some level of functional homology. Is this an example of gene duplication and divergence, and therefore indicative of a common evolutionary bond between *dpn* and *h*?

Supporting this, similarities in sequence, exon–intron structure, and patterns of expression for *h* and *dpn* in different developmental pathways led Bier *et al.* to propose that *h* and *dpn* arose from a common ancestor (Bier *et al.*, 1992). If so, does any redundancy exist between *dpn* and *h* in these different pathways or have their roles completely diverged? Finally, does the use of *runt, dpn,* and *h* in these different pathways represent an example of the conservation of molecular mechanisms in distinct developmental pathways?

Indirect evidence for redundancy in sex determination comes from the discovery that high levels of ectopic *h* expression during sex determination inhibit the activation of *Sxl* (Parkhurst *et al.*, 1990). However, unlike *dpn*, loss of *h* activity does not lead to ectopic activation of *Sxl* in males (Duffy, 1992). This indicates that if *h* is involved in sex determination, it is as a redundant element. Direct evidence that *h* may have such a redundant role comes from two genetic experiments. First, reduced *h* activity can suppress the effect of reduced *runt* and *sis-b* activity and second, slight feminizing effects are seen in triploid intersexes with reduced *h* activity (Duffy, 1992). Thus, in sex determination, redundancy may only exist for *dpn* and *h* under near-threshold conditions.

A similar type of developmental redundancy also appears to exist for *sis-b* through the *AS-C* genes *achaete* (T5) and *lethal of scute* (T3). Both have significant sequence similarity to *sis-b* and have weak feminizing activities under near-threshold conditions (Cline, 1988; Parkhurst *et al.*, 1990, 1993). These types of weak redundancy may represent evolutionary remnants of gene duplication and divergence that currently act only as a buffering system for molecular decisions.

A second and more mechanistic question involves the regulation of *Sxl* in the presegmental region and the roles of *runt* and *dpn*. While *sis-a* and *sis-b* are absolutely required in all somatic cells for *Sxl* expression, the complete absence of *runt* activity reduces *Sxl* expression only in the presegmental region (Bopp *et al.*, 1991; Duffy and Gergen, 1991; Duffy, 1992). One possible explanation is that *runt* functions to amplify the ability of other numerator elements to activate *Sxl* transcription. In this scenario *runt* provides a necessary level of refinement in the transduction of the X/A ratio and other gene(s) must exist to provide this refinement in the terminal regions of the embryo. Alternatively, since the presegmental expression of *runt* has other functions (Tsai and Gergen, submitted), did *dpn* evolve simply to prevent *runt* from disrupting sex determination (or vice versa)? According to this model, additional genes are not required at the termini since *runt* and *dpn* are not direct regulators of *Sxl* and function in a mutually antagonistic fashion. If this model is correct in its simplest form, *Sxl* should be properly regulated in the presegmental region of embryos lacking the expression of both *runt* and *dpn*.

Regardless of the answers to such questions, analysis of this pathway has clearly demonstrated that one of the first cell fate choices in *Drosophila*, the determination of somatic sex, is initiated through combinatorial and concentration-dependent interactions among a number of distinct and related transcription factors (Table 1 and Fig. 1). With the knowledge that roles in other pathways had been previously characterized for some of these factors (*T4, runt, da,* and *emc*), the discovery of their involvement in sex determination demonstrates the

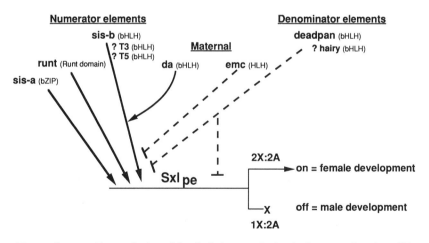

FIG. 1. Sex-specific regulation of *Sex-lethal* transcription in the central region of *Drosophila* embryos. For a full description of this model, see the text. The *Sxl* embryonic promoter responds to the X/A ratio; 2X:2A = ON, 1X:2A = OFF. This regulation utilizes the positive input of the genes: *sis-a, sis-b, runt,* and *da,* which is depicted by solid lines and arrows. Dotted lines depict the inhibitory effects of the genes *dpn* and *emc*. Putative direct effects are shown by a line drawn to the promoter. Effects mediated through a putative interaction (+ or −) are shown by directing a line toward its putative partner. For a discussion of the possible redundant roles of T3, T5, and *hairy,* see the text.

reutilization of transcription factors in diverse developmental pathways. These examples of reutilization, coupled with the similarities of sequences of these proteins to proteins involved in other pathways, initially support the use of similar mechanisms of cell fate determination in different pathways. In Sections III and IV we review more evidence of reutilization and redundancy that further support this notion of a common theme of cell fate determination for diverse developmental pathways.

III. Determination of Metameric Units: Segmentation of the Anterior–Posterior Axis

A. OVERVIEW: GRADIENTS TO PERIODICITY

Shortly after the determination of sexual fate, the embryonic body is organized into a metameric pattern along the anterior–posterior (A–P) axis. The genetic and molecular characterization of genes required for this process has led to a model in which patterns of gene expression are refined in a hierarchical fashion (for review see St. Johnston and Nüsslein-Volhard, 1992; Driever, 1993; Martinez Arias 1993; Pankratz and Jackle, 1993; Sprenger and Nüsslein-Volhard, 1993; St Johnston, 1993; and references therein). Global coordinates set by the maternal anterior, posterior, and terminal group genes serve to define large overlapping domains of zygotic gap gene expression. These overlapping gap gene domains then generate the first signs of periodicity through the regulation of the zygotic pair-rule genes. In turn, the pair-rule genes serve to sharpen their own domains of expression and initiate expression of the segment-polarity genes. In this manner the embryo is progressively subdivided and cell fates established in a parasegmental or segmental array.

Consistent with the notion that transcriptional regulation controls the early stages of this process, all of the gap and pair-rule genes currently identified encode transcriptional motifs (Laughon and Scott, 1984; McGinnis et al., 1984; Frigerio et al., 1986; Kilcherr et al., 1986; Macdonald et al., 1986; Rosenberg et al., 1986; Nauber et al., 1988; Tautz et al., 1987; Rushlow et al., 1989; Coulter et al., 1990; Finkelstein et al., 1990; Kania et al., 1990; Pignoni et al., 1990; Capovilla et al., 1992; Grossniklaus et al., 1992). Thus, as we have described for sex determination, the determination of cell fates in a metameric pattern along the A–P axis also relies heavily on transcriptional control. In the following section we review data showing that this process involves combinatorial and concentration-dependent interactions among transcription factors, many of which are reutilized in other developmental pathways.

B. GENERATING PERIODICITY: GAP TO PAIR-RULE

A key step in defining the metameric pattern of the embryo is the transition from aperiodic (gap gene) patterns to periodic (pair-rule gene) patterns. Molecularly, this step is initiated through stripe-specific cis-regulatory elements located in the regulatory regions of some of the pair-rule genes (Harding *et al.*, 1989; Goto *et al.*, 1989; Pankratz *et al.*, 1990; Howard and Struhl 1991; Riddihough and Ish-Horowicz 1991; M. Klingler, J. Soong, and J. P. Gergen, personal communication. These initial patterns are then refined and maintained through pair-rule response cis-regulatory elements (discussed in Section III,C).

Of the stripe-specific cis-regulatory sequences, the *eve* stripe 2 element is the most well characterized and its regulation is an elegant paradigm for the combinatorial and concentration-dependent control of gene expression (reviewed in Ip *et al.*, 1992). Analysis of the *eve* cis-regulatory region has identified a 480-bp region capable of establishing expression of *eve* stripe 2 (Goto 2 *et al.*, 1989; Harding *et al.*, 1989). This element responds in a concentration-dependent fashion to the transcriptional regulators *bicoid* (*bcd*, homeodomain motif), *hunchback* (*hb*, zinc finger motifs), *giant* (*gt*, bzip motif) and *Kruppel* (*Kr*, zinc finger motifs) to establish *eve* stripe 2 expression (Goto *et al.*, 1989; Harding *et al.*, 1989; Stanojevic *et al.*, 1991; Small *et al.*, 1991, 1992).

Genetically, the maternal gene *bcd* and the gap gene *hb* were identified as activators of this stripe (Small *et al.*, 1991). There are five *bcd* and one *hb* binding sites in the stripe 2 element (Stanojevic *et al.*, 1989, 1991; Small *et al.*, 1991, 1992). It has been shown that these sites are necessary for the activation of *eve* stripe 2 *in vivo* and suggested that cooperativity between *bcd* and *hb* plays a role in activation (Small *et al.*, 1991, 1992; Stanojevic *et al.*, 1991). Characterization of the temporal and spatial ontogeny of this stripe suggested that the anterior gradients of *bcd* and *hb* lead to a broad anterior domain of *eve* expression, while other genes are required to set the sharp borders (Stanojevic *et al.*, 1989, 1991; Small *et al.*, 1991, 1992).

Supporting the notion that the gap genes *gt* and *Kr* set these borders, their expression appears concomitantly with the anterior and posterior boundaries, respectively, of *eve* stripe 2 (Stanojevic *et al.*, 1989; Small *et al.*, 1991). The anterior border is set through direct repression by the binding of *gt* to any of the three sites within this 480-bp region (Stanojevic *et al.*, 1991; Small *et al.*, 1991, 1992). The posterior border appears to be set in response to decreasing levels of the activators *bcd* and *hb*, and increasing levels of the repressor *Kr*. Like *gt*, regulation of this element by *Kr* also involves direct repression through any of the three *Kr* binding sites (Stanojevic *et al.*, 1989, 1991; Small *et al.*, 1991, 1992). Cooperativity between *bcd* and *hb* means that *gt* and *Kr* need to block only a single activator site to repress *eve* transcription and effectively generate sharp on/off boundaries for *eve* stripe 2 expression (reviewed in Ip *et al.*, 1992).

The combined use of genetic, molecular, and biochemical approaches to dissect the regulation of this stripe has provided one of the most elegant examples of combinatorial and concentration-dependent control of cell fate determination. The analysis of other pair-rule stripe-specific elements has suggested that similar mechanisms utilizing other combinations and concentrations of gap proteins are likely to be directing expression from these other elements (Pankratz *et al.*, 1990; Howard and Struhl, 1991; Riddihough and Ish-Horowicz, 1991).

C. Maintaining Periodicity: Pair-Rule Interactions

After stripe initiation by gap genes, a complex network of interactions among pair-rule genes serves to refine and maintain these metameric patterns. As in the regulation of stripe-specific elements by gap genes, combinatorial and concentration-dependent interactions also appear to govern this network. Distinct cis-regulatory elements responsible for governing these interactions have been identified (Hiromi *et al.*, 1985; Hiromi and Gehring, 1987; M. Klingler, J. Soong, and J. P. Gergen, personal communication). The *ftz* zebra element is one of the most noted of these elements and its regulation illustrates some key features of the pair-rule network.

First, a hierarchy exists among these interactions. Primary pair-rule genes (*eve, h,* and *runt*) are at the top of this hierarchy and affect the regulation of each other and of downstream pair-rule genes (i.e., *ftz*) but are not regulated by these downstream genes (Carroll and Scott, 1986; Frasch and Levine, 1987; Ingham and Gergen, 1988; Carroll and Vavra, 1989; Hooper *et al.*, 1989; Klingler and Gergen, 1993).

Second, antagonistic relationships have been defined. For example, loss-of-function mutations in *h* and *runt* have opposing effects on the expression of *ftz; h* mutations result in an expansion, while *runt* mutations result in a reduction of *ftz* expression (Carroll and Scott, 1986;

Howard and Ingham, 1986; Ingham and Gergen, 1988). Similar effects are seen with a β-galactosidase reporter gene under the control of the zebra element, indicating that these effects are transcriptional and mediated at least in part through the zebra element (Hiromi and Gehring, 1987). Ectopic expression studies using the heat shock promoter to control the expression of *runt* and *h* have supported these initial observations and shown that these effects are concentration-dependent (Ish-Horowicz and Pinchin, 1987; C. Tsai and J. P. Gergen, personal communication).

Further characterization of the zebra element led to the identification of a small region, termed the fDE (*ftz* dual element), that is capable of responding to cues from both activators and repressors (Dearolf *et al.*, 1989; Topol *et al.*, 1991). Mutation of binding sites within the fDE for a DNA-binding protein of the steriod receptor superfamily—FTZ-F1—results in a strong reduction of *ftz* expression in stripes 1, 2, 3, and 6, and a weaker reduction in stripes 4, 5, and 7 (Ueda *et al.*, 1990). This and the observation that *ftz* stripes 3 and 6 are preferentially lost in a hypomorphic *runt* allele led Ueda *et al.*, to speculate that FTZ-F1 may mediate the effects of *runt* on *ftz*. Support for the notion that FTZ-F1 mediates the effects of *runt* and also of *h* on *ftz* has recently been obtained through ectopic expression studies with *runt* and *h,* and their effects on a β-galactosidase reporter gene under the control of oligomerized FTZ-F1 sites (C. Tsai and J. P. Gergen, personal communication).

This relationship between *runt* and *h* is reminiscent of the relationship between *runt* and *dpn*. In both cases *runt* has regulatory effects that oppose those of a specific bHLH protein—*dpn* in sex determination and *h* in segmentation. A further intriguing molecular link between *runt, dpn,* and *h* is the presence of related tetrapeptide sequences at their C terminal. This motif, WRPW, is the last four amino acids in *h, dpn,* and the bHLH proteins of the *Enhancer of split* complex (Bier *et al.*, 1992). The sequence of *h* mutations in *D. melanogaster* and the conservation of this motif in an *h* homolog from *D. virilis* indicates that this motif serves an important functional role (Wainwright and Ish-Horowicz, 1992).

While *runt* does not have this exact sequence, its last four amino acids, WRPY, are very similar (Kania *et al.*, 1990). This related motif may also serve an important function in *runt*, since it is conserved in two vertebrate homolog of *runt*, the murine genes *PEBP2α* and mouse *AML1*, (Bae *et al.*, 1993; Ogawa *et al.*, 1993). What is the significance of this motif and the relationships among *runt, dpn* and *h?* The molecular link between *runt* and *dpn* for sex determination is not clear. In segmentation, however, the effects of *runt* and *h* involve a common regulatory element and potentially a common factor, FTZ-F1. Given this observation, one is tempted to speculate that the WRPW/Y motifs are involved in regulating protein–protein interactions and that their similarity may provide a molecular basis for the opposing effects of *h* and *runt* on FTZ-F1. Regardless of the exact function of this motif, its conservation in *dpn* and *h* suggests that similar molecular scenarios for *dpn* and *runt*, and *h* and *runt* are likely to exist. Curiously, although this tetrapeptide motif is present in an ortholog of *h* from the flour beetle *Tribolium castaneum*, the hexapeptide sequence VVETVM is found C terminal to the WRPW sequence (Sommer and Tautz, 1993). The continued identification of homologs of these genes (*dpn*, *h*, and *runt*) may shed further light on the functional importance of the WRPW/Y motifs and the evolutionary links among these genes.

IV. Determination of Cell Fates in the Central Nervous System

A. OVERVIEW: ECTODERM TO NEURON

In the introduction we asked the following question: Given finite resources, how does the embryo reproducibly generate enormous cellular diversity? After the early determinative events of sex determination and segmentation, the embryo embarks upon what may be its greatest challenge with respect to this question: the determination of cell fates in the central nervous system (CNS). In an effort to efficiently generate such enormous diversity, many of the transcription factors used in sex determination and segmentation are reexpressed and most likely reutilized during neural development (for reviews see Doe, 1992; Goodman and Doe, 1993; see also Table 1). While prior roles introduce a greater level of difficulty in addressing the function of these factors in determining neural fates, they also provide an opportunity to further explore the links between developmental promiscuity, redundancy, and the determination of cell fates.

The development of the *Drosophila* nervous system starts with the establishment of neural equivalence groups within the ectoderm. While all cells within a particular group have the capability to generate equivalent neural identities, only one cell per group adopts a neural fate (Doe, 1992; Goodman and Doe, 1993). That cell, now called a neuroblast (NB), enlarges and delaminates from the ectoderm. Each NB then goes through successive asymmetric divisions, with each division producing a stem cell NB and a smaller ganglion mother cell (GMC). Each GMC then divides once to produce two sibling neurons.

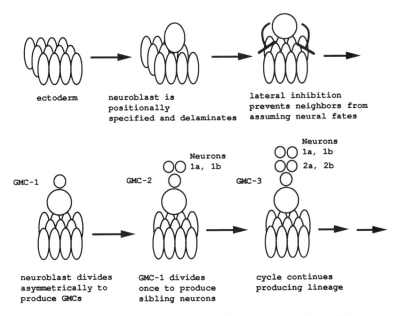

FIG. 2. Schematic diagram of neural development in *Drosophila*.

This series of cell divisions results in the generation of a neuronal lineage specific for each NB (see Fig. 2).

B. NEURAL VS. ECTODERMAL: PRONEURAL GENES AND THE DETERMINATION OF A NEURAL FATE

CNS development begins with the generation of a reproducible NB pattern that is controlled through the combined action of the zygotic pair-rule and dorsal–ventral genes (Skeath *et al.*, 1992; for reviews see Doe, 1992; Goodman and Doe, 1993). These two classes of transcription factors act in concert to establish specific expression domains of a group of genes termed the "proneural" genes (Skeath *et al.*, 1992). Because the absence or increase of proneural gene activity leads to corresponding reductions or expansions in the number of cells adopting neural fates, the proneural genes appear to provide the competence to adopt a neural fate (Brand and Campos-Ortega, 1988; Jimenez and Campos-Ortega, 1990). Thus, these clusters of proneural expression most likely correspond to equivalence groups. Within each proneural cluster, only one cell maintains proneural expression and develops into an NB. While the exact details of this process are still being elucidated, part of this restriction to a single cell requires lateral and mutual inhibition through cell–cell communication involving a class of genes termed the neurogenics (Lehmann *et al.*, 1981). In loss-of-function mutations in these genes, all cells of the proneural cluster appear to adopt equivalent neural identities (thus the term equivalence group) (Kania *et al.*, 1990; see also Doe, 1992; Goodman and Doe, 1993).

This result indicates that neurogenics are not required for the assignment of a cell to a specific neural pathway, but instead function to inhibit cells from adopting neural fates. Because this section is concerned with mechanisms directing cells into a specific neural pathway, we refer readers to excellent reviews by Campos-Ortega (1993) and Ghysen *et al.*, (1993, and references therein) for a discussion of the establishment of a single NB from a proneural cluster and the role of cell–cell communication.

Genetic analysis has identified some but not all of the proneural genes (for reviews see Doe, 1992; Cabrera, 1992; Goodman and Doe, 1993). Support for the notion that uncharacterized proneural genes still exist comes from the observation that some NBs are still present in embryos lacking all known proneural genes (Jimenez and Campos-Ortega, 1990). The currently identified proneural genes include: *ventral nervous system condensation defective (vnd), da,* and the members of the *AS-C: achaete (ac;* T5), *scute (sc;* T4; *sis-b), lethal of scute (l'sc;* T3), and *asense (ase;* T8/T1a). As discussed in Section II,B, the proneural genes *da* and *sc* encode factors previously used during sex determination.

As mentioned in Section II, one of the best examples of gene duplication and divergence and its links to developmental redundancy is seen with the proneural genes of the *AS-C*. During neuroblast determination, *ac, sc,* and *l'sc* are expressed in overlapping domains (the precise register of *ase* has not yet been reported) (Cabrera, 1990; Martin-Bermudo *et al.*, 1991; Skeath and Carroll, 1991; Skeath *et al.*, 1992). While their sequence similarity and co-expression is suggestive, direct evidence for redundancy has been provided through the genetic analysis of these genes. Deletion of these genes leads to no detectable (*ac, sc,* and *ase)* or only mild (*l'sc)* neural hypotrophy within the CNS (Jimenez and Campos-Ortega, 1979, 1990; Cabrera *et al.*, 1987; Martin-Bermudo *et al.*, 1991). However, when all members of the complex are simultaneously removed, the extent of neural hypotrophy is dramatically increased (Jimenez and Campos-Ortega, 1979). This neural hypotrophy is correlated with a reduction in NBs (Cabrera *et al.*, 1987; Jimenez and Campos-Ortega, 1990; Martin-Bermudo *et al.*, 1991). Based on these observations, a functional redundancy appears to exist among the *AS-C* in the determination of NBs within the CNS.

In addition to a role for the *AS-C* in the determination of NB fate, other roles in neural development also seem likely. First, the reduction in NBs seen in *AS-C* deficiencies does not directly account for the amount of neural hypotrophy seen (Jimenez and Campos-Ortega, 1990). Second, members of the complex continue to be expressed in specific patterns during later stages of neural development (Cabrera, 1990; Martin-Bermudo *et al.*, 1991; Skeath and Carroll, 1991; Skeath *et al.*, 1992).

C. ESTABLISHMENT OF A NEURAL LINEAGE: DETERMINATION OF NB, GMC, AND NEURONAL IDENTITIES

After acquiring neural competence, each NB is assigned a specific identity that appears to be related to its position along the anterior–posterior and dorsal–ventral axes (for reviews see Doe, 1992; Goodman and Doe, 1993). Doe *et al.* have proposed a model in which the A–P and D–V patterning molecules direct the expression of "NB identity genes" that are directly responsible for controlling the identity of the NB throughout its lineage. NB identity genes also control the fate of newly born GMCs by establishing the expression of "GMC identity genes" that act to control GMC and neuronal identities. As we will discuss later, some of the segmentation genes are reutilized in these later stages and serve specific functions as NB or GMC identity genes.

Many of the genes expressed in patterns consistent for NB and GMC identity genes encode transcription factors (for reviews see Doe, 1992; Goodman and Doe, 1993). This led Doe *et al.* to propose that the determination of these neural fates is under combined control. Taking this one step further, we propose that the reutilization of many of the regulatory factors from sex determination and segmentation dictates that the determination of neural fates be accomplished through combinatorial and concentration-dependent mechanisms very similar to those described for these other pathways. Although many of these regulatory molecules are expressed in patterns suggestive of roles as NB or GMC identity genes, the corresponding functional analysis has been extremely limited. One obvious reason for this is the need to bypass their earlier requirements to directly ascertain their specific roles in neural development.

Here we focus on the roles of three of these promiscuous regulators, *runt*, *eve* and *ftz*, because they have been shown to function in the determination of neural fates as putative NB and GMC identity genes, respectively (Doe *et al.*, 1988a,b; Duffy *et al.*, 1991). *runt* is expressed in NBs, GMCs, and neurons, while *eve* and *ftz* are expressed only in GMCs and neurons (Doe *et al.*, 1988a,b; Duffy *et al.*, 1991). Consistent with this, *runt* has been classified as an NB identity gene, while *eve* and *ftz* have been classified as GMC identity genes (for reviews see Doe, 1992; Goodman and Doe, 1993).

During neural development, *runt* expression is first detected in ~ 5 NBs per hemisegment and evolves in a temporally and spatially complex manner (Kania *et al.*, 1990; Duffy *et al.*, 1991). Whether *runt* continues to be expressed in all of the progeny (GMCs and neurons) of these 5 NBs or has a more complex evolution is not clear. A straightforward analysis of the importance of this expression to neural development is obscured because of *runt*'s prior role in segmentation as a pair-rule gene. Simple loss-of-function mutations in *runt* produce drastic segmentation defects that result in a grossly malformed nervous system (Nüsslein-Volhard and Wieschaus, 1980; Gergen and Wieschaus, 1986; Perrimon *et al.*, 1989). To circumvent this problem, a combination of approaches was used to bypass *runt*'s earlier requirement and specifically address its function during neural development (Duffy *et al.*, 1991; Duffy, 1992). One of these approaches had been used earlier for *ftz* and another for *eve* (Doe *et al.*, 1988a,b).

In the first approach, a transposon that lacked elements necessary for *runt* neural expression, but provided *runt* expression during segmentation, was used. In this *runt* background, a subset of *eve*-expressing CNS cells, the EL (*eve*-lateral) GMCs, and neurons failed to express *eve*. This result indicated that *runt* was required for the normal development of this subset of identified CNS cells. The second approach utilized a temperature-sensitive (ts) allele to provide *runt* activity during segmentation and modulate the activity of *runt* during neural development. This approach confirmed the earlier observation and also determined the temporal requirement for *runt*'s function in the determination of EL GMCs and neurons. These results indicated that *runt* was required during the stages of NB delamination and stem cell divisions, and not for maintaining *eve* expression in GMCs and neurons that had already formed.

More recently, hypomorphic alleles that preferentially affect *runt* neural activity have been used to support these observations (Duffy, 1992). Based on these results, it appears that *runt* is required for maintaining NB identity and for regulating downstream GMC identity genes. An alternative hypothesis in which *runt* functions as a GMC identity gene that is required in all of the EL GMCs to maintain their identity and thereby control GMC and neuronal fates has not been ruled out. However, the temporal requirement for *runt* and the fact that *runt* activity is no longer required once *eve* expression is initiated, provide stronger support for the first hypothesis.

As mentioned above, similar approaches had been previously used to characterize the roles of *ftz* and *eve* in neural development and led to their classification as GMC identity genes (Doe *et al.*, 1988a,b). By removing a cis-regulatory element responsible for *ftz* neural expression from a transposon capable of rescuing *ftz* mutants, Doe *et al.* were able to specifically address *ftz*'s role in neural development. The reduction of *ftz* activity during neural development led to the absence of *eve* expression in GMC-1 of neuroblast 4–2. Strong morphological evidence for a switch in cell fate was provided by filling this neuron with dye in the mutant background and observing its axonal morphology. The change in axonal morphology of the RP2 neuron seen in these experiments suggests that the absence of *ftz* activity changes the fate of GMC-1 to that of its sibling GMC-2, which gives rise to RP1 and RP3. These results demonstrated that *ftz* acts as a GMC identity gene. As expected from these results, the use of a ts allele to reduce *eve* activity during neural development also resulted in a change of GMC-1 to a GMC-2 fate, as indicated by a change in RP2 axon morphology. However, it is not known if *eve* acts to control GMC-1 or RP2 fate. In addition, reducing *eve* activity with the ts allele also led to a change in the axon trajectory of the aCC neuron. Thus, *eve* is also required for determining neuronal identity.

The inability to distinguish among the alternative roles for *runt* in specifying the identities of NBs vs. GMCs and for *eve* and *ftz* in GMCs vs. neurons highlights some of the difficulties of analyzing the mechanisms of neural specification. First, the activity of NB identity genes may be continuously required in the NB stem cell to maintain its identity, as opposed to being required in its GMC progeny. Second, the lack of markers for specific NBs, GMCs, and neurons has also made it difficult to determine if it is the NB, GMCs, or neurons that have adopted an alternative fate. Finally, the absence of a complete description of lineages has also proved burdensome.

In part owing to these difficulties, *runt, eve,* and *ftz* are currently the only segmentation genes for which direct roles in CNS development have been demonstrated. However, indirect evidence for the involvement of other segmentation genes has also been obtained (Patel *et al.*, 1989). Recent advances, including the identification of more markers, lineage descriptions, the ability to target gene expression, and the ability to selectively mark mutant clones, should allow these difficulties to be overcome and the roles of these and other genes to be more accurately described (Doe, 1992; Brand and Perrimon, 1993; Harrison and Perrimon, 1993).

Even with the limited data presented in Sections IV,B and IV,C, a few conclusions can be drawn. First, transcription factors used in other pathways are involved at all stages of neural determination. Second, to generate different fates, novel regulatory hierarchies are used in neural determination. In segmentation, *eve* and *ftz* are not coexpressed and *eve* negatively regulates *ftz* (Carroll and Scott, 1986; Frasch and Levine, 1987; Ingham and Gergen, 1988; Carroll and Vavra, 1989; Hooper *et al.*, 1989). In contrast, during neural development, *eve* and *ftz* are coexpressed, and *ftz* positively regulates *eve* expression (in the NB4–2 lineage) (Doe *et al.*, 1988a,b). Finally, the neural defects ascribed to each of the genes discussed above are relatively mild compared with their extensive expression during neural development. While it has yet to be proved, it is possible that the relatively minor effects of these genes are due to functional redundancy (Doe *et al.*, 1988a,b). Given the task of reproducing the enormous complexity of the nervous system, the continued analysis of this pathway may reveal that more redundancy exists here than for any other pathway.

V. Speculation

Although there is currently little direct evidence for combinatorial and concentration-dependent control of the successive steps of neural determination, the reexpression and reutilization of many of the sex determining and patterning molecules provides strong circumstantial evidence supporting such a model. Thus, the characterization of sex determination, segmentation, and neurogenesis has demonstrated that by altering the combinations and concentrations of these promiscuous developmental regulators, distinct programs of gene expression can be generated. As such, the use of these molecules in different developmental pathways strongly suggests that these pathways exploit common molecular mechanisms to determine diverse cell fates.

While developmental approaches will continue to reveal important information on the mechanisms of cell fate determination, evolutionary approaches may also shed light on these mechanisms and their relationships (for discussions see Wolpert, 1992; Raff, 1992). Consider the following observations. All of the genes discussed here encode transcription factors, some of which appear to have arisen from a common ancestral gene. They all act in a variety of pathways, where they have distinct, overlapping, or homologous roles in determining cell fates. Finally, specific molecular ties to other transcription factors have been retained throughout the evolution of these molecules and their roles in these developmental pathways. The coordinated reutilization of many of these genes may be indicative of certain molecular interactions that are functionally crucial (i.e., certain molecules only work in the context of others). Thus, the evolution of these genes may not be completely independent, but may occur in an interrelated fashion as gene networks.

Viewing this from an evolutionary standpoint, the needs that reutilization and redundancy serve in meeting the demands of efficiency and reproducibility may dictate that these two processes evolve in an interdependent fashion. It will be interesting to see if developmental pathways of diverse species will provide further support for this premise.

While space has limited our discussion to a small number of genes, there are many other examples of transcription factors that function in more than one developmental pathway. We look forward to seeing if the continued analysis of these genes supports a common theme of cell fate determination and its evolutionary links to reutilization and redundancy. Finally, we believe that a combination of developmental and evolutionary approaches to understanding these transcriptional molecules will prove beneficial in elucidating key molecular relationships and mechanisms involved in cell fate determination, not only in diverse developmental pathways, but in other organisms as well. Furthermore, the identification of putative mammalian homologs of some of the genes discussed in this chapter and the implication of at least one of these homologs (*AML-1*) in oncogenic transformation (myelogenous leukemia) leads to the hope that the analysis of these invertebrate genes will provide important insights into the understanding and treatment of such diseases.

ACKNOWLEDGMENTS

We thank Jeff Axelrod, Rich Binari, Martin Klingler, Michele Musacchio, and Esther Siegfried for fruitful discussions and comments.

REFERENCES

Alonso, M. C., and Cabrera, C. V. (1988). The *achaete–scute* gene complex of *Drosophila melanogaster* comprises four homologous genes. *EMBO J.* **7**, 2585–2591.

Bae, S. C., Yamaguchi-Iwai, Y., Ogawa, E., Maruyama, M., Inuzuka, M., Kagoshima, H., Shigesada, K., Satake, M., and Ito, Y. (1993). Isolation of *PEBP2aβ* cDNA representing the mouse homolog of human acute myeloid leukemia gene, *AML1*. *Oncogene* **8**, 809–814.

Bell, L. R., Maine, E. M., Schedl, P., and Cline, T. W. (1988). *Sex-lethal*, a *Drosophila* sex determination switch gene, exhibits sex-specific RNA splicing and sequence similarity to RNA binding proteins. *Cell* **55**, 1037–1046.

Bell, L. R., Horabin, J. I., Schedl, P., and Cline, T. W. (1991). Positive autoregulation of *Sex-lethal* by alternative splicing maintains the female determined state in *Drosophila*. *Cell* **65**, 229–239.

Bengal, E., Ransone, L., Scharfmann, R., Dwarki, V. J., Tapscott, S. J., Weintraub, H., and Verma, I. M. (1992). Functional antagonism between c-Jun and MyoD proteins: a direct physical association. *Cell* **68**, 507–519.

Bier, E., Vaessin, H., Younger-Shepherd, S., Jan, L. Y., and Jan, Y. N. (1992). *deadpan*, an essential pan-neural gene in *Drosophila*, encodes a helix–loop–helix protein similar to the *hairy* gene product. *Genes Dev.* **6**, 2137–2151.

Bopp, D., Bell, L. R., Cline, T. W., and Schedl, P. (1991). Developmental distribution of female-specific *Sex-lethal* proteins in *Drosophila melanogaster*. *Genes Dev.* **5**, 403–415.

Botas, J., Moscoso del Prado, J., and Garcia-Bellido, A. (1982). Gene dose titration analysis in the search of trans-regulatory genes in *Drosophila*. *EMBO J.* **1**, 307–310.

Brand, A., and Perrimon, N. (1993). Targeted gene expression as a means of altering cell fates and generating dominant phenotypes. *Development* **118**, 401–415.

Brand, M., and Campos-Ortega, J. A. (1988). Two groups of interrelated genes regulate early neurogenesis in *Drosophila melanogaster*. *Wilhelm Roux's Arch. Dev. Biol.* **197**, 457–470.

Bridges, C. B. (1921). Triploid intersexes in *Drosophila melanogaster*. *Science* **54**, 252–254.

Bridges, C. B. (1925). Sex in relation to chromosomes and genes. *Am. Nat.* **59**, 127–137.

Cabrera, C. V. (1990). Neuroblast determination and segregation in *Drosophila*, the interactions between *scute, Notch*, and *Delta*. *Development* **109**, 733–742.

Cabrera, C. V. (1992). The generation of cell diversity during early neurogenesis in *Drosophila*. *Development* **115**, 893–901.

Cabrera, C. V., Martinez-Arias, A., and Bate, M. (1987). The expression of three members of the *achaete–scute* gene complex correlates with neuroblast segregation in *Drosophila*. *Cell* **50**, 425–433.

Campos-Ortega, J. A. (1993). Early neurogenesis in *Drosophila melanogaster*. In "The development of *Drosophila melanogaster*" (M. Bate and A. Martinez-Arias, Eds.), pp. 1091–1129. Cold Spring Harbor Press, Cold Spring Harbor, New York.

Campuzano, S. L., Carramolino, L., Cabrera, C. V., Ruiz-Gomez, M., Villares, R., Boronat, A., and Modelell, J. (1985). Molecular genetics of the *achaete–scute* gene complex of *D. melanogaster*. *Cell* **40**, 327–338.

Capovilla, M., Eldon, E., and Pirrotta, V. (1992). The *giant* gene of *Drosophila* encodes a b-ZIP DNA-binding protein that regulates the expression of other segmentation genes. *Development* **114**, 99–112.

Carroll, S., and Scott, M. (1986). Zygotically active genes that affect the spatial expression of the *fushi tarazu* segmentation gene during early *Drosophila* embryogenesis. *Cell* **45**, 113–126.

Carroll, S. B., and Vavra, S. H. (1989). The zygotic control of *Drosophila* pair-rule expression II. Spatial repression by gap and pair-rule gene products. *Development* **107**, 673–683.

Caudy, M., Grell, E. H., Dambly-Chaudiere, C., Ghysen, A., Jan, L. Y., and Jan, Y. N. (1988a). The maternal sex determination gene *daughterless* has zygotic activity necessary for the formation of peripheral neurons in *Drosophila*. *Genes Dev.* **2**, 843–852.

Caudy, M., Vaessin, H., Brand, M., Tuma, R., Jan, L. Y., and Jan, Y. N. (1988b). *daughterless*, a *Drosophila* gene essential for both neurogenesis and sex determination, has sequence similarities to *myc* and the *achaete–scute* complex. *Cell* 55, 1061–1067.

Cline, T. W. (1978). Two closely linked mutations in *Drosophila melanogaster* that are lethal to opposite sexes and interact with *daughterless*. *Genetics* 90, 683–698.

Cline, T. W. (1983). The interaction between *daughterless* and *Sex-lethal* in Triploids: A lethal sex-transforming maternal effect linking sex determination and dosage compensation in *Drosophila melanogaster*. *Dev. Biol.* 95, 260–274.

Cline, T. W. (1984). Autoregulatory functioning of a *Drosophila* gene product that establishes and maintains the sexually determined state. *Genetics* 107, 231–277.

Cline, T. W. (1986). A female-specific lethal lesion in an X-linked positive regulator of the *Drosophila* sex determination gene, *Sex-lethal*. *Genetics* 113, 641–663.

Cline, T. W. (1988). Evidence that *sisterless-a* and *sisterless-b* are two of several discrete "numerator elements" of the X/A sex determination signal in *Drosophila* that switch *Sxl* between two alternative stable expression states. *Genetics* 119, 829–862.

Coulter, D. E., Swaykus, E. A., Beran-Koehn, A., Goldberg, D., Wieschaus, E., and Schedl, P. (1990). Molecular analysis of *odd-skipped*, a zinc finger encoding segmentation gene with a novel pair-rule expression pattern. *EMBO J.* 8, 3795.

Cronmiller, C., Schedl, P., and Cline, T. W. (1988). Molecular characterization of *daughterless*, a *Drosophila* sex determination gene with multiple roles in development. *Genes Dev.* 2, 1666–1676.

Cubas, P., de Celis, J. F., Campuzano, S., and Modelell, J. (1991). Proneural clusters of *achaete–scute* expression and the generation of sensory organs in the *Drosophila* imaginal wing disc. *Genes Dev.* 5, 996–1008.

Daga, A., Tighe, J. E., and Calabi, F. (1992). Leukemia/*Drosophila* homology. *Nature (London)* 356, 484.

Dearolf, C. R., Topol, J., and Parker, C. S. (1989). Transcriptional control of *Drosophila fushi tarazu* zebra stripe expression. *Genes Dev.* 3, 384–398.

Doe, C. Q. (1992). The generation of neuronal diversity in the *Drosophila* embryonic central nervous system. In "Determinants of Neuronal Identity." (M. Shankland and E. Macagno, eds.), pp. 119–154. Academic Press, San Diego.

Doe, C. Q., Hiromi, Y., Gehring, W. J., and Goodman, C. S. (1988a). Expression and function of the segmentation gene *fushi tarazu* during *Drosophila* neurogenesis. *Science* 239, 170–175.

Doe, C. Q., Smouse D., and Goodman, C. S. (1988b). Control of neuronal fate by the *Drosophila* segmentation gene *even-skipped*. *Nature (London)* 333, 376–378.

Driever, W. (1993). Maternal control of anterior development in the *Drosophila* embryo. In "The development of *Drosophila melanogaster*" (M. Bate and A. Martinez-Arias, Eds.), pp. 301–324. Cold Spring Harbor Press. Cold Spring Harbor, New York.

Duffy, J. B. (1992). A genetic characterization of *runt* activity during *Drosophila* embryogenesis. Thesis. Univ. of Texas Health Sci. Cent. Grad. Sch. of Biomed. Sci., Houston.

Duffy, J. B., and Gergen, J. P. (1991). The *Drosophila* segmentation gene *runt* acts as a position-specific numerator element necessary for the uniform expression of the sex-determining gene *Sex-lethal*. *Genes Dev.* 5, 2176–2187.

Duffy, J. B., Kania, M. A., and Gergen, J. A. (1991). Expression and function of the *Drosophila* gene *runt* in early stages of neural development. *Development* 113, 1223–1230.

Ellis, H. M., Spann, D. R., and Posakony, J. W. (1990). *extramacrochaetae*, a negative regulator of sensory organ development in *Drosophila*, defines a new class of helix–loop–helix proteins. *Cell* 61, 27–38.

Erickson, J., and Cline, T. W. (1991). The molecular nature of the *Drosophila* sex determination signal and its link to neurogenesis. *Science* **251**, 1071–1074.

Erickson, J., and Cline, T. W. (1993). A bZIP protein, Sisterless-a, collaborates with bHLH transcription factors early in *Drosophila* development to determine sex. *Genes Dev.* **9**, 1688–1702.

Erickson, P., Gao, J., Chang, K. S., Look, T., Whisenant, E., Raimondi, S., Lasher, R., Trujillo, J., Rowley, J., and Drabkin, H. (1992). Identification of breakpoints in t(8;21) acute myelogenous leukemia and isolation of a fusion transcript, *AML1/ETO* with similarity to *Drosophila* segmentation gene *runt*. *Blood* **80**, 1825–1831.

Finkelstein, R., Smouse, D., Capaci, T. M., Spradling, A. C., and Perrimon, N. (1990). The *orthodenticle* gene encodes a novel homeo domain protein involved in the development of the *Drosophila* nervous system and ocellar visual structures. *Genes Dev.* **4**, 1516–1527.

Frasch, M. and Levine, M. (1987). Complementary patterns of *even-skipped* and *fushi tarazu* expression involve their differential regulation by a common set of segmentation genes in *Drosophila*. *Genes Dev.* **1**, 981–995.

Frigerio, G., Burri, M., Bopp, D., Baumgartner, S., and Noll, M. (1986). Structure of the segmentation gene *paired* and the *Drosophila* PRD gene set as part of a gene network. *Cell* **47**, 735–746.

Garcia-Alonso, L. A., and Garcia-Bellido, A. (1988). *Extramacrochaetae*, a trans-acting gene of the *achaete-scute* complex of *Drosophila* involved in cell communication. *Wilhelm Roux's Arch. Dev. Biol.* **197**, 328–338.

Garrell, J., and Modolell, J. (1990). The *Drosophila extramacrochaetae* locus, an antogonist of proneural genes that, like these genes, encodes a helix–loop–helix protein. *Cell* **61**, 39–48.

Gergen, J. P. (1987). Dosage compensation in *Drosophila*: evidence that *daughterless* and *Sex-lethal* control X chromosome activity at the blastoderm stage of embryogenesis. *Genetics* **117**, 477–485.

Gergen, J. P., and Wieschaus, E. (1986). Dosage requirements for *runt* in the segmentation of *Drosophila* embryos. *Cell* **45**, 289–299.

Ghysen, A., and Dambly-Chaudiere, C. (1988). From DNA to form: the *achaete-scute* complex. *Genes Dev.* **2**, 495–501.

Ghysen, A., Dambly-Chaudiere, C., Jan, L. Y., and Jan, Y. N. (1993). Cell interactions and gene interactions in peripheral neurogenesis. *Genes Dev.* **7**, 723–733.

Goodman, C. S., and Doe, C. Q. (1993). Embryonic development of the *Drosophila* nervous system. In "The development of Drosophila melanogaster" (M. Bate and A. Martinez Arias, Eds.), pp. 1131–1206. Cold Spring Harbor Laboratory Press, Cold Spring Harbor, New York.

Goto, T., Macdonald, P., and Maniatis, T. (1989). Early and late periodic patterns of *even-skipped* expression are controlled by distinct regulatory elements that respond to different spatial cues. *Cell* **57**, 413–422.

Grossniklaus, U., Kurth-Pearson, R., and Gehring, W. J. (1992). The *Drosophila sloppy paired* locus encodes two proteins involved in segmentation that show homology to mammalian transcription factors. *Genes Dev.* **6**, 1030–1051.

Harding, K., Hoey, T., Warrior, R., and Levine, M. (1989). Autoregulatory and gap response elements of the *even-skipped* promoter of *Drosophila*. *EMBO J.* **8**, 1205–1212.

Harrison, D., and Perrimon, N. (1993). Simple and efficient generation of marked clones in *Drosophila*. *Curr. Biol.* **3**, 424–433.

Hiromi, Y., and Gehring, W. J. (1987). Regulation and function of the *Drosophila* segmentation gene *fushi tarazu*. *Cell* **50**, 963–974.

Hiromi, Y., Kuroiwa, A., and Gehring, W. J. (1985). Control elements of the *Drosophila* segmentation gene *fushi tarazu*. *Cell* **43**, 603–613.

Hooper, K. L., Parkhurst, S. M., and Ish-Horowicz, D. (1989). Spatial control of *hairy* protein expression during embryogenesis. *Development* **102**, 489–504.

Howard, K., and Ingham, P. W. (1986). Regulatory interactions between the segmentation genes *fushi tarazu*, *hairy* and *engrailed* in the *Drosophila* blastoderm. *Cell* **44**, 949–957.

Howard, K., and Struhl, G. (1991). Decoding positional information: regulation of the pair-rule gene *hairy*. *Development* **110**, 1223–1231.

Ingham, P. W., and Gergen, J. P. (1988). Interactions between the pair-rule genes *runt, hairy, even-skipped* and *fushi tarazu* and the establishment of periodic pattern in the *Drosophila* embryo. *Development* **104**, Suppl., 51–60.

Ip, Y. T., Levine, M., and Small, S. (1992). The *bicoid* and *dorsal* morphogens use a similar strategy to make stripes in the *Drosophila* embryo. *J. Cell Sci.* **16**, Suppl., 33–38.

Ish-Horowicz, D., and Pinchin, S. M. (1987). Pattern abnormalities induced by ectopic expression of the *Drosophila* gene *hairy* are associated with repression of *ftz* transcription. *Cell* **51**, 405–415.

Jan, Y. N., and Jan, L. Y. (1993). The peripheral nervous system. In "The development of *Drosophila melanogaster*" (M. Bate and A. Martinez-Arias, Eds.), pp. 1207–1244. Cold Spring Harbor Press, Cold Spring Harbor, New York.

Jimenez, F., and Campos-Ortega, J. A. (1979). On a region of the *Drosophila* genome necessary for central nervous system development. *Nature (London)* **282**, 310–312.

Jimenez, F., and Campos-Ortega, J. A. (1990). Defective neuroblast commitment in mutants of the *achaete–scute* complex and adjacent genes of *D. melanogaster*. *Neuron* **5**, 81–89.

Kagoshima, H., Shigesada, K., Satake, M., Ito, Y., Miyoshi, H., Ohki, M., Pepling, M., and Gergen, P. (1993). The Runt-domain identifies a new family of heteromeric DNA-binding transcriptional regulatory proteins. *Trends Genet.* **9**, 338–341.

Kania, M. A., Bonner, A. S., Duffy, J. B., and Gergen, J. P. (1990). The *Drosophila* segmentation gene *runt* encodes a novel nuclear regulatory protein that is also expressed in the developing nervous system. *Genes Dev.* **4**, 1701–1713.

Keyes, L. N., Cline, T., and Schedl, P. (1992). The primary sex determination signal of *Drosophila* acts at the level of transcription. *Cell* **68**, 933–943.

Kilchherr, F., Baumgartner, S., Bopp, D., Frei, E., and Noll, M. (1986). Isolation of the *paired* gene of *Drosophila* and its spatial expression during early embryogenesis. *Nature (London)* **321**, 493–499.

Klingler, M., and Gergen, J. P. (1993). Regulation of *runt* transcription by *Drosophila* segmentation genes. *Mech. Dev.* **43**, 3–19.

Laughon, A., and Scott, M. P. (1984). Sequence of a *Drosophila* segmentation gene: Protein structure homology with DNA-binding proteins. *Nature (London)* **310**, 25–31.

Lehmann, R., Dietrich, U., Jimenez, F., and Campos-Ortega, J. A. (1981). Mutations of early neurogenesis in *Drosophila*. *Wilhelm Roux's Arch. Dev. Biol.* **190**, 226–229.

Lucchesi, J. C., and Skripsky, T. (1981). The link between dosage compensation and sex differentiation in *Drosophila melanogaster*. *Chromosoma* **82**, 217–227.

Macdonald, P., Ingham, P. W., and Struhl, G. (1986). Isolation structure and expression of *even-skipped*: a second pair-rule gene of *Drosophila* containing a homeobox. *Cell* **47**, 721–734.

Maine, E. M., Salz, H. K., Schedl, P., and Cline, T. W. (1985). *Sex-lethal*, a link between sex determination and sexual differentiation in *Drosophila melanogaster*. *Cold Spring Harbor Symp. Quant. Biol.* **50**, 596–604.

Martin-Bermudo, M.D., Martinez, C., Rodriguez, A., and Jimenez, F. (1991). Distribution and function of the *lethal of scute* gene product during early neurogenesis in *Drosophila. Development* 113, 445–454.

Martinez Arias, A. (1993). Development and patterning of the larval epidermis of *Drosophila.* In "The development of *Drosophila melanogaster*" (M. Bate and A. Martinez-Arias, Eds.), pp. 517–608. Cold Spring Harbor Press, Cold Spring Harbor, New York.

McGinnis, W., Levine, M., Hafen, E., Kuroiwa, A., and Gehring, W. (1984). A conserved DNA sequence in homeotic genes of the *Drosophila Antennapedia* and bithorax complexes. *Nature (London)* 308, 428–433.

Moscoso del Prado, J., and Garcia-Bellido, A. (1984a). Genetic regulation of the *achaete–scute* complex of *Drosophila melanogaster. Wilhelm Roux's Arch. Dev. Biol.* 193, 242–245

Moscoso del Prado, J., and Garcia-Bellido, A. (1984b). Cell interactions in the generation of chaetae pattern in *Drosophila. Wilhelm Roux's Arch. Dev. Biol.* 193, 246–251.

Murre, C., McCaw, P. S., and Baltimore, D. (1989a). A new DNA binding and dimerization motif in immunoglobulin enhancer binding, *daughterless, MyoD,* and *myc* proteins. *Cell* 56, 777–783.

Murre, C. Mccaw, P. S., Vaessin, H., Caudy, M., Jan, L. Y., Jan, Y. N., Cabrera, C., Buskin, J. N., Hauschka, S. D., Lassar, A. B., Weintraub, H., and Baltimore, D. (1989b). Interactions between heterologous helix–loop–helix proteins generate complexes that bind specifically to a common DNA sequence. *Cell* 58, 537–544.

Nauber, U., Pankratz, M. J., Kienlin, A., Seifert, E., Klemm, U., and Jäckle, H. (1988). Abdominal segmentation of the *Drosophila* embryo requires a hormone receptor-like protein encoded by the gap-gene *knirps. Nature (London)* 336, 489–492.

Nüsslein-Volhard, C., and Wieschaus, E. (1980). Mutations affecting segment number and polarity in *Drosophila. Nature (London)* 287, 795–801.

Ogawa, E., Maruyama, M., Kagoshima, H., Inuzuka, M., Lu, J., Satake, M., Shigesada, K., and Ito, Y. (1993). PEBP2/PEA2 represents a new family of transcription factors homologous to the products of the *Drosophila runt* and the human AML1. *Proc. Nat'l. Acad. Sci. U.S.A.* 90, 6859–6863.

Pankratz, M. J., and Jackle, H. (1993). Blastoderm segmentation. In "The development of *Drosophila melanogaster*" (M. Bate and A. Martinez-Arias, Eds.), pp. 467–516. Cold Spring Harbor Press, Cold Spring Harbor, New York.

Pankratz, M. J., Seifert, E., Gerwin, N., Billi, B., Nauber, U., and Jäckle, H. (1990). Gradients of *Krüppel* and *knirps* gene products direct pair-rule gene stripe patterning in the posterior region of the *Drosophila* embryo. *Cell* 61, 309–317.

Parkhurst, S. M., Bopp, D., and Ish-Horowicz, D. (1990). X:A ratio, the primary sex-determining signal in *Drosophila,* is transduced by helix–loop–helix proteins. *Cell* 63, 1179–1191.

Parkhurst, S. M., Lipshitz, H. D., and Ish-Horowicz, D. (1993). *achaete–scute* feminizing activities and *Drosophila* sex determination. *Development* 117, 737–749.

Patel, N. H., Schafer, B., Goodman, C. S., and Holmgren, R. 1989). The role of segment polarity genes during *Drosophila* neurogenesis. *Genes Dev.* 3, 890–904.

Perrimon, N., Smouse, D., and Gabor Miklos, G. L. (1989). Developmental genetics of loci at the base of the X chromosome of *Drosophila melanogaster. Genetics* 121, 313–331.

Pignoni, F., Baldarelli, R. M., Steingrimsson, E., Diaz, R. J., Patapoutian, A., Merriam, J. R., and Lengyel, J. A. (1990). The *Drosophila* gene *tailless* is expressed at the embryonic termini and is a member of the steroid receptor superfamily. *Cell* 62, 151–163.

Raff, R. A. (1992). Evolution of developmental decisions and morphogenesis: the view from two camps. *Development* 1 , Supp., 15–22.

Riddihough, G., and Ish-Horowicz, D. (1991). Individual stripe regulatory elements in the *Drosophila hairy* promoter respond to maternal, gap, and pair-rule genes. *Genes Dev.* 5, 840–854.

Rosenberg, U. B., Schroder, C., Preiss, A., Kienlin, A., Cote, S., Riede, I., and Jäckle, H. (1986). Structural homology of the product of the *Drosophila Krüppel* gene with Xenopus transcription factor IIIA. *Nature (London)* 319, 336–339.

Rushlow, C. A., Hogan, A., Pinchin, S. M., Howe, K. M., Lardelli, M., and Ish-Horowicz, D. (1989). The *Drosophila hairy* protein acts both in segmentation and bristle patterning and shows homology to N-*myc*. *EMBO J.* 8, 3095–3103.

Salz, H. K., Maine, E. M., Keyes, L. N., Samuels, M. E., Cline, T. W., and Schedl, P. (1989). The *Drosophila* female-specific sex-determination gene, *Sex-lethal*, has stage-, tissue-, and sex-specific RNAs suggesting multiple modes of regulation. *Genes Dev.* 3, 708–719.

Sanchez, L., and Nothiger, R. (1983). Sex determination and dosage compensation in *Drosophila melanogaster:* production of male clones in XX females. *EMBO J.* 2, 485–491.

Skeath, J. B., and Carroll, S. (1991). Regulation of *achaete–scute* gene expression and sensory organ pattern formation in the *Drosophila* wing. *Genes Dev.* 5, 984–995.

Skeath, J. B., Panganiban, G., Selegue, J., and Carroll, S. B. (1992). Gene regulation in two dimensions: the proneural *achaete* and *scute* genes are controlled by combinations of axis-patterning genes through a common intergenic control region. *Genes Dev.* 6, 2602–2619.

Small, S., Kraut, R., Hoey, T., Warrior, R., and Levine, M. (1991). Transcriptional regulation of a pair-rule stripe in *Drosophila*. *Genes Dev.* 5, 827–839.

Small, S., Blair, A., and Levine, M. (1992). Regulation of *even-skipped* stripe 2 in the *Drosophila* embryo. *EMBO J.* 11, 4047–4057.

Sommer, R. J., and Tautz, D. (1993). Involvement of an orthologue of the *Drosophila* pair-rule gene *hairy* in segment formation of the short germ-band embryo of *Tribolium* (Coleoptera). *Nature (London)* 361, 448–450.

Sprenger, F., and Nüsslein-Volhard, C. (1993). The terminal system of axis determination in the *Drosophila* embryo. In "The development of *Drosophila melanogaster*" (M. Bate and A. Martinez-Arias, Eds.), pp. 365–386. Cold Spring Harbor Lab. Press, Cold Spring Harbor, New York.

Stanojevic, D., Hoey, T., and Levine, M. (1989). Sequence-specific DNA-binding activities of the gap proteins encoded by *hunchback* and *Krüppel* in *Drosophila*. *Nature (London)* 341, 331–335.

Stanojevic, D., Small, S., and Levine, M. (1991). Regulation of a segmentation stripe by overlapping activators and repressors in the *Drosophila* embryo. *Science* 254, 1385–1387.

St. Johnston, D. (1993). Pole plasm and the posterior group genes. In "The development of *Drosophila melanogaster*" (M. Bate and A. Martinez-Arias, Eds.), pp. 325–363. Cold Spring Harbor Lab. Press, Cold Spring Harbor, New York.

St. Johnston, D., and Nüsslein-Volhard, C. (1992). The origin of pattern and polarity in the *Drosophila* embryo. *Cell* 68, 201–219.

Tautz, D., Lehmann, R., Schnürch, H., Schuh, R., Seifert, E., Kienlin, A., Jones, K., and Jäckle, H. (1987). Finger protein of novel structure encoded by *hunchback*, a second member of the gap class of *Drosophila* segmentation genes. *Nature (London)* 327, 383–389.

Topol, J., Dearolf, C. R., Prakash, K., and Parker, C. S. (1991). Synthetic oligonucleotides recreate *Drosophila fushi tarazu* zebra-stripe expression. *Genes Dev.* **5,** 855–867.

Torres, M., and Sanchez, L. (1989). The *scute* (T4) gene acts as a numerator element of the X:A signal that determines the state of activity of *Sex-lethal* in *Drosophila*. *EMBO J.* **8,** 3079–3086.

Torres, M., and Sanchez, L. (1992). Gap gene properties of the pair-rule gene *runt* during Drosophila segmentation. *Genet. Res.* **59,** 189–198.

Tsai, C., and Gergen, J. P. (1993). Zygotic expression of the *Drosophila* segmentation gene *runt* antagonizes the activity of the maternal morphogen *bicoid*. Submitted.

Ueda, H., Sonoda, S., Brown, J. L., Scott, M. P., and Wu, C. (1990). A sequence-specific DNA-binding protein that activates *fushi tarazu* segmentation gene expression. *Genes Dev.* **4,** 624–635.

Villares, R., and Cabrera, C. V. (1987). The *achaete–scute* gene complex of *D. melanogaster:* conserved domains in a subset of genes required for neurogenesis and their homology to *myc. Cell* **50,** 415–424.

Wainwright, S. M., and Ish-Horowicz. D. (1992). Point mutations in the *Drosophila hairy* gene demonstrate *in vivo* requirements for basic, helix–loop–helix, and WRPW domains. *Mol. Cell. Biol.* **12,** 2475–2483.

Wolpert, L. (1992). Gastrulation and the evolution of development. *Development* (suppl.), 7–13.

Younger-Shepherd, S., Vaessin, H., Bier, E., Jan, L. Y., and Jan, Y. N. (1992). *deadpan*, an essential pan-neural gene encoding an HLH protein, acts as a denominator in *Drosophila* sex determination. *Cell* **70,** 911–922.

NEUROGENESIS IN *Xenopus:*
A MOLECULAR GENETIC PERSPECTIVE

Beatriz Ferreiro[1] and William A. Harris

Department of Biology and Center for Molecular Genetics, University of California at San Diego, La Jolla, California 92093-0357

I. Introduction

Embryological work in amphibians over the past century taught us that the nervous system originates from a series of inductive interac-

---

[1]Present address: La Jolla Cancer Research Foundation, La Jolla, CA 92037.

ADVANCES IN GENETICS, Vol. 31

tions. In the past few years, molecular genetic approaches to the generation of the nervous system, particularly in *Xenopus*, have revealed a number of genes that may be involved in specific aspects of neurogenesis. It is our intention to review both embryological and molecular studies and to provide a framework within which it may be possible to imbue the genes with embryological significance. One may be struck when reading this chapter with how many of the genes that appear to be involved are homologs of developmentally significant genes that were first identified in *Drosophila* as developmental mutants. This, of course, is no accident, but part of a strategy to move directly into the molecular basis of *Xenopus* development, shortcutting some of the biochemical purifications that haunted the embryologists of the previous era. It should also become clear to the reader that we are still far away from even a moderately complete molecular understanding of neural induction.

II. The Organizer

Spemann and colleagues showed that the dorsal lip of the blastopore has the ability to induce and pattern the surrounding ectoderm and mesoderm into an entire embryo (Spemann and Mangold, 1924). Thus, an embryo with an extra dorsal lip transplanted to its ventral side produces a secondary embryo from host tissue. Because of the ability of this tissue to organize surrounding tissues, Spemann called it "the organizer."

A. INDUCTION OF THE ORGANIZER

The organizer itself was shown to be induced by what is now called the Nieuwkoop center, a dorsovegetal region of the early blastula in a *Xenopus* embryo (Nieuwkoop, 1973). Nieuwkoop exploited a powerful assay of induction which is now extensively used for molecular analysis of induction (Figs. 1A, 2). By itself, isolated animal cap tissue forms epidermis (Figs. 1A, 2A), but when it is combined with dorsal vegetal

---

FIG. 1. From blastula to neurula, views of the *Xenopus* embryo. (A) A blastula embryo with relevant axes, animal–vegetal and dorsal–ventral. The dorsal involuting marginal zone (DIMZ), which is equivalent to the dorsal mesoderm and contains the organizer, is lightly stippled. The prospective neural plate, also known as the dorsal noninvoluting marginal zone (DNIMZ), is shown darkly stippled. Also shown in this figure (to the right) is the animal cap explant that is dissected off the late blastula or early gastrula. Left by itself, this tissue will form ectoderm, but when it is combined

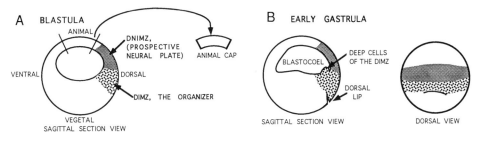

A BLASTULA

ANIMAL
DNIMZ,
(PROSPECTIVE
NEURAL PLATE)   ANIMAL CAP
VENTRAL   DORSAL

DIMZ, THE ORGANIZER
VEGETAL
SAGITTAL SECTION VIEW

B EARLY GASTRULA

BLASTOCOEL   DEEP CELLS
                    OF THE DIMZ

DORSAL
LIP
SAGITTAL SECTION VIEW   DORSAL VIEW

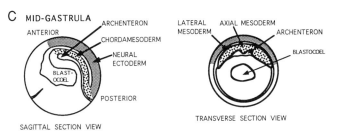

C MID-GASTRULA

ANTERIOR   ARCHENTERON

CHORDAMESODERM

NEURAL
ECTODERM
BLAST-
OCOEL

POSTERIOR

SAGITTAL SECTION VIEW

LATERAL AXIAL MESODERM
MESODERM   ARCHENTERON

BLASTOCOEL

TRANSVERSE SECTION VIEW

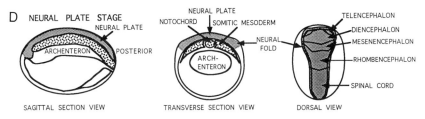

NEURAL PLATE
D NEURAL PLATE STAGE   NOTOCHORD SOMITIC MESODERM   TELENCEPHALON
      NEURAL PLATE   DIENCEPHALON
ARCHENTERON   POSTERIOR   NEURAL   MESENENCEPHALON
                                   ARCH-   FOLD
                                   ENTERON   RHOMBENCEPHALON

                                                               SPINAL CORD

SAGITTAL SECTION VIEW   TRANSVERSE SECTION VIEW   DORSAL VIEW

with dorsal lip tissue, it will turn on a neural program (see Fig. 2). In (B), gastrulation is just beginning; the dorsal lip is forming and the DIMZ beginning to involute. An arrow points to the deep cells of the DIMZ that crawl along the blastocoel roof and express *goosecoid*. Another arrow points to the dorsal lip of the blastopore. To the right is a dorsal surface view of the embryo. Here one can appreciate how compressed the fated neural plate and dorsal mesoderm are in the anterior–posterior axis. In (C), the DIMZ has involuted, converged, and extended, and so become the chordamesoderm. The fur thest point of involution is the anterior of the embryo; the blastopore is posterior, although there is evidence that A–P positional identity is not fixed by this stage either in the neural plate or the chordamesoderm (Saha and Grainger, 1992). By midgastrulation, the neural plate has been partially induced and it has also converged and extended along the A–P axis. A transverse view through a midgastrula embryo is shown to the right, emphasizing the neural ectoderm (dark stipple) and the subjacent axial (chorda) mesoderm. (D) Three views of a neural plate embryo. The sagittal section (left) shows the extended neural plate with the elongated chordamesoderm subjacent to it. The transverse section (middle) shows that the chordamesoderm has broken up into a distinct notochord and somitic mesoderm, with unlabeled lateral mesoderm at the lateral edges of the somitic mesoderm. In this view, one can also appreciate the thickening of the neural plate, especially at the edges where it is raised into neural folds. To the right is a dorsal surface view showing the fate map of the neural plate. By this stage A–P values in the mesoderm and neural plate are relatively fixed (Eagleson and Harris, 1990; Saha and Grainger, 1992).

31

ANIMAL CAP

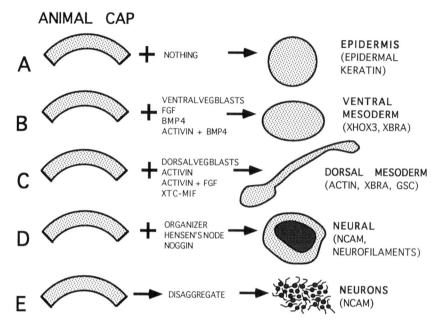

A   + NOTHING ⟶ EPIDERMIS (EPIDERMAL KERATIN)

B   + VENTRALVEGBLASTS FGF BMP4 ACTIVIN + BMP4 ⟶ VENTRAL MESODERM (XHOX3, XBRA)

C   + DORSALVEGBLASTS ACTIVIN ACTIVIN + FGF XTC-MIF ⟶ DORSAL MESODERM (ACTIN, XBRA, GSC)

D   + ORGANIZER HENSEN'S NODE NOGGIN ⟶ NEURAL (NCAM, NEUROFILAMENTS)

E   ⟶ DISAGGREGATE ⟶ NEURONS (NCAM)

FIG. 2. The induction of animal caps. Animal caps when cultured by themselves with no additional factors as in (A) turn into ectoderm and express epidermal genes such as epidermal keratin. When cultured with FGF or BMP-4 (B), or when combined with ventral vegetal blastomeres, animal cap tissue makes ventral mesoderm and expresses genes such as *Xhox3* and *Xbra,* as well as producing blood cells which are characteristic of ventral mesoderm (Slack, 1990). When cultured in the presence of picomolar concentrations of activin or activin plus bFGF, or activin containing *Xenopus* tissue culture-mesodermal-inducing factor (XTC-MIF) (C), the tissue converges and extends and expresses dorsal mesodermal markers such as muscle-specific actin and *goosecoid* (GSC), and pan-mesodermal markers such as *Xbra* (Green *et al.,* 1990). When BMP-4 is combined with activin, however, it has a dominant ventralizing effect and leads to the expression of ventral mesodermal fate (B) (Dale *et al.,* 1992; Jones *et al.,* 1992). Animal cap tissue can also be induced to turn on neural genes and tissue (dark area) when (D) combined with the organizer, with Hensen's node from a chick embryo (Kintner and Dodd, 1991), cultured in the presence of the noggin protein, or (E) simply when disaggregated and cultured as individual cells (Grunz and Tacke, 1989; Godsave and Slack, 1989).

cells, cap tissue is induced to form mesoderm and an organizer (Fig. 2C) (Nieuwkoop, 1969).

Using the animal cap assay, several growth factors and secreted molecules have been implicated in mesodermal induction (Fig. 2) (see review in Green and Smith, 1991). Activin, a member of the TGF-β family, induces dorsal mesoderm and organizer properties (Eijnden-Van Raaij *et al.,* 1990; Smith *et al.,* 1990; Thomsen *et al.,* 1990). The secreted proteins noggin and wnt-8 enhance this process (Christian and Moon, 1993; Smith and Harland, 1991, 1992; Sokol *et al.*, 1991; Steinbeisser *et al.*, 1993). bFGF induces ventral mesoderm (Kimelman *et al.*, 1988; Slack *et al.*, 1988), but when combined with activin, produces mesoderm of dorsal character (Cooke, 1989). Activin and bFGF in various doses and combinations can induce a wide array of mesodermal cell types (Green *et al.*, 1992). Bone morphogenetic protein-4 (BMP-4), another TGF-β family member, also induces ectoderm to become ventral mesoderm, but it acts dominantly with respect to activin so that the two together produce ventral rather than dorsal mesoderm (Fig. 2B) (Dale *et al.*, 1992; Jones *et al.*, 1992; Koster *et al.*, 1991). These gene products (and others yet to be discovered), released from vegetal cells, acting on their various receptors in the equatorial region of the blastula, set up the dorsal–ventral axis of the marginal zone and the organizer (see reviews in Kimelman *et al.*, 1992; Moon and Christian, 1992; Sive, 1993).

B. MOLECULAR BIOLOGY OF THE ORGANIZER

Spemann's organizer, also known as the dorsal involuting marginal zone (DIMZ) (Fig. 1A,B) in *Xenopus* embryos, involutes and forms the notochord (Keller, 1985), which continues to be a potent inducer. The DIMZ was originally imagined to be the source of one or two organizer substances (see review in Gilbert and Saxen, 1993). Recent studies, however, suggest that the DIMZ may turn out to be very complex, both functionally and molecularly (see Table 1). This tissue has several distinguishable inductive roles in early development (Spemann and Mangold, 1924). Some involve neural induction and patterning of the nervous system, and others involve induction and patterning of the dorsal mesoderm. As shown in this section, several genes of the organizer encode transcription factors whose expression is highly localized to the dorsal lip and the involuting notochord. These transcription factors lie downstream of the Nieuwkoop center inducers and some lie upstream of the inducing signals released by the organizer. Some must also regulate the molecules governing movements and cellular structure of the DIMZ and notochord.

At the beginning of gastrulation, the marginal zone cells around the dorsal half circumference of the embryo converge toward the dorsalmost side, extending by the intercalation of these cells as they involute (Keller and Danilchik, 1988), (see Fig. 1C). The tissue that extends the most during this time, driving the embryo from the spherical blastula to the elongated neurula, is the notochord (Keller *et al.*, 1989), a rigid

TABLE 1
Genes of the Organizer

Gene	Temporal and spatial expression	Inducers	Protein characteristics	In ventral effect of misexpression	Reference
goosecoid	Blastopore lip, deep mesodermal cells, head mesoderm	Activin	Homeodomain (*bicoid* and *gooseberry*)	In ventral blastomeres produces 2nd axis formation	Blumberg *et al.* (1991); Cho *et al.* (1991a); Niehrs *et al.* (1993)
Xlim-1	Low maternal, early gastrula in dorsal lip and notochord, tadpole brain	Activin and RA	LIM homeodomain (*lin-11*, *isl-1*, *mec-3*)	Untested	Taira *et al.* (1992)
Pintallavis XForkhead	Organizer, axial midline, all germ layers. Low in tailbud	Activin	DNA binding protein (*forkhead*, *HNF-3α*)	Expansion of posterior neural tube, and repression of anterior and dorsal neurons	Ruiz i Altaba and Jessel (1992); Dirksen and Jamrich (1992); Knochel *et al.* (1992)
Xnot	Uniform at MBT, organizer, notochord and floor plate	Activin and bFGF	Homeodomain	Untested	von Dassow *et al.* (1993)
Xbra	Presumtive mesodermal cells around the blastopore, notochord. Low after neurula	Activin	Intracellular protein. *N*-glycosylation sites	Ectopic mesoderm formation with posterior character	Smith *et al.* (1991); Cunliffe and Smith (1992)

34

noggin	Maternal uniform, organizer, notochord and head mesoderm	Unknown	Secreted protein. N-glycosylation site	Dorsalization of mesoderm and neural induction of ectoderm	Smith and Harland (1992); Smith et al. (1993); Lamb et al. (1993)
FGF3/int-2	Around blastopore lip, disappears as the lip involutes	Activin and bFGF	FGF family	Untested	Tannahill et al. (1992)
Xcad-1, -2	Blastopore lip	Unknown	Homeodomain (*caudal*)	Untested	Blumberg et al. (1991)
Xlab-1, -2	Enriched in blastopore lip	Unknown	Homeodomain (*labial*)	Untested	Blumberg et al. (1991); Sive and Cheng (1991)
Xtwist	Early gastrula mesoderm, tailbud in non-somitic mesoderm	Vegetal tissue	bHLH protein (*twist*)	Untested	Hopwood et al. (1989)
Xsnail	MBT (uniform), mesoderm	Activin and bFGF	Zn finger protein (*snail*)	Untested	Sargent and Bennett (1990)
Xhox-3	Axial mesoderm with posterior anterior gradient. Tailbud and tadpole; anterior CNS and tailbud	bFGF, less TGF-β	Homeodomain, pair rule (*even skipped*)	Anterior defects, headless embryos. No homeotic transformation	Ruiz i Altaba and Melton (1989a,b)

structure that is circular in cross section and composed of highly vacu-
olated cells shaped like pizza slices (Fig. 1D). The strength of the notochord must derive from the genes for the structural proteins that it
expresses in its cells. There has been no molecular analysis of the ri-
gidity of the notochord, but the following cytoskeletal and extracellular matrix proteins, expressed in the notochord, could be involved: collagen type-2 (Bieker and Yazdani, 1992), tenascin (Umbhauer *et al.*,
1992), Xk-endo-B (a type-1 keratin) (LaFlamme and Dawid, 1990),
keratin sulfate proteoglycan (Smith and Watt, 1985), and Xfibrillin
(P. Skoglund, personal communication).

1. Transcription Factors of the Organizer

a. Goosecoid. Goosecoid was originally found in a screen for homeobox genes specifically expressed in the dorsal lip (Blumberg *et al.*,
1991). It derives its name from two *Drosophila* genes, *gooseberry* and
bicoid, with which it shares considerable homology in the homeodomain. Both of these *Drosophila* genes are also involved in early pattern
formation (Lindsley and Zimm, 1992). *Goosecoid* is induced in animal
caps by activin even in the presence of inhibitors of protein synthesis,
but not by bFGF (Cho *et al.,* 1991a). This indicates that it is part of the
early response to dorsal mesoderm induction. It is also induced by
Xwnt-8, but only in the marginal zone (Steinbeisser *et al.,* 1993). Since
its discovery in *Xenopus,* a *goosecoid* homolog has been found in mouse
(Blum *et al.,* 1992), expressed in the anterior end of the primitive
streak, the equivalent of the *Xenopus* organizer.

During gastrulation, *goosecoid* is expressed predominantly in the
migratory deep cells of the dorsal lip (Cho *et al.,* 1991a). These cells do
not intercalate but instead crawl along the blastocoel roof, and, as the
leading edge of the mesoderm (Fig. 1B,C), eventually give rise to head
mesoderm and endoderm (Shih and Keller, 1992). The expression of
goosecoid message in the dorsal lip diminishes as gastrulation is completed (Cho *et al.,* 1991a), which suggests that it has a transient function in development.

Injection of *goosecoid* RNA into ventral blastomeres of the zygote
can lead to partial and sometimes full secondary axis formation (Cho
et al., 1991a; Steinbeisser *et al.,* 1993), showing that the *goosecoid* gene
product may function early in the induction hierarchy of the organizer;
yet animal caps from embryos injected with *goosecoid* do not turn on
neural genes (Niehrs *et al.,* 1993). It has recently been demonstrated
that ectopic expression of *goosecoid* leads to increased involution and
the anterior migration of the misexpressing and neighboring cells

(Niehrs *et al.*, 1993), indicating that *goosecoid* is involved in certain aspects of gastrulation movements. When *goosecoid* is misexpressed in the marginal zone, it changes the behavior of cells that would normally converge and extend to that of the migratory deep cells (Niehrs *et al.*, 1993). This suggests that *goosecoid* might regulate the expression of cell surface adhesion molecules that are involved in the migration process.

b. Xcaudal-1 and 2, Xlabial-1 and 2. *Xenopus* homeobox genes, *Xcaudal-1* and *-2* and *Xenopus labial-1* were cloned in the same screen in which *goosecoid* was found (Blumberg *et al.*, 1991), and *Xlabial-1* and *-2* were isolated in a search for *labial* genes (Sive and Cheng, 1991). In flies, *caudal* is a posteriorly expressed maternal gene needed for the development of posterior structures, and *labial* is a member of the *Antennapedia* complex necessary for the formation of head structures (Lindsley and Zimm, 1992). These *Xenopus* homologs are expressed at stage 10.5, the beginning of gastrulation, until stage 20, neural tube closure. Both *Xcaudal* genes are specifically expressed in the blastopore lip whereas *Xlabial* genes are only enriched in this region and are expressed in other parts of the gastrula embryo (Blumberg *et al.*, 1991; Sive and Cheng, 1991). The function of these genes has not yet been explored by induction or misexpression studies.

c. Xnot. *Xnot* is another gene encoding a transcription factor expressed in the organizer (von Dassow *et al.*, 1993). *Xnot* contains a homeodomain that represents a novel class of homeobox gene. Originally this gene turns on broadly in the embryo at the mid blastula transition (MBT). It becomes restricted to the organizer region and then to the notochord, and the floor plate. Both bFGF and activin induce *Xnot* expression in animal caps. In the absence of protein synthesis, *Xnot* never becomes restricted to the organizer region, which suggests that another protein, very likely BMP-4, inhibits its ventral expression (von Dassow *et al.*, 1993).

d. Xlim-1. *Xlim-1* is a homeodomain gene (Taira *et al.*, 1992) of the lim family, indicating homology to the *Caenorhabditis elegans* genes *lin-11* (Freyd *et al.*, 1990) and *mec-3* (Way and Chalfie, 1988), and to the rat gene *isl-1* (Karlsson *et al.*, 1990), all three implicated in cell-lineage determination. Lim homeodomain proteins contain, in addition to a homeobox, two tandemly repeated cysteine–histidine motifs. *Xlim-1* turns on strongly in the early gastrula, predominantly in the dorsal marginal zone. It is also an immediate response to dorsal meso-dermal induction because it is induced by activin, even in the absence of protein synthesis, while it is not induced by bFGF. Interestingly, retinoic acid (RA) also induces *Xlim-1* in animal caps and enhances *Xlim-1* expression in these caps in response to activin (Taira *et al.*, 1992). RA has been shown to have a morphogenetic and patterning effect in anterior–posterior axis signaling, inhibiting an-terior development in a dose-dependent manner (Durston *et al.*, 1989; Ruiz i Altaba and Jessell, 1991; Sive *et al.*, 1990). The induction of *Xlim-1* by RA shows that RA may also be involved in regulating early dorsal mesodermal genes, which first turn on at the portion of the blas-topore fated to become the posterior pole of the embryo. Interestingly, Hensen's node (the chicken and mammalian equivalent of the organ-izer) is a place where RA is synthesized (Hogan *et al.*, 1992; Chen *et al.*, 1992). *Xlim-1* expression is reduced by late neurula stages, but it comes back again during later embryogenesis in the brain, where it is expressed at high levels into adulthood (Taira *et al.*, 1992).

    e. Xforkhead. Xenopus forkhead (*XFD-1* or *XFKH*1) (Dirksen and Jamrich, 1992; Knochel *et al.*, 1992) and *Pintallavis* (Ruiz i Altaba and Jessell, 1992) encode transcription factors homologous to mammalian hepatocyte nuclear factor-3α (HNF-3α) (Lai *et al.*, 1991) and *Dro-sophila forkhead* protein (Weigel *et al.*, 1989). In *Drosophila, forkhead* is a homeotic gene which is expressed in the terminal regions of the embryo, and which may repress thoracic development in the head (Jur-gens and Weigel, 1988). *XFD-1* and *Pintallavis* are only very slightly different from each other at the nucleotide sequence level and probably therefore represent two copies of the same gene in the pseudotetraploid *Xenopus*. The expression patterns match almost exactly. We will there-fore refer to this gene(s) as *Xforkhead*.

    Xforkhead first turns on after the mid blastula transition in the dor-sal marginal zone. By mid tailbud stages, the levels are low. In animal caps it is induced by activin but not by bFGF, even in the presence of cyclohexamide, which suggests that *Xforkhead* is also an immediate early response to dorsal mesodermal induction. During gastrulation, *Xforkhead* expression becomes restricted to the organizer and espe-cially the involuting notochord. It is also induced in the central strip of the neural plate, a region destined to give rise to the floor plate. This expression is dependent on mesodermal involution, as it does not turn on in the ectoderm of exogastrula.

    Injection of synthetic *Xforkhead* mRNA into embryos results in the suppression of anterior structures of the CNS and the expansion of posterior parts, that is, the hindbrain and spinal cord (Ruiz i Altaba and Jessell, 1992). This is analogous to the proposed function of *fork-head* in flies, which is to promote terminal and repress segmental development in the *Drosophila* embryo. This consequence of ectopic *Xforkhead* may be reflective of its normal domain of expression, which does not extend more anterior than the midbrain. In addition, *Xfork-head*-injected embryos show a decreased number of cell types of the dorsal spinal cord, for example, Rohon-Beard neurons (Ruiz i Altaba and Jessell, 1992). This suggests that *Xforkhead* may be responsible for generating a patterning element localized to the notochord and/or the floor plate which antagonizes a dorsalizing neural influence emanating from the roof plate. Animal caps from *Xforkhead*-injected embryos do not turn on neural markers (Ruiz i Altaba and Jessell, 1992), which suggests that this gene is not directly implicated in neural induction.

f. Xenopus brachyury. *Xbra* is a *Xenopus* homolog to the *brachyury* gene (Smith *et al.*, 1991) already described in mice (Herrmann *et al.*, 1990). *Xbra* is expressed at the beginning of gastrulation in a wide band around the embryo coincident with the presumptive mesoderm (Smith *et al.*, 1991). As gastrulation proceeds, *Xbra* becomes localized around the blastopore and is concentrated in the involuting cells of the presumptive notochord. Brachyury is an intracellular gene product with regions of similarity to NFκB, rel, dorsal (Perrin, 1992) and the proposed DNA binding domain of optomotor blind (omb) (Pflugfelder *et al.*, 1992), which suggests that it is a transcription factor that may be capable of regulating gene expression by translocating to the cytoplasm or the nucleus.

Unlike *goosecoid, Xlim-1,* and *Xforkhead,* and like *Xnot, Xbra* is induced by both activin and bFGF. Like all of them, however, the message is turned on in response to these growth factors even in the absence of protein synthesis (Smith *et al.*, 1991). In mouse, the *brachyury* gene is coded for by the t-locus, and mutants in this gene lack posterior mesoderm, including the notochord (Wilkinson *et al.*, 1990). Zebrafish *brachyury* mutants, called notail (*ntl*), also lack a notochord (Schulte *et al.*, 1992; B. Melby and C. B. Kimmel, unpublished observations). Overexpression of *Xbra* in *Xenopus* embryos drives the formation of ectopic posterior mesoderm in the misexpressing cells (Cunliffe and Smith, 1992). These results, like those in mouse chimeras (Rashbass *et al.*, 1991) and zebrafish mosaics (B. Melby and C. B. Kimmel, unpublished observations), suggest that *brachyury* function is autonomous: only the cells that express this gene form dorsal posterior mesoderm. Isolated animal caps from embryos injected with *Xbra* express various muscle markers, such as muscle-specific actin (Cunliffe and Smith, 1992). *Xbra* expression in animal caps also leads to the expression of *Xsnail* and *Xhox3* genes (Cunliffe and Smith, 1992), two downstream transcription factors. Interestingly, *Drosophila snail,* which we discuss again in the next section, is essential for mesoderm formation because it represses the expression of neural genes in the ventral side of the embryo (Kosman *et al.*, 1991).

   g. Xtwist. Xtwist codes for a helix–loop–helix transcription factor that begins to be expressed in the involuted nonsomitic mesoderm, that is, the notochord and lateral plate (Hopwood *et al.*, 1989). *Xtwist* is a *Xenopus* homolog of the *Drosophila* gene *twist* (Thisse *et al.*, 1988), which is also expressed predominantly in the mesoderm and seems to be essential for mesodermal differentiation. Vegetal cells are able to induce *Xtwist* expression in animal cap recombinants, though direct activation with bFGF or activin has not been shown. The first appearance of *Xtwist* transcripts (Hopwood *et al.*, 1989) follows the initial involution of some mesodermal cells at gastrulation (Keller, 1985). This suggests that *Xtwist* is involved in events somewhat later than the primary response to mesodermal induction. *Xtwist* is also induced in the lateral folds of the anterior neural plate during neurulation. These latter *Xtwist*-expressing cells will go on to form the anterior neural crest, giving rise to mesoderm-like derivatives, that is, the connective and skeletal tissues of the head (Hopwood *et al.*, 1989).
   In *Drosophila, twist* expression is upregulated by the dorsal protein (Kosman *et al.*, 1991), which is predominantly expressed in the nuclei of ventral cells of the blastula. *Twist* also has a role in neural development in *Drosophila* since it is a regulator of the proneural *achaete-scute* complex (*AS-C*) of genes in the lateroventral neurogenic region of the embryo. *AS-C* genes are not expressed in the most ventral region of the *Drosphila* embryo because *snail,* a zinc finger type of transcription factor, which is also induced by *dorsal,* is a strong repressor of *AS-C* gene expression (Kosman *et al.*, 1991). In *twist*-defective mutants, the level of *snail* expression also decreases (Ip *et al.*, 1992); therefore the neuroectodermal genes normally repressed by *snail* become derepressed. Thus the effect of *twist* on the *AS-C* genes seems to be mediated by *snail.* It is possible that *Xtwist* has a similar function in vertebrate development, preventing neurogenesis in the prospective mesodermal region of the embryo.

2. Secreted Proteins of the Organizer

a. Noggin. Ultraviolet light directed on the vegetal pole of the early zygote inhibits the microtubule-dependent rotation of the fertilized egg cortex with respect to the cytoplasmic core (Gerhart *et al.*, 1989). Because this rotation is necessary to specify the dorsal–ventral embryonic axis, embryos that develop after UV treatment consist of an endodermal mass surrounded by ciliated epithelia. *Noggin* was isolated, along with *Xwnt-8,* by the ability of its mRNA to rescue dorsal structures in UV-ventralized embryos in an expression screening strategy (Smith and Harland, 1991, 1992; Sokol *et al.*, 1991). From this, it seems that *noggin* is able to make a dorsalizing or Nieuwkoop center or a Spemann organizer, showing that it has the ability to act as an inducer and perhaps a patterner of nervous tissue. *Noggin* is a maternally expressed gene that codes for a novel secreted glycosylated protein (Smith and Harland, 1992). *Xwnt-8* has been shown to mimic this early effect of the *noggin* gene (Sokol *et al.*, 1991; Smith and Harland, 1991) but it is not expressed maternally, suggesting that another related *wnt* protein, not yet identified, may have this function naturally.

Noggin mRNA level increases dramatically at the MBT when it starts to become localized in the dorsal marginal zone, particularly the organizer (Smith and Harland, 1992). As the embryo develops, *noggin* expression continues to be restricted to the notochord, the head mesoderm, and the roof plate of the neural tube (Smith and Harland, 1992). When the noggin protein is applied to animal caps, it does not induce dorsal mesoderm, but when it is applied to ventral marginal explants from gastrula stage embryos, it can lead to the expression of muscle from this tissue (Smith *et al.*, 1993). This suggests that noggin is mimicking the role of the organizer in its ability to recruit and organize mesodermal tissue to become dorsal muscle. Can it mimic another aspect of the Spemann organizer, the ability to induce neural tissue? In recent very exciting experiments, the application of noggin protein to animal caps induced NCAM expression and anterior neural structures directly without inducing dorsal mesoderm. Noggin induces neural tissue in gastrula stages, when normal neural induction occurs. This suggests that noggin may also have a role in neural induction (Lamb *et al.*, 1993) (Fig. 2D).

b. Activin and Follistatin. If animal caps are cultured in saline, the tissue expresses only epidermal genes. If, however, the same cells are disaggregated and cultured in the same medium, many of the isolated animal cap cells express neuronal markers (Grunz and Tacke, 1989; Godsave and Slack, 1989) (Fig. 2E). This suggests that these cells may

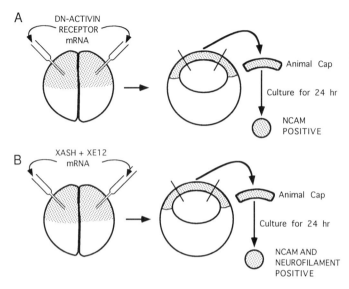

FIG. 3. Genes in the animal cap that turn on neural pathways. In (A) a dominant negative (DN) form of the activin receptor was injected into embryos. Animal cap tissue from such embryos no longer responded to activin by turning on mesodermal genes. Yet, even in the absence of inducing factors, these injected caps turned on NCAM, suggesting that the complete lack of an activin signal may push this tissue toward a neural fate (Hemmati and Melton, 1992). In (B), embryos were injected with mRNAs encoding an *XASH* gene and *XE12*. Animal caps from these embryos expressed both NCAM and neurofilament genes, which suggests that *XASH* acts upstream of neural differentiation (B. Ferreiro, C. R. Kintner, and W. A. Harris, unpublished data).

have a natural tendency to become neural and that it is the interaction among the cells that inhibits them from accepting this fate. One possible source of this inhibition is activin, the inducer of dorsal mesoderm. This suggestion comes from experiments in which mRNA encoding a dominant negative activin receptor was injected into embryos (Hemmati and Melton, 1992). In animal caps from these embryos, there were no mesoderm genes (*Xbra* and muscle *actin*) turned on by activin, yet, surprisingly, these caps turned on NCAM, even when they were not challenged with activin (Fig. 3A). Since animal cap cells produce low levels of activin, this result is consistent with the idea that low levels of activin lead to epidermal development, while the complete absence of activin might lead to neural development. If this is an important *in vivo* aspect of neurogenesis, then there must be some way that the embryo keeps activin away from the prospective neural plate.

Follistatin is an extracellular protein that is one of a growing number of molecules that bind activin (Nakamura *et al.*, 1990). Mammalian P19 cells can be transformed to neurons by retinoic acid, and this transformation can be inhibited by activin (Hashimoto *et al.*, 1990). Follistatin blocks this inhibition (Hashimoto *et al.*, 1992).

In *Xenopus*, there are several *follistatin* transcripts; they are expressed maternally and their concentration increases dramatically at the mid blastula transition (Tashiro *et al.*, 1991). Information about the spatial localization of *follistatin* transcripts at early gastrula stages is limited, but if they were restricted to the most dorsal aspect of the DIMZ, they could limit the field of activin action. By mid to late gastrula stages *follistatin* does become restricted to the chordamesoderm (Hemmati and Melton, 1994), which is consistent with follistatin having a role in neural induction. In support of this idea, recent experiments show that when certain *follistatin* messages are overexpressed in embryos, this leads to massive neural induction (Hemmati and Melton, 1994).

c. FGF3/int-2. *FGF3,* otherwise known as *int-2,* is first expressed around the blastopore lip (Tannahill *et al.*, 1992). It is induced by bFGF and activin. Its role in the organizer is unclear because it turns off as the mesoderm involutes. *FGF3/int-2* becomes restricted to the posterior third of the neural plate and mesoderm and to subregions of the anterior third of the neuroectoderm, suggesting a role in the formation of axial specific structures. Xenoplastic transplantations of animal cap from *Xenopus* into the axolotl neural plate can lead to the expression of FGF3/*int-2* in the transplanted tissue as it is induced to become neural, showing that FGF3/*int-2* is expressed in response to two different inductive events, mesodermal and neural. Knockout of the FGF3/*int-2* gene, either by culturing chick embryo fragments in the presence of antisense oligonucleotides (Represa *et al.*, 1991) or by homologous recombination in mice (Mansour *et al.*, 1993), leads to defects in the inner ear. Thus, FGF3/*int-2* has a crucial role in the development of specific structures in the regions where it is normally expressed.

C. PERSPECTIVE

Many of the genes that have been discovered to be expressed in the organizer and notochord, especially those isolated by homology, encode transcription factors. This may not seem too unusual given the bias of the screens. What seems remarkable is that a good number of these genes are turned on directly by mesodermal inducers and in that sense are acting as parallel members of an induction hierarchy. Not only do we have to continue to elucidate the different embryological roles of these different transcription factors, but we also must find their downstream targets. Do they work cooperatively on some of these targets and individually upon others? What do the target genes do? Are any of them secreted growth factors or inducers that have particular roles in neural induction? Do they, for instance, regulate *noggin* or *follistatin* expression?

The discovery of *noggin*, using an expression rescue rather than an homology screen, and the work with *follistatin* and other genes not yet described in *Drosophila* show that an understanding of vertebrate molecular neurogenesis clearly involves more than known *Drosophila* homologs. Noggin is a candidate for an anterior neural induction factor of the kind suggested by certain embryological experiments (see later discussion). These studies also show that there must be additional signals involved in neural induction. Follistatin may be one such molecule, and there are possibly others which have yet to be identified.

III. Inductive Signaling and Neural Origins: Three Pathways

In order to assess the roles that particular gene products might play in neural induction, we review some of the experimental embryological work that constitutes the basis of our knowledge of this process.

A. VERTICAL SIGNALING

Spemann and Mangold's original experiments (Spemann and Mangold, 1924) did not distinguish between two distinct routes of neural induction, one through the plane of the embryonic epithelium, especially across the boundary between the DIMZ and the DNIMZ (dorsal noninvoluting marginal zone), and the other a vertical signal from the involuting chordomesoderm to the overlying ectoderm. Mangold discovered, using einsteck experiments (i.e., transplant of a piece of inducing tissue into the blastocoel of an early gastrula, as in the legend to Fig. 7A) that neural induction was possible by vertical signaling alone (Mangold, 1933), and Holtfretter observed that the ectoderm of exogastrula in which there is no mesodermal involution and no possibility of vertical signaling showed little neural cytoarchitecture (Holtfreter, 1933). So it was thought for many years that vertical signaling was indeed the only route for natural neural induction. Indeed, the search for the vertical signal that could pass through a filter was the source of a great deal of biochemical work over the past few decades (see reviews in Hamburger, 1988; Nakamura and Toivonen, 1978).

B. PLANAR SIGNALING

Recent experiments, however, clearly show that in *Xenopus,* the inducing signal(s) from the Spemann organizer can act in the plane of the ectodermal-mesodermal boundary as in exogastrula (Kintner and Melton, 1987) or Keller sandwiches (Keller and Danilchik, 1988) (Fig. 4A). This route of signaling is sufficient to turn on neural markers such as NCAM (Kintner and Melton, 1987; Keller *et al.,* 1992b) and cause convergent extension of the neural plate (Keller and Danilchik, 1988; Keller *et al.,* 1992b) (Fig. 4A).

Either vertical (Sharpe and Gurdon, 1990) or planar (Doniach *et al.,* 1992; Ruiz i Altaba, 1992) signaling alone can lead to the correct anteroposterior patterns of expression of many neural-specific homeobox genes and other anterior–posterior (A–P) pattern genes.

C. COMBINED SIGNALING

Neither Keller sandwiches (planar signaling) nor recombinants (vertical signaling) induce anteroventral neural structures such as eyes, yet when one allows both planar and vertical signals as in a "Keller sandwich recombinant," the resultant embryoids not only have more NCAM expression compared with normal levels, but they also have eyes (Dixon and Kintner, 1989; Ruiz i Altaba, 1992) (Fig. 4E) This suggests that to get normal patterning of a full set of neural structures, both vertical and planar signaling, acting synergistically, may be necessary (Dixon and Kintner, 1989). In accord with this, older studies in *Triturus* show that pieces of chordamesoderm induce ectoderm to express neural fates characteristic of a more anterior position than that of the inducing piece (Horst, 1948), as reviewed by Slack and Tannahill (1992). This suggests that within the neural plate there may be additional planar signals that are needed to induce the neural structures that usually form at specific locations. Each signaling component may in addition have a specific role. Vertical signaling from the notochord appears to be critical for some aspects of dorso-ventral (D-V) patterning of the neural tube because animals without notochords lack ventral structures in the spinal cord (Clarke *et al.,* 1991).

The notoplate is formed by the cell population of the DNIMZ closest to the organizer which extends within the ectodermal sheet along the

46   BEATRIZ FERREIRO AND WILLIAM A. HARRIS

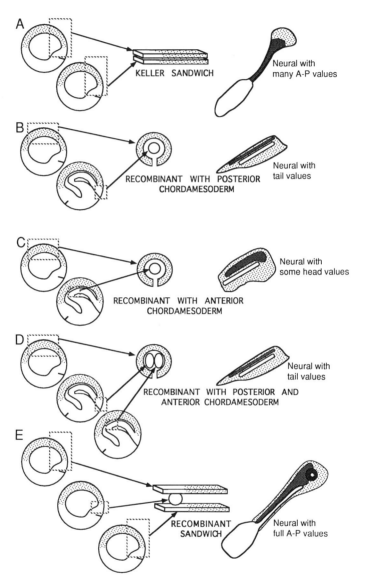

FIG. 4. Sandwiches and recombinants. A Keller sandwich (A) is made by combining the DIMZ and the DNIMZ and adjacent animal tissue from two early gastrula. The result is a piece of tissue in which the mesoderm (DIMZ) extends in one direction and the ectoderm extends in the other. Sandwiches like this therefore allow only planar signaling, yet the ectodermal tissue turns on neural genes and expresses many regionally distinct A–P markers, such as homeobox genes, in a correct order with respect to each other (Keller *et al.*, 1992b; Doniach *et al.*, 1992). Classic recombinants are shown in (B) and

midline of the prospective neural plate during gastrulation and neu-rulation (Keller, 1985). The notoplate will transform into the floor plate of the neural tube in a process that requires vertical signals from the notochord. As demonstrated in the chick by grafting the notochord to ectopic places of the neural tube and removing it, the floor plate has a key role in the D-V patterning of the neural tube, as reviewed by Jessell and Dodd (1992). Planar signaling, on the other hand, may be crucial for the normal convergent extension of the neural plate since vertical signaling in Keller sandwiches seems to inhibit rather than accentuate this process (Keller *et al.*, 1992b).

D. HOMEOGENETIC INDUCTION

In addition to inducing signals that pass from the mesoderm to the prospective neural ectoderm, neuralizing signals can also pass within the plane of the neural plate. This neural-to-neural pathway is called homeogenetic induction, and was first described by Mangold and Spemann (1927). The capacity of the neural tissue to induce competent ectoderm (assayed by antibodies that recognize NCAM and 2G9 neural antigen) was restudied recently in transplants and recombinant experiments (Servetnick and Grainger, 1991). When competent gastrula ectoderm (stage 10) labeled with fluorescein–dextran was transplanted to head ectoderm or ventral ectoderm of an early neural plate embryo (stage 14), neural tissue was formed only when the transplant was located close to the host neural plate. To dissect out the source of the neuralizing signal, the ectoderm was recombined separately with either anterior neural plate or dorsolateral head mesoderm, and it was found that the neuralizing signal comes primarily from the neural tissue (Servetnick and Grainger, 1991).

(C). In (B), cap tissue is combined with posteriorly fated mesoderm at the dorsal lip of a midlate gastrula, and the result is an embryoid consisting of organized neural (spinal cord) and mesodermal components with posterior or tail characteristics, while in (C) when anterior mesoderm from the same stage embryo is wrapped in cap tissue, the result is an embryoid with anterior or head structures, including brain elements. In (D), we see the dominant posteriorizing effect in recombinant induction, for in this case anterior and posterior mesoderm together, when wrapped in cap tissue, induce posterior nervous system and tail structures. One argument that both planar and vertical signaling is used in normal development is outlined in the recombinant sandwich experiment shown in (E). In this experiment, a piece of inducer tissue is used as a filling between the two layers of the Keller sandwich. The result is a more complete embryoid with ventral forebrain structures such as eyes and a stronger induction of neural markers (Dixon and Kintner, 1989).

In Keller sandwiches and exogastrula, the signal that proceeds up the axis of the induced neural plate may have a homeogenetic character, and it is unclear whether the homeogenetic, planar, or even vertical signaling work by distinct molecular pathways.

E. PERSPECTIVE

The post-Spemann einsteck and recombinant studies, including the biochemical studies (Saxen and Toivonen, 1962), all pointed to a vertical signaling mechanism. Now, starting from the work of Kintner and Melton (1987) and Keller and Danilchik (1988), it is clear that neural induction in real embryos probably consists of a combination of planar and vertical signaling pathways. Moreover, the fact that a great deal of anterior–posterior pattern can arise from a planar induction signal(s) undercuts the long-held viewpoint that regionalization in the neural plate arises primarily from local signals in a patterned chordomesoderm. Now it appears that the patterning and induction of the neural plate may arise during gastrulation, and that interactions both between germ layers and homeogenetically within the tissues help to coordinate positional identity. The three neural induction pathways we have described may well be partially redundant mechanisms that the vertebrate embryo has enlisted to ensure the formation of a perfectly organized nervous system.

IV. Competence to Respond to Induction

Whatever the neural-inducing pathways and molecules are, they must act on competent ectodermal tissue, that is, they must be capable of responding to the signals by turning on neural pathways of development. At a molecular level, we can understand that at least the receptors and the transduction apparatuses for the inducing signals must be present, as well as the nuclear machinery for interpreting the cascade and turning on neural genes. In addition, factors that may inhibit this process must be turned off or neutralized. In this section, we review the embryological and molecular studies of competence.

A. TEMPORAL AND SPATIAL ASPECTS OF COMPETENCE

1. Temporal Competence

The temporal competence of the ectoderm to respond to neural induction decreases during gastrulation and is lost by early neurula stages. This was demonstrated by (1) transplanting ectoderm over inducing chordomesoderm (Lehmann, 1926); (2) transplanting the organizer to the ventral ectodermal region (Machemer, 1932; Schechtman, 1938); and (3) combining uninduced ectoderm with Hensen's node (Kintner and Dodd, 1991). When three different markers (*XIF3*, *XIF6*, and *XlHbox6*) expressed in different A–P regions of the neural plate were examined (Sharpe and Gurdon, 1990), it was found in recombinant experiments that the competence of the ectoderm to respond by expressing each of these genes decreased at about the same stage (stages 13–14). These three neural markers were also used to show that the duration of contact between the mesoderm and ectoderm is critical in the neural induction process. There is a dramatic increase in the expression of these neural markers when the contact is maintained for 8.5 hr which rises to a maximal effect at about 11 hr. The different markers all behaved about the same, indicating that the anterior versus posterior specification of the neural plate is not simply a matter of the time over which induction occurs (Sharpe and Gurdon, 1990).

It is intriguing that the point at which the ectoderm loses its competence to begin responding, stage 14, is considerably earlier than the end of the period of time that is necessary for the ectoderm to be in contact with the mesoderm to express these neural markers in a stable fashion. This suggests that the mesoderm may be necessary not only to induce neural tissue but also to maintain its induced state until it begins to differentiate.

2. Spatial Competence

If a tissue responds differently to the same inducing signal, depending on its position, this indicates differential spatial competence and is strongly suggestive of prepattern. Such a prepattern has been found in the animal cap tissue with respect to its ability to respond to activin (Sokol and Melton, 1991). For example, ventral ectoderm is less likely than dorsal to form notochord (Bolce *et al.*, 1992; Sokol and Melton, 1991) and to induce neural tissue in response to a given dose of activin (Sokol and Melton, 1991) (Fig. 5A). It is possible that BMP-4, which antagonizes the dorsal effects of activin, or noggin or a wnt-8-like molecule which enhances activin induction, may be the cause of this prepattern. This prepattern of activin sensitivity may be an important part of normal development, especially since small changes in the dosage of activin can dramatically affect the fate of ectodermal cells (Green *et al.*, 1992).

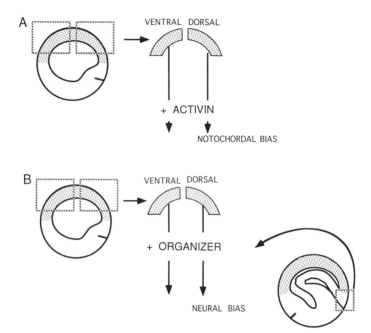

FIG. 5. Prepattern in the animal cap. In these two experiments, animal caps were separated along their dorsoventral boundaries. In (A), when both halves were treated with equal doses of activin, the dorsal half was more likely to form distinct notochord, indicating a predisposition of this tissue to form dorsal mesoderm (Sokol and Melton, 1991). In (B), it is shown that dorsal cap tissue also has a predisposition to become neural. In this experiment, the two halves were individually combined with the organizer, and the dorsal half turned on neural genes more strongly than the ventral half (Sharpe et al., 1987).

The dorsal–ventral prepattern of the animal ectoderm, as revealed by exposure to mesodermal induction with activin, may also influence its competence to respond to neural induction. This was shown in a recombinant experiment in which stage 11 dorsal mesoderm induced NCAM and *XlHbox6* expression preferentially in dorsal rather than ventral animal cap tissue from stage 10 early gastrula (Sharpe *et al.*, 1987). This shows that the tissue fated to become neural is somehow predisposed to respond to neural induction (Fig. 5B). Whether this prepattern in the ectoderm is significant in normal development is again difficult to determine, especially since ventral ectoderm can give rise to a fully ordered neural plate when it is induced by a transplanted organizer (Spemann and Mangold, 1924).

B. Factors Affecting the Competence of Ectoderm to Respond to Neural Induction

1. Protein Kinase C

PKC signal transduction pathways are activated during neural induction (Otte *et al.*, 1988, 1989, 1991). PKC comes in at least two isoforms, α and β (Chen *et al.*, 1989), and the α form is expressed at much higher levels in the dorsal ectoderm of the early gastrula (Otte and Moon, 1992). While injection of the mRNA encoding these two isoforms into the zygote does not by itself induce animal caps to turn on neural genes, it does make overexpressing ectoderm more responsive to neural induction in recombinants with dorsal mesoderm, expressing higher amounts of *NCAM, XIF3,* and *XIF6* than controls. Moreover, when the α form is overexpressed, then both ventral and dorsal ectoderm respond equally well to neural induction (Otte and Moon, 1992), which suggests that the distribution of this gene product could be biasing the competence of animal cap tissue to neural induction, and may in part account for the spatial prepattern of competence.

2. Xotch

Xotch, the *Xenopus* homolog of *Drosophila Notch,* encodes a transmembrane receptor (Coffman *et al.*, 1990). It is expressed maternally, and continues to be expressed through development, though later it becomes restricted to portions of the embryo where cell fates are being decided, such as proliferative zones in the neuroepithelium (Coffman *et al.*, 1990, 1993). In the gastrula and early neurula, *Xotch* is expressed predominantly in the neural plate and neighboring dorsal ectoderm, and in the mesoderm. As tissues decide their fates, *Xotch* seems to diminish, such that in midneurula, there is little or no expression in the notochord or lateral epidermis, which have by this time committed to these fates (Coffman *et al.*, 1993). Thus *Xotch* could be a factor in establishing temporal competence.

Studies in *Drosophila* and nematodes have shown that *Notch* and *Notch*-like genes function in equivalence groups where cell–cell interactions limit the number of cells that follow the primary fate (see reviews in Artavanis and Simpson, 1991; Campos and Knust, 1990). These molecules appear to act as receptors for an inhibitory signal from cells that follow the primary fate. As a result, cells receiving the *Notch*-like signal are inhibited from following this fate, and instead choose a secondary fate. In *Xenopus,* the overexpression of an extracellular deletion mutant of *Xotch,* a dominant activated form of the

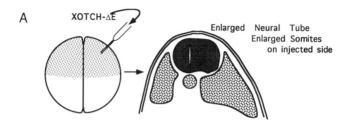

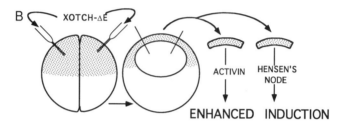

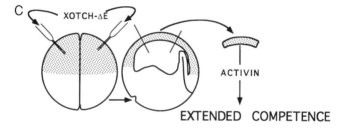

FIG. 6. *Xotch* and competence. (A) When embryos at the two-cell stage were injected in one blasomere with an mRNA that codes for a *Xotch* molecule with an extracellular deletion (*Xotch*-DE) that acts as dominant activated form of the *Xotch* molecule, then the resultant embryos at the neural tube stage showed an enlarged neural tube and somites on the injected side. One explanation for this effect is that the ability of embryonic tissue to respond to mesodermal and neural induction is enhanced in these embryos. The experiment shown in (B) demonstrated that cap tissue from embryos injected with *Xotch*-DE mRNA showed enhanced mesodermal and neural induction, in terms of the expression of tissue-specific markers such as actin (mesoderm) and NCAM (neural). In (C), it is shown that animal cap tissue taken from stage 11.5 embryos still responded to activin. Similar tissue from uninjected or control injected embryos no longer responded to activin (not shown) (Coffman *et al.*, 1993).

gene, alters cell fates in the gastrula, resulting in the overproduction of neural and muscle tissue and the underproduction of other tissues (Coffman *et al.*, 1993) (Fig. 6A). Animal cap assays show that tissue expressing dominant activated *Xotch* has enhanced competence to respond to both mesodermal and neural induction, and this appears to be the result of extended competence (Coffman *et al.*, 1993) (Fig. 6B,C). Thus, the normal activation of Xotch may keep cells in an undetermined and therefore competent state. In neurogenesis, Xotch may function to limit the size of the induced neural plate by its activity at the borders. Subsequently in the tadpole, when Xotch is expressed in the proliferative zones of the neuroepithelium, it may act in the continued determination of cells in the CNS.

C. PERSPECTIVE

The competence to respond to neural induction is biased spatially and extends only through a window of development around gastrulation and neurulation. The molecular correlates of this competence have only begun to be studied. There are likely to be other inhibitors of differentiation than Xotch. Xotch appears to be a receptor, but its ligands and its downstream machinery in vertebrates are unknown. They probably involve a Delta-like molecule (one of the Notch ligands in *Drosophila* (Fehon *et al.*, 1990), and (anti-)transcription factors of the E(split) (Campos and Knust, 1990), or id families (Benezra *et al.*, 1990) that inhibit differentiation.

V. Origin of Anterior–Posterior Axis Formation

During gastrulation, the chordamesoderm becomes patterned in the anterior–posterior axis, and this pattern can be reflected in the ability of subdivisions to act as regionally specific neural inducers in einsteck or recombinant experiments (Mangold, 1933; Hemmati *et al.*, 1990). Yet we have seen that planar induction can also allow the generation of an anterior–posterior axis on the neural plate. In the animal, this axis of the neural plate probably comes from a combination of these two routes, and now the question arises about how this axis emerges in the first place, and what its molecular nature is. By homology with *Drosophila* genes of the homeotic complex, a series of *Xenopus Hox* genes have been found, and there is evidence that in frogs there is a similar strategy to their function in setting up the anterior-posterior identity in the mesoderm and the neural plate.

A. PREPATTERN AND PATTERN LABILITY

It is unlikely that an anterior-to-posterior prepattern exists in the fated dorsal invaginating mesoderm before it has invaginated. Because of intercalation, the DIMZ cells are randomized during gastrulation so that it is impossible to map the fate of anterior versus posterior notochord (Keller *et al.*, 1989; Hama *et al.*, 1985; Slack and Tannahill, 1992). Therefore, it is likely that A–P pattern arises during gastrulation.

Is there any prepattern in the ectoderm which is fated to become neural tissue in its competence to express anterior versus posterior neural genes? Since the presumptive neural plate at the early gastrula stage is also rather a thin strip of cells just above the involuting marginal zone (Fig. 1B), and it is also induced to go through extensive convergence toward the midline and intercalation (Keller *et al.*, 1992a), reproducible fates would be again impossible to assign. It is therefore likely, as with the mesoderm, that A–P positional cues are not assigned or prepatterned onto the presumptive neuroectoderm at the blastula or early gastrula stage of development, and that A–P patterning in the embryo may be determined at later, gastrula or neurula stages. In agreement with this idea are the results of a recent study (Saha and Grainger, 1992) of the regional responsiveness of the prospective neuroectoderm and inductive capacities of the mesoderm. Using an anterior (*opsin*), a posterior (*XlHbox6*), and general (*tubulin*) neural markers, they showed that A–P properties are only fixed at early neurula stages. During gastrulation the dorsal mesoderm as well as the prospective neuroectoderm activated all three markers along the entire A–P axis as assayed using ectoderm and mesoderm recombinants and explants, respectively (Saha and Grainger, 1992). At neural plate stages, however, the inducing capacities of dorsal mesoderm as well as the regional character of the neural plate become restricted, such that in each recombinant or explant only the regionally appropriate markers are expressed.

B. INDUCTION OF PATTERN

The induction of the anterior–posterior axis of the neural tube has been examined by various manipulations of the dorsal lip. The first experiments done by Spemann and Mangold (1924) showed that when the dorsal lip was transplanted to the ventral marginal zone, an entire secondary embryo could be induced, while the transplanted tissue only gave rise to the chordamesoderm. These embryos thus had all A–P values in the secondary neural tube. Other experiments in *Cynops,* in which dorsal lip tissue is grafted to the ventral side of another embryo, suggest that posterior neural plate can be induced by both early and late dorsal lip tissue, as long as this tissue is taken from the region above the invaginating lip (Fig. 4B). Once early lip tissue has invaginated, however, and has moved to a more anterior position inside the embryo, it is capable of inducing only anterior plate structures in the ectoderm (Okada and Takaya, 1942a,b) (Fig. 4C). It has also been found that early dorsal lip tissue, if it is kept in culture for several hours, loses its potency to induce posterior structures in ectoderm, but can induce anterior ones (Okada and Takaya, 1942a,b). This suggests that the same tissue, depending on context and age, can induce anterior values, posterior values, or both.

The extent of anterior induction appears to depend on the size or quantity of the marginal zone organizer as well, for when various amounts of organizer tissue were combined with UV-ventralized embryos, there was a correlation between the amount of organizer tissue present and the extent of anterior structures (Stewart and Gerhart, 1990). Blocking gastrulation at different times also compromises anterior development (Gerhart *et al.*, 1989). Thus it is possible to imagine that the dorsal lip tissue needs to be in the embryo in sufficient quantity and to be able to migrate sufficiently anteriorly to induce anterior plate.

Recent experiments (S. Zoltewicz and J. Gerhart, personal communication) have shown regional neural inducing differences in the early (stage 10) gastrula organizer in *Xenopus*. When the animal half of the dorsal lip is implanted into the blastocoele as an einsteck experiment, it induces in the host a secondary trunk and tail, whereas the vegetal half induces a secondary head. The same results when these two pieces are transplanted in the ventral side of the embryo. The animal half of the dorsal lip is fated to become notochord and somites, and the vegetal half to become head mesoderm and endoderm.

Because specific areas of the dorsal involuting marginal zone have distinct inductive capacities, it may be that this region is, in fact, composed of two or more organizers, as was suggested by Spemann (Hamburger, 1988). It may also be that there is a sequential induction, with the Niewkoop center inducing an anterior organizer, and the anterior organizer inducing a posterior one.

C. POSTERIOR DOMINANCE IN INDUCTION

While anterior fated mesoderm induces head structures and posterior fated mesoderm produces tail when wrapped in ectoderm, when both were placed together, the recombinant gives rise only to tail structures (Okada and Takaya, 1942a,b) (Fig. 4D). Other experiments with unnatural inducers showed basically the same result, giving rise to the notion that posterior induction is dominant to anterior induction (Saxen and Toivonen, 1961). From the results above, it can be argued that neural induction may take place in several stages. In the first, anterior structures are induced by prechordamesoderm, and in the second, these anterior structures are suppressed by posteriorizing factor(s) that spread throughout the mesoderm and ectoderm, which is why the posterior mesoderm or notochord is capable of inducing only posterior structures. Noggin could, for example, act as the activating inducer of anterior neural structures, for in animal caps to which noggin is applied, mostly anterior neural structures and cement glands are formed (Lamb *et al.*, 1993).

D. ROLE OF RETINOIC ACID AND THE *Hox* GENES

Retinoic acid may be a morphogen in the vertebrate limb and other systems, affecting the positional identity (Brockes, 1989, 1990). The anterior development of *Xenopus* embryos is inhibited by RA applied in a dose-dependent fashion (Durston *et al.*, 1989). One explanation for this is that RA acts by activating *Hox* gene activity differentially along the *Hox* complex, being most effective on the more posterior or distal genes (Papalopulu *et al.*, 1991). The effects of RA on *Xenopus* embryos are reviewed by Papalopulu and Kintner (1992).

The dominant posteriorizing influences seen embryologically might thus come in the form of genes activated in the *Xenopus Hox* gene cluster that turn on during gastrulation and neurulation in successively more posterior regions of the embryo. As in flies, this cluster of genes is arranged along the chromosome in an order which parallels their order of spatial expression (McGinnis and Krumlauf, 1992).

Xhox-3 (homolog to the pair rule *even skippped* gene) (Ruiz i Altaba and Melton, 1989a) is a posteriorly expressed homeobox gene which is maximally expressed in the mesoderm at the early neurula stages. When this gene is misexpressed anteriorly, embryos without heads develop (Ruiz i Altaba and Melton, 1989b), as if the expression of this gene is incompatible with anterior development.

Xhox-1 (homolog of the *Antennapedia* complex genes) is expressed at more anterior levels than *Xhox-3* as a narrow band extending through the CNS, neural crest, and mesoderm; the anterior border of its expression is in the rostral spinal cord. When antibodies to the Xhox-1 protein are injected into the zygote or when a short form of the transcript is injected, the rostral cord appears to be transformed into caudal hindbrain (Wright *et al.*, 1989). These results suggest that these *XHox* genes are involved in posterior specification of the *Xenopus* A–P axis and anterior suppression or transformation.

This suggestion is also supported by experiments with *Xhox-6,* which is yet more posteriorly expressed than *Xhox-1* (Sharpe *et al.,* 1987; Wright *et al.*, 1990). When *Xhox-6* is misexpressed more anteriorly, brain cells expressing this RNA fail to differentiate (Niehrs and De Robertis, 1991) and when *Xhox-6* is misexpressed in animal caps, then these are capable of inducing only tail structures when used in an einsteck experiment (Fig. 6) (Cho *et al.*, 1991b). Thus this gene also seems to posteriorize the A–P axis, inhibiting anterior development and supporting more posterior development. There are several other *XHox* genes that have been found, and although the functions of most of them have not been studied, their expression pattern in the nervous system and mesoderm suggests that they may also contribute to posteriorizing the A–P axis of the embryo.

In *Drosophila,* there are two clusters of homeotic genes of this class, *BX-C* which, like the ones discussed, serve to posteriorize segments, and the *ANT-C* genes, which serve to anteriorize segments. *Xenopus* also has homeobox genes that are expressed in anterior head regions, such as *distalless*. There is a series of *distalless* transcripts, *X-dll1–4,* that are expressed in ventral forebrain, cranial neural crest, cement gland, olfactory placode, and developing eye (Papalopulu and Kintner, 1993). These genes may have a role in anteriorizing the nervous system, such as the *ANT-C* genes in *Drosophila,* although no functional studies have yet been attempted with *Xdll*. *XANF-1* is another homeobox gene that is expressed in the anterior neural fold region (Zaraisky *et al.*, 1992). It is first expressed throughout the entire gastrula ectoderm. When this ectoderm is combined with anterior axial mesoderm, the expression of *XANF-1* is increased, while it is repressed by posterior axial mesoderm (Zaraisky *et al.*, 1992). This suggests that the posterior mesoderm, at least, has repressive as well as inductive activities on the axial patterning of ectoderm.

E. PERSPECTIVE

As discussed earlier, recent experiments using Keller sandwiches argue that vertical signaling of the types mentioned is not the sole determiner of anterior–posterior position in the neural plate, but rather that an inductive signal at the posterior end can spread or be propagated along the A–P axis and induce regionally specific markers (Doniach *et al.*, 1992) as far anteriorly as the forebrain (Papalopulu and Kintner, 1993). In these Keller sandwich explants, it would be useful to investigate whether some of the findings from transplant experiments apply, such as whether anterior neural markers turn on only when there is a sufficient quantity of DIMZ attached, and whether dorsal lip at different ages induces neural plate with A–P values reminiscent of what the same piece would do when transplanted to a ventral marginal zone.

The specification of anterior parts of the neural plate depends upon inductive signals from the anterior chordamesoderm. More posterior chordamesoderm is capable of inducing more posterior neural plate and it is also capable of posteriorizing tissue that is coinduced with an anterior piece of chordamesoderm. The A–P pattern also appears to self-propagate within the plane of the neural plate although it is not clear whether this is the dominant mode of regionalization along this axis *in vivo*. The homeobox genes are turned on or off in a pattern along the A–P axis that reflects their chromosomal order, as in flies and other vertebrates (McGinnis and Krumlauf, 1992), and as in these other animals, more posteriorly expressed genes of this class transform anterior-to-posterior structures or impede anterior development.

This molecular finding is reminiscent of the dominant posteriorizing influence that has been studied embryologically and may therefore offer a molecular explanation. Moreover, the mentioned morphogenetic effects of retinoic acid are likely to work via the action of homeobox-containing genes since many vertebrate homeobox genes are RA responsive (Arcioni *et al.*, 1992; Cho and De Robertis, 1990; Dekker *et al.*, 1992; Sive and Cheng, 1991). Key questions that remain are: (1) How is the original expression of the *Hox* genes controlled at a molecular level in the vertebrate embryo? (2) How do *Hox* genes regulate the identity of cells in that part of the nervous system, that is, what are the downstream targets of these genes in the neural plate?

VI. Response of the Neural Plate to Induction

When competent ectoderm is induced, it switches on a program of neural development. Obviously, since the nervous system is an extremely complex tissue, one cannot hope to review all the genes that are expressed in this tissue in response to neural induction. Rather, we will mention a few that seem to turn on early in response to the inductive signal and that may have key roles in the response of ectoderm to the inducing signals, (see Table 2). Some of the genes whose transcription is activated following induction are the *Hox* genes that we have already described (see also Table 2), and we do not do so again in this section. In any case, it is unclear whether the role of *Hox* genes in axis formation is dependent or independent of the machinery of neural induction.

A. TRANSCRIPTION FACTORS

1. Helix–Loop–Helix

The bHLH transcription factors, XASH-1 and XASH-3 (Ferreiro *et al.*, 1993; Zimmerman *et al.*, 1993), are *Xenopus* homologs of the *Drosophila* achaete-scute complex (AS-C) proteins and the mammalian *achaete-scute* homolog *MASH*-1 (Johnson *et al.*, 1990). In *Drosophila*, this complex is involved in the proneural decision of undifferentiated ectodermal cells; embryos with deletions at the locus have a reduced number of neuroblasts, and embryos with extra copies of *AS-C* show neural hyperplasia (Jimenez and Campos, 1990). *AS-C* encodes four proteins of the neural bHLH family of transcription factors, which are localized in the neurogenic regions (Alonso and Cabrera, 1988; Gonzalez *et al.*, 1989; Murre *et al.*, 1989; Villares and Cabrera, 1987). AS-C proteins bind proneural bHLH daughterless (*da*) protein through the HLH domain, and the complex binds DNA-E2 box sequences through the basic region (Van Doren *et al.*, 1991; Davis *et al.*, 1990). Antineural HLH emc protein forms complexes with bHLH proteins but there is no DNA binding (Van Doren *et al.*, 1991). *Id* is the vertebrate homolog of *emc* (Benezra *et al.*, 1990).

XASH-3 is the earliest neural marker described; it appears midway through gastrulation (stage 11.5) as two broad stripes along the mid-hemineural plate (Zimmerman *et al.*, 1993). XASH-1 is expressed at the end of neural plate stages in specific domains of the anterior CNS (Ferreiro *et al.*, 1993). At tadpole stages, its expression is concentrated in the ventricular areas of the neural tube where neuroblasts continue to be born. Both of these genes, like their *Drosophila* homologs, have been proposed to be involved in early differentiation of subsets of neuroblasts. Hyperplasia of the brain and spinal cord has been observed in recent misexpression experiments of *XASH* mRNAs with or without *XE12* (B. Ferreiro *et al.*, unpublished observations; D. Turner and H. Weintraub, personal communication), the *Xenopus* homolog of *da* (Rashbass *et al.*, 1992). Animal caps isolated from these injected embryos express different neural markers (*NCAM* and *NF3*)

TABLE 2
Early Neural Genes

Gene	Type	Temporal expression	Spatial expression	Suggested function	Reference
Part 1. Transcription Factors					
XASH3	Helix–loop–helix	stage 11.5—tadpole	Two broad stripes along the mid-neural plate	Early response to neural induction, turns on neural differentiation. Involved in dorsoventral pattern or neural plate or tube	Zimmerman et al. (1993)
XASH1	Helix–loop–helix	Neural tube—tadpole	Specific areas in anterior CNS	Involved in early differentiation of specific neural cells	Ferreiro et al. (1993)
Xtwist	Helix–loop–helix	Neural plate—tadpole	Cephalic neural crest	Crest differentiation factor for cells that make tissues of mesodermal type: connective and skeleton tissues of head	Hopwood et al. (1989)
Xkrox-20	Zn finger domain	Late gastrula (12.5-13)—tadpole	Rhombomeres 3 and 5; neural crest	Pattern of hindbrain, neural crest	Bradley et al. (1993)
Xdll	Homeobox	Neural plate—tadpole	Anterior	Establishment of the structures in the anterior part of the embryo	Asano et al. (1992)
Xdll-3	Homeobox	stage 10.5—tadpole	Ventral forebrain, cranial neural crest, cement gland, olfactory and otic placode	Determines the rostral-most part of the neural plate, which gives rise to ventral forebrain	Papalopulu and Kintner (1993)

60

Xdll-4	Homeobox	Neural tube—tadpole	Ventral forebrain, cranial neural crest and cement gland, developing eye	Determines anterior identities	Papalopulu and Kintner (1993)
en-2	Homeobox	stage 14—tadpole	Midbrain-hindbrain junction	Regional identity	Hemmati *et al.* (1991)
XlHbox6	Homeobox	Late gastrula (stage 13)—tailbud	Posterior neural plate, spinal cord, and lateral plate mesoderm	Specifies cell fate in position-specific and cell-autonomous manner	Sharpe *et al.* (1987); Wright *et al.* (1990); Niehrs and De Robertis (1991)
XlHbox1	Homeobox	Late gastrula—tadpole	Narrow band across the cervical region of the CNS, neural crest, and mesoderm	Involved in the transformation of hindbrain to spinal cord	Carrasco *et al.* (1984); Oliver *et al.* (1988); Wright *et al.* (1989)
Xhox7.1	Homeobox	Early gastrula—neural tube	Dorsal mesoderm, dorsal neural tube and neural crest, dorsal hindbrain	Involved in morphogenesis	Su *et al.* (1991)
XeNK-2	Homeobox	Neurula—tadpole	Dorsoventral midline area of forebrain, midbrain and hindbrain	Dorsoventral patterning	Saha *et al.* (1993)
XANF1	Homeobox	Early gastrula—tadpole	Uniform in ectoderm of gastrula, anterior part of developing nervous system	Involved in early neurogenesis	Zaraisky *et al.* (1992)
XLPOU-1, -2	POU domain	Neural plate—tadpole	Skin and anterior CNS (POU-1), brain and kidney (POU-2)	Specifications of neuronal phenotypes	Agarwal and Sato (1991)

(*continues*)

TABLE 2—*Continued*

Gene	Type	Temporal expression	Spatial expression	Suggested function	Reference
Part 1. Transcription Factors—*Continued*					
XBrn-3	POU domain	Neural tube—tadpole	Brain and subset of neural crest cells. Later in retinal ganglion cells and tectum	Differentiation factor for certain neural cells	N. Hirsch (personal communication)
PAX-6	PAX domain	Early neural tube—tadpole	Eye and forebrain	Patterning and positional identity	N. Hirsch (personal communication)
Part 2. Differentiation Products					
XIF3	Intermediate filament	Low maternal—tadpole	Ectoderm, cranial ganglia, motor neurons of hindbrain	Neuronal cytoskeleton	Sharpe *et al.* (1989)
XIF6	Intermediate filament	Neural tube closure—tadpole	General neural	Neuronal cytoskeleton	Sharpe (1988)
NCAM	Cell adhesion molecule	Late gastrula—tadpole	Neural plate, neural tube	Regulates cell adhesion. Misexpression causes no effect in neurogenesis	Kintern and Melton (1987); Kintner (1988)
NCad	Cell adhesion molecule	Late gastrula—tadpole	Neural plate and dorsal mesoderm	Involved in morphogenetic changes of neural plate or tube	Detrick *et al.* (1990); Kintner (1992); Simonneau *et al.* (1992)
FCad	Cell adhesion molecule	Late gastrula—tadpole	Specific regions of neural plate or tube; associated with flexures of plate or tube	Involved in dorsoventral pattern of neural plate or tube	C. R. Kintner (personal communication)
XSha2	Potassium channel	Neural fold—tadpole	Thoughout nervous system	Physiological activity of neurons	Ribera (1990)

62

Part 3. Secreted Factors and Patterning Factors

	Family	Expression	Location	Function	Reference
FGF3 (int-2)	FGF family	Late blastula—neural tube	Anterior third of neuroectoderm. Tadpole: optic cups, hypothalamus, mid-hindbrain junction	Involved in development of brain-derived structures	Tanahill et al. (1992)
Xwnt-1	wnt family	Neurula—tadpole	Midbrain and hindbrain boundary, dorsal and ventro/lateral midline of midbrain, dorsal midline of hindbrain	Involved in pattern of neural ectoderm	Noordermeer et al. (1989); McGrew et al. (1992); Wolda et al. (1993)
Xwnt-3A	wnt family	Neurula—tadpole	Anteriormost neural fold, dorsal midline of developing brain and dorsal surface of otic vesicle	Patterning CNS. Misexpression can lead to the formation of a 2nd axis	Wolda et al. (1993)
Xwnt-4	wnt family	Late gastrula—tadpole	Inner layer of neuroectoderm, lateral ridge of neural plate. Specific regions of CNS (dorsal midline midbrain and hindbrain, floor plate of hindbrain and spinal cord)	Patterning CNS. Misexpression has no effect on development	McGrew et al. (1992)
Xwnt-7A	wnt family	Tadpole	Ventral regions of neural tube	Patterning CNS	Wolda and Moon (1992)
Xwnt-10	wnt family	Tadpole	Dorsal region of hindbrain	Patterning CNS	Wolda and Moon (1992)

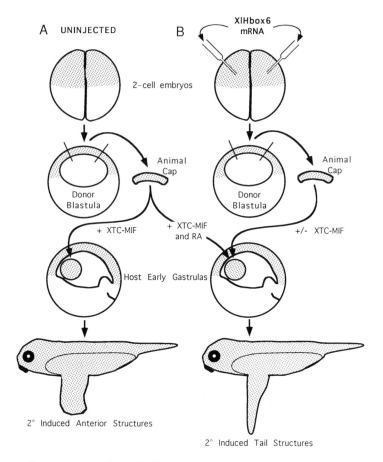

FIG. 7. Posteriorizing effects of a Hox gene. In this experiment, cap tissue from blas-
tula-stage embryos was taken from uninjected or control injected animals (A) or animals
injected with *XlHbox6* mRNA (B) at the two-cell stage. This cap tissue was then in-
cubated in activin containing XTC-MIF to induce it with a dorsal mesoderm character.
The tissue was then implanted into the ventral blastocoel of an early gastrula in an
einsteck experiment. The mesodermally induced control cap tissue induced secondary
heads (A) while the cap tissue that expressed *XlHbox6* (B) induced tails. Interestingly,
this *XlHbox6*-expressing cap tissue no longer required XTC-MIF induction to show organizing ability in this assay. If retinoic acid was added with XTC-MIF to the incubation
medium of the uninjected caps, the caps gained posterior induction capacity. The interpretation of these experiments is that posteriorly expressed *Hox* genes, possibly turned
on by RA, act dominantly to posteriorize tissue that would normally express and induce
anterior values (Cho *et al.*, 1991b).

(B. Ferreiro *et al.*, unpublished observations) (Fig. 3B). These results
suggest that the *XASH* genes are in the pathway of early neural
differentiation.

2. Others

a. Zn-Finger. *Xkrox-20* (Bradley *et al.*, 1993) encodes a protein that belongs to the Zn finger family of transcription factors; this gene was first described in mice (Wilkinson *et al.*, 1989). The early expression pattern of *Xkrox-20* is restricted to the anterior neuroectoderm region of the late gastrula (stage 12.5) that will become rhombomeres 3 and 5 of the hindbrain and the adjacent neural crest cells. In mouse, *Krox-20* regulates the restricted expression of the *HoxB2* gene during hindbrain segmentation (Sham *et al.*, 1993).

b. Paired Domain. *Pax* genes encode a family of transcription factors that appear to be involved in the regionalization of the early nervous system in vertebrates (Goulding, 1992). They are homologous to the paired-box genes of *Drosophila,* and like them, contain both a homeobox and a paired domain. *Pax* genes in vertebrates are expressed in restricted A–P and D–V patterns in the brain and spinal cord (Goulding, 1992; Gruss and Walther, 1992). In mice there are several mutants in which the expression of particular *Pax* genes is affected, and this leads to regional defects in the nervous system. For instance, the *Pax-6* gene is heavily expressed in the eye and the diencephalon of the forebrain (Walther and Gruss, 1991). Heterozygotes for a *Pax-6* mutation (Small Eye) lead to reduced eye size, while homozygotes have a complete absence of the visual pathway (Hill *et al.*, 1992). Recently a *Xenopus* homolog of *Pax-6* has been identified, and is expressed in a pattern similar to the mouse gene (N. Hirsch, personal communication).

c. POU Domain. POU domain transcription factors contain a homeobox and a specific POU domain (Treacy and Rosenfeld, 1992). Work in nematodes and mammals has shown that these genes appear to be involved in controlling the differentiation of particular cells or groups of cells. In *Xenopus,* a few POU domain genes have been identified that are expressed primarily in parts of the developing nervous system. *XLPOU-1* is expressed in the anterior neural plate (Agarwal and Sato, 1991). In tailbud stages, it is expressed in the eye and the brain, with weak expression along the length of the nerve cord and the skin. *XLPOU-2* is expressed in the brain and the kidney (Agarwal and Sato, 1991). *Xbrain-3* is first expressed at neural tube stages in the anterior neural crest, dorsal midbrain, and the ganglion cells of the retina (N. Hirsch, personal communication).

B. *WNT* GENES

The *Xwnt* genes encode cysteine-rich, secreted proteins that belong to the *wingless/int-1* family. *Drosophila wingless* is a segment polarity gene (Lindsley and Zimm, 1992). In mice, targeted disruption of *wnt-1*, a gene normally expressed in various regions of the developing brain, leads to severe deformities in the midbrain and cerebellum (McMahon and Bradley, 1990). In *Xenopus*, members of this family are expressed in different areas of the CNS, suggesting therefore a role in its patterning. *Xwnt-1* is first expressed at the neurula stage (Noordermeer *et al.*, 1989; Wolda *et al.*, 1993); later, it is expressed at the dorsal and lateral midbrain and dorsal hindbrain. *Xwnt-3A* is expressed at stage 16 in the anteriormost neural fold as a consequence of neural induction (Wolda *et al.*, 1993). Later, it is expressed in the dorsal midline of the developing CNS and dorsal surface of the otic vesicles. *Xwnt-4* is expressed at stage 12.5 in the lateral ridge of the neural plate, and in the inner layer of the neuroectoderm (McGrew *et al.*, 1992). Its expression is dependent on neural induction. Later in development, it is expressed in the dorsal midline of the mid- and hindbrain, and in the floor plate of the spinal cord. The overexpression of *Xwnt-4* mRNA in the zygote does not have an apparent effect on the embryo (McGrew *et al.*, 1992). *Xwnt-7A* and *10* start being expressed at tadpole stages, the first in ventral regions of the neural tube and the second in the dorsal region of the hindbrain (Wolda and Moon, 1992). Their functional roles have not been assayed.

C. STRUCTURAL AND PHYSIOLOGICAL PROTEINS OF THE NERVOUS SYSTEM

1. Cell Adhesion Molecules

a. NCAM. NCAM is a Ca^{2+}-independent cell adhesion molecule of the immunoglobulin gene family. NCAM is a general neural marker (Kintner and Melton, 1987) which has been suggested to mediate homophilic cell adhesion in the nervous system. Its expression has a very low maternal component and a higher zygotic one starting at stage 12 (Kintner and Melton, 1987; Krieg *et al.*, 1989). Expression of NCAM is dependent on neural induction as assayed by recombinants of animal caps and dorsal lip tissue and by the apposition of Hensen's node to animal ectoderm (Kintner and Melton, 1987; Kintner and Dodd, 1991). Overexpression of this molecule has no appreciable effect on morphogenesis of the neural plate or tube (Kintner, 1988).

b. Cadherins. Cadherins are another group of homophilic cell adhesion molecules that are Ca^{2+}-dependent. *NCAD* is expressed at late gastrula (stage 12–13) in the neural plate and dorsal mesoderm (Detrick *et al.*, 1990). The expression of this molecule in the neuroecto-derm also depends upon neural induction as assayed with Hensen's node-blastula ectoderm recombinants. Ectopic expression of *NCAD* produces formation of novel cell boundaries and morphological defects in the embryo (Detrick *et al.*, 1990). This suggests that NCAD has an active role in the morphogenetic changes of the neural plate and tube. Expression of *NCAD* mutant that lacks the extracellular domain in *Xenopus* embryos causes a dramatic inhibition of cell adhesion (Kint-ner, 1992). Analysis of the mutant phenotype shows that mutant cad-herin can inhibit binding of E-cadherin to the intracellular protein ca-tenin (Kintner, 1992). These results suggest that cadherin-mediated adhesion can be regulated by cytoplasmic interactions and that this regulation may contribute to morphogenesis when emerging tissues coexpress several cadherin types.

Different tissues appear to express different cadherins, giving rise to the notion that cadherins have a role in organogenesis and regionali-zation. Thus, for example, in addition to *NCAD,* which is expressed throughout the neural plate, there is also *RCAD,* which in chickens is expressed specifically on the retina (Inuzuka *et al.*, 1991a,b), and *FCAD* which in *Xenopus* is expressed in flexure regions of the neural plate or tube in a pattern that may help divide basal and alar regions of the CNS (C. Kintner, personal communication).

2. Intermediate Filaments

In *Xenopus* there are several neural-specific intermediate filaments. XIF3 and XIF6 are intermediate filament proteins of type III and IV, respectively. *XIF3* expression has two components, one low mater-nal and a second, higher zygotic one. The second one depends on neu-ral induction as assayed by ectoderm–mesoderm recombinants. The neural-inducing signals restrict its expression to the anterior neuro-ectoderm (Sharpe *et al.*, 1989). XIF6 is expressed generally in the neu-roectoderm at the end of neurulation (stage 20), also as a consequence of neural induction (Sharpe, 1988).

3. Channels

The normal function of neurons as information carriers in the ner-vous system depends on their ability to fire action potentials, and to receive and transmit synaptic activity. To this end, all neurons must express an array of voltage, ion, or ligand-gated channels. While the physiological development of *Xenopus* neurons has been studied in some detail, making it clear which channels are developmentally ex-pressed in neurons (Spitzer, 1991), molecular work on these channels is just beginning. A potassium channel encoding gene, *XSha-2*, homologous to the *Drosophila shaker* gene, has been described in *Xenopus* (Ribera, 1990). Expression studies show that it is turned on soon after neural induction, which suggests that some of these genes may be directly activated by the neural induction process.

D. PERSPECTIVE

The genes mentioned in this section are clearly just a very few of those that turn on in response to neural induction. Some, like the *XASH* genes, may be a critical early response driving a proneural pathway. Others, like *NCAM* and *N-cadherin,* are panneural adhesion molecules and may act to add cohesiveness to neural tissue. The *wnts, Pax, Krox,* and *POU* genes and other *cadherins* are more restricted in their expression and may be important in regionalization of the CNS. Yet others, like the channels and neurofilaments, are products that need to be expressed in neurons for their proper structure and function.

Given the incredible molecular complexity of the nervous system, it was difficult for us to define a natural stopping point for this review. It is clear that there is a chain of inductive signals that continues along with the maturation of the nervous system. In the spinal cord, the floor and the roof plate appear to be involved in generating local dorsoventral patterns and thus cellular identity (Jessell and Dodd, 1992). In other parts of the nervous system, as in the retina, there are likely to be cascades of sequential inductions to generate the host of cell types that populate the nervous system (Harris and Messersmith, 1992). Clearly, most of these interactions occur after neural induction is completed. At issue now are the questions of: (1) What genes are directly activated by the inducing signals? (2) Is there a positional dependence of this direct response? (3) How many of these early response genes are transcription factors and what genes do these factors regulate to establish neural identity and commitment?

VII. Conclusions

From the establishment of the organizer to the beginnings of the neural tube there are various inductive events and responses. As this chapter shows, an impressive collection of genes is already known to be involved in this process. Perhaps, therefore, it is fair to say that the search for a single simple molecular solution to the problem of neural induction, considered a holy grail by a previous generation of embryologists, is turning out to uncover a rich molecular network that will not resolve into a simple or solitary pathway.

ACKNOWLEDGMENTS

We thank Chris Kintner for offering useful suggestions on the manuscript. Stubbornly, we did not take all his excellent advice. We thank John Gerhart, Susie Zoltewicz, Nick Hirsch, Paul Skoglund, David Turner, and Chris Kintner for allowing us quote their unpublished findings. We also thank Clark Coffman and Sarah McFarlane for reading parts of the manuscript. Finally, we acknowledge the support of the National Institutes of Health, the McKnight Foundation, and the Spanish Ministerio de Educación y Ciencia.

REFERENCES

Agarwal, V. R., and Sato, S. M. (1991). *XLPOU 1* and *XLPOU 2,* two novel *POU* domain genes expressed in the dorsoanterior region of *Xenopus* embryos. *Dev. Biol.* **147,** 363–373.

Alonso, M. C., and Cabrera, C. V. (1988). The *achaete–scute* gene complex of *Drosophila melanogaster* comprises four homologous genes. *Embo J.* **7,** 2585–2591.

Arcioni, L., Simeone, A., Guazzi, S., Zappavigna, V., Boncinelli, E., and Mavilio, F. (1992). The upstream region of the human homeobox gene *HOX3D* is a target for regulation by retinoic acid and *HOX* homeoproteins. *Embo J.* **11,** 265–277.

Artavanis, T. S., and Simpson, P. (1991). Choosing a cell fate: a view from the *Notch* locus. *Trends Genet.* **7,** 403–408.

Asano, M., Emori, Y., Saigo, K., and Shiokawa, K. (1992). Isolation and characterization of a *Xenopus* cDNA which encodes a homeodomain highly homologous to *Drosophila Distal-less*. *J. Biol. Chem.* **267,** 5044–5047.

Benezra, R., Davis, R. L., Lockshon, D., Turner, D. L., and Weintraub, H. (1990). The protein Id: a negative regulator of helix–loop–helix DNA binding proteins. *Cell* **61,** 49–59.

Bieker, J. J., and Yazdani, B. M. (1992). Distribution of type II collagen mRNA in *Xenopus* embryos visualized by whole-mount *in situ* hybridization. *J. Histochem. Cytochem.* **40,** 1117–1120.

Blum, M., Gaunt, S. J., Cho, K. W., Steinbeisser, H., Blumberg, B., Bittner, D., and De Robertis, E. (1992). Gastrulation in the mouse: the role of the homeobox gene *goosecoid*. *Cell* **69,** 1097–1106.

Blumberg, B., Wright, C. V., De Robertis, E., and Cho, K. W. (1991). Organizer-specific homeobox genes in *Xenopus laevis* embryos. *Science* **253,** 194–196.

Bolce, M. E., Hemmati, B. A., Kushner, P. D., and Harland, R. M. (1992). Ventral ectoderm of *Xenopus* forms neural tissue, including hindbrain, in response to activin. *Development* **115,** 681–688.

Bradley, L. C., Snape, A., Bhatt, S., and Wilkinson, D. G. (1993). The structure and expression of the *Xenopus Krox-20* gene: conserved and divergent patterns of expression in rhombomeres and neural crest. *Mech. Dev.* **40,** 73–84.

Brockes, J. P. (1989). Retinoids, homeobox genes, and limb morphogenesis. *Neuron* **2,** 1285–1294.

Brockes, J. P. (1990). Retinoic acid and limb regeneration. *J. Cell Sci., Suppl.* **13,** 191–198.

Campos, O. J., and Knust, E. (1990). Molecular analysis of a cellular decision during embryonic development of *Drosophila melanogaster:* epidermogenesis or neurogenesis. *Eur. J. Biochem.* **190,** 1–10.

Carrasco, A. E., McGinnis, W., Gehring, W. J., and De Robertis, E. M. (1984). Cloning of an *X. laevis* gene expressed during early embryogenesis coding for a peptide region homologous to *Drosophila* homeotic genes. *Cell* **37,** 409–414.

Chen, K., Peng, Z., Lavu, S., and Kung, H. (1989). Molecular cloning and sequence analysis of two distinct types of *Xenopus laevis* protein kinase C. *Second Messengers Phosphoproteins* **12,** 251–260.

Chen, Y., Huang, L., Russo, A. F., and Solursh, M. (1992). Retinoic acid is enriched in Hensen's node and is developmentally regulated in the early chicken embryo. *Proc. Natl. Acad. Sci. U.S.A.* **89,** 10056–10059.

Cho, K. W. Y., and De Robertis, E. M. (1990). Differential activation of *Xenopus* homeo box genes by mesoderm-inducing growth factors and retinoic acid. *Genes Dev.* **4,** 1910–1916.

Cho, K. W. Y., Blumberg, B., Steinbeisser, H., and De Robertis, E. M. (1991a). Molecular nature of Spemman's organizer: the role of the *Xenopus* homeobox gene *goosecoid. Cell* **67,** 1111–1120.

Cho, K. W. Y., Morita, E. A., Wright, C. V., and De Robertis, E. M. (1991b). Overexpression of a homeodomain protein confers axis-forming activity to uncommitted *Xenopus* embryonic cells. *Cell* **65,** 55–64.

Christian, J. L., and Moon, R. T. (1993). Interactions between *Xwnt-8* and Spemann organizer signaling pathways generate dorsoventral pattern in the embryonic mesoderm of *Xenopus. Genes Dev.* **7,** 13–28.

Clarke, J. D., Holder, N., Soffe, S. R., and Storm, M. J. (1991). Neuroanatomical and functional analysis of neural tube formation in notochordless *Xenopus* embryos; laterality of the ventral spinal cord is lost. *Development* **112,** 499–516.

Coffman, C. R., Harris, W. A., and Kintner, C. R. (1990). *Xotch,* the *Xenopus* homolog of *Drosophila Notch. Science* **249,** 1438–1441.

Coffman, C. R., Skoglund, P., Harris, W. A., and Kintner, C. R. (1993). An extracellular deletion of *Xotch* diverts cell fate in *Xenopus* embryos. *Cell* **73,** 659–671.

Cooke, J. (1989). Mesoderm inducing factors and the Spemann's organizer phenomenon in amphibian development. *Development* **101,** 229–241.

Cunliffe, V., and Smith, J. C. (1992). Ectopic mesoderm formation in *Xenopus* embryos caused by widespread expression of a *Brachyury* homologue. *Nature (London)* **358,** 427–430.

Dale, L., Howes, G., Price, B. M., and Smith, J. C. (1992). Bone morphogenetic protein 4: a ventralizing factor in early *Xenopus* development. *Development* **115,** 573–585.

Davis, R. L., Cheng, P. F., Lassar, A. B., and Weintraub, H. (1990). The *MyoD* DNA binding domain contains a recognition code for muscle-specific gene activation. *Cell* **60,** 733–746.

Dekker, E. J., Pannese, M., Houtzager, E., Timmermans, A., Boncinelli, E., and Durston, A. (1992). *Xenopus Hox-2* genes are expressed sequentially after the onset of gastrulation and are differentially inducible by retinoic acid. *Development* (Suppl.), 195–202.

Detrick, R. J., Dickey, D., and Kintner, C. R. (1990). The effects of *N-cadherin* misexpression on morphogenesis in *Xenopus* embryos. *Neuron* **4,** 493–506.

Dirksen, M., and Jamrich, M. (1992). A novel, activin-inducible, blastopore lip-specific gene of *Xenopus laevis* contains a *fork-head* DNA-binding domain. *Genes Dev.* **6,** 599–608.

Dixon, J. E., and Kintner, C. R. (1989). Cellular contacts required for neural induction in *Xenopus* embryos: evidence for two signals. *Development* **106**, 749–757.

Doniach, T., Phillips, C. R., and Gerhart, J. C. (1992). Planar induction of anteroposterior pattern in the developing central nervous system of *Xenopus laevis*. *Science* **257**, 542–545.

Durston, A. J., Timmermans, J. P., Hage, W. J., Hendriks, H. F., de Vries, N. J., Heideveld, M., and Nieuwkoop, P. D. (1989). Retinoic acid causes an anteroposterior transformation in the developing central nervous system. *Nature (London)* **340**, 140–144.

Eagleson, G. W., and Harris, W. A. (1990). Mapping of the presumptive brain regions in the neural plate of *Xenopus laevis*. *J. Neurobiol.* **21**, 427–440.

Eijnden-Van Raaij, van den A., Zoelent, van E., Nimmen, van K., Koster, C., Snoek, G., Durston, A. J., and Huylebroeck, D. (1990). Activin-like factor from a *Xenopus laevis* cell line responsible for mesoderm induction. *Nature (London)* **345**, 732–734.

Fehon, R. G., Kooh, P. J., Rebay, I., Regan, C. L., Xu, T., Muskavitch, M. A., and Artavanis, T. S. (1990). Molecular interactions between the protein products of the neurogenic loci *Notch* and *Delta*, two EGF-homologous genes in *Drosophila. Cell* **61**, 523–534.

Ferreiro, B., Skoglund, P., Bailey, A., Dorsky, R., and Harris, W. A. (1993). *XASH1*, a *Xenopus* homolog of *achaete–scute:* a proneural gene in anterior regions of the vertebrate CNS. *Mech. Dev.* **40**, 25–36.

Freyd, G., Kim, S. K., and Horvitz, H. R. (1990). Novel cysteine-rich motif and homeodomain in the product of the *Caenorhabditis elegans* cell lineage gene *lin-11. Nature (London)* **344**, 876–879.

Gerhart, J. C., Danilchik, M., Doniach, T., Roberts, S., Rowning, B., and Stewart, R. (1989). Cortical rotation of the *Xenopus* egg: consequences for the anteroposterior pattern of embryonic dorsal development. *Development* **107** (Suppl.), 37–51.

Gilbert, S. F., and Saxen, L. (1993). Spemann's organizer: Models and molecules. *Mech. Dev.* **41**, 73–89.

Godsave, S. F., and Slack, J. M. (1989). Clonal analysis of mesoderm induction in *Xenopus laevis. Dev. Biol.* **134**, 486–490.

Gonzalez, F., Romani, S., Cubas, P., Modolell, J., and Campuzano, S. (1989). Molecular analysis of the asense gene, a member of the *achaete–scute* complex of *Drosophila melanogaster*, and its novel role in optic lobe development. *Embo J.* **8**, 3553–3562.

Goulding, M. (1992). Paired box genes in vertebrate neurogenesis. *Semin. Neurosci.* **4**, 327–335.

Green, J. B., and Smith, J. C. (1991). Growth factors as morphogens: do gradients and thresholds establish body plan? *Trends Genet.* **7**, 245–250.

Green, J. B., Howes, G., Symes, K., Cooke, J., and Smith, J. C. (1990). The biological effects of XTC-MIF: quantitative comparison with *Xenopus* bFGF. *Development* **108**, 173–184.

Green, J. B., New, H. V., and Smith, J. C. (1992). Responses of embryonic *Xenopus* cells to activin and FGF are separated by multiple dose thresholds and correspond to distinct axes of the mesoderm. *Cell* **71**, 731–739.

Grunz, H., and Tacke, L. (1989). Neural differentiation of *Xenopus laevis* ectoderm takes place after disaggregation and delayed reaggregation without inducer. *Cell Differ. Dev.* **28**, 211–217.

Gruss, P., and Walther, C. (1992). *Pax* in development. *Cell* **69**, 719–722.

Hama, T., Tsujimura, H., Kaneda, T., Takata, K., and Ohara, A. (1985). Inductive capacities for the dorsal mesoderm of the dorsal marginal zone and pharyngeal endoderm in the very early gastrula of the newt, and presumptive pharingeal endoderm as an initiatior of the organization center. *Dev. Growth Differ.* **27**, 419–433.

Hamburger, V. (1988). "The Heritage of Experimental Embryology: Hans Spemann and the Organizer." Oxford Univ. Press, New York.

Harris, W. A., and Messersmith, S. L. (1992). Two cellular inductions involved in photoreceptor determination in the *Xenopus* retina. *Neuron* **9**, 357–372.

Hashimoto, M., Kondo, S., Sakurai, T., Etoh, Y., Shibai, H., and Muramatsu, M. (1990). Activin/EDF as an inhibitor of neural differentiation. *Biochem. Biophys. Res. Commun.* **173**, 193–200.

Hashimoto, M., Nakamura, T., Inoue, S., Kondo, T., Yamada, R., Eto, Y., Sugino, H., and Muramatsu, M. (1992). *Follistatin* is a developmentally regulated cytokine in neural differentiation. *J. Biol. Chem.* **267**, 7203–7206.

Hemmati, B. A., and Melton, D. A. (1992). A truncated activin receptor inhibits mesoderm induction and formation of axial structures in *Xenopus* embryos. *Nature (London)* **359**, 609–614.

Hemmati, B. A., Stewart, R. M., and Harland, R. M. (1990). Region-specific neural induction of an engrailed protein by anterior notochord in *Xenopus*. *Science* **250**, 800–802.

Hemmati, B. A., de la Torre, J. R., Holt, C., and Harland, R. M. (1991). Cephalic expression and molecular characterization of *Xenopus* En-2. *Development* **111**, 715–724.

Hemmati, B. A., Kelly, O., and Melton, D. A. (1994). Follistatin, an antagonist of activin, is expressed in the Spemann organizer and displays direct neuralizing activity. *Cell* (in press).

Herrmann, B. G., Labeit, S., Poustka, A., King, T. R., and Lehrach, H. (1990). Cloning of the T gene required in mesoderm formation in the mouse. *Nature (London)* **343**, 617–622.

Hill, R. E., Favor, J., Hogan, B. L., Ton, C. C., Saunders, G. F., Hanson, I. M., Prosser, J., Jordan, T., Hastie, N. D., and van Heyningen, V. (1992). Mouse Small eye results from mutations in a paired-like homeobox-containing gene. *Nature (London)* **355**, 750.

Hogan, B. L., Thaller, C., and Eichele, G. (1992). Evidence that Hensen's node is a site of retinoic acid synthesis. *Nature (London)* **359**, 237–241.

Holtfreter, J. (1933). Die totale exogastruation, eine selbstrablosung des ektoderms von entomesoderm: entwicklung und funktionelles verhalten nervenloser organe. *Wilhelm Roux's Arch. Entwicklungsmech. Org.* **129**, 669–793.

Hopwood, N. D., Pluck, A., and Gurdon, J. B. (1989). A *Xenopus* mRNA related to *Drosophila* twist is expressed in response to induction in the mesoderm and the neural crest. *Cell* **59**, 893–903.

Horst, J. (1948). Differenzierungs und Induktionsleistungen verschiedener Abschnitte der Medullarplatte und des Urdarmdaches von Triton im Kombinat. *Wilhelm Roux's Arch. Entwicklungsmech. Org.* **143**, 275–303.

Inuzuka, H., Miyatani, S., and Takeichi, M. (1991a). R-cadherin: a novel Ca(2+)-dependent cell–cell adhesion molecule expressed in the retina. *Neuron* **7**, 69–79.

Inuzuka, H., Redies, C., and Takeichi, M. (1991b). Differential expression of R- and N-cadherin in neural and mesodermal tissues during early chicken development. *Development* **113**, 959–967.

Ip, Y. T., Park, R. E., Kosman, D., Yazdanbakhsh, K., and Levine, M. (1992). dorsal–twist interactions establish *snail* expression in the presumptive mesoderm of the *Drosophila* embryo. *Genes Dev.* **6**, 1518–1530.

Jessell, T., and Dodd, J. (1992). Midline signals that control the dorso-ventral polarity of the neural tube. *Semin. Neurosci.* **4**, 317–325.

Jimenez, F., and Campos, O. J. (1990). Defective neuroblast commitment in mutants of the *achaete–scute* complex and adjacent genes of *D. melanogaster. Neuron* **5**, 81–89.

Johnson, J. E., Birren, S. J., and Anderson, D. J. (1990). Two rat homologues of *Drosophila achaete–scute* specifically expressed in neuronal precursors. *Nature (London)* **346**, 858–861.

Jones, C. M., Lyons, K. M., Lapan, P. M., Wright, C. V., and Hogan, B. L. (1992). DVR-4 (bone morphogenetic protein-4) as a posterior-ventralizing factor in *Xenopus* mesoderm induction. *Development* **115**, 639–647.

Jurgens, G., and Weigel, D. (1988). Terminal versus segmental development in the *Drosophila* embryo: the role of the homeotic gene *forkhead. Wilhelm Roux's Arch. Dev. Biol.* **197**, 345–354.

Karlsson, O., Thor, S., Norberg, T., Ohlsson, H., and Edlund, T. (1990). Insulin gene enhancer binding protein Isl-1 is a member of a novel class of proteins containing both a homeo- and a Cys-His domain. *Nature (London)* **344**, 879–882.

Keller, R. E. (1985). The cellular basis of amphibian gastrulation. *Dev. Biol.* **2**, 241–327.

Keller, R. E., and Danilchik, M. (1988). Regional expression, pattern and timing of convergence and extension during gastrulation of *Xenopus laevis. Development* **103**, 193–209.

Keller, R. E., Cooper, M. S., Danilchik, M., Tibbetts, P., and Wilson, P. A. (1989). Cell intercalation during notochord development in *Xenopus laevis. J. Exp. Zool.* **251**, 134–154.

Keller, R. E., Shih, J., and Sater, A. (1992a). The cellular basis of the convergence and extension of the *Xenopus* neural plate. *Dev. Dyn.* **193**, 199–217.

Keller, R. E., Shih, J., Sater, A. K., and Moreno, C. (1992b). Planar induction of convergence and extension of the neural plate by the organizer of *Xenopus. Dev. Dyn.* **193**, 218–234.

Kimelman, D., Abraham, J. A., Haaparanta, T., Palisi, T. M., and Kirschner, M. W. (1988). The presence of fibroblast growth factor in the frog egg: its role as a natural mesoderm inducer. *Science* **242**, 1053–1056.

Kimelman, D., Christian, J. L., and Moon, R. T. (1992). Synergistic principles of development: overlapping patterning systems in *Xenopus* mesoderm induction. *Development* **116**, 1–9.

Kintner, C. R. (1988). Effects of altered expression of the neural cell adhesion molecule, N-CAM, on early neural development in *Xenopus* embryos. *Neuron* **1**, 545–555.

Kintner, C. R. (1992). Regulation of embryonic cell adhesion by the cadherin cytoplasmic domain. *Cell* **69**, 225–236.

Kintner, C. R., and Dodd, J. (1991). Hensen's node induces neural tissue in *Xenopus* ectoderm. Implications for the action of the organizer in neural induction. *Development* **113**, 1495–1505.

Kintner, C. R., and Melton, D. M. (1987). Expression of *Xenopus* N-CAM RNA is an early response of ectoderm to induction. *Development* **99**, 311–325.

Knochel, S., Lef, J., Clement, J., Klocke, B., Hille, S., Koster, M., and Knochel, W. (1992). Activin A induced expression of a *fork head* related gene in posterior chordamesoderm (notochord) of *Xenopus laevis* embryos. *Mech. Dev.* **38**, 157–165.

Kosman, D., Ip, Y. T., Levine, M., and Arora, K. (1991). Establishment of the mesoderm–neuroectoderm boundary in the *Drosophila* embryo. *Science* **254**, 118–122.

Koster, M., Plessow, S., Clement, J. H., Lorenz, A., Tiedemann, H., and Knochel, W. (1991). Bone morphogenetic protein 4 (BMP-4), a member of the TGF-beta family, in early embryos of *Xenopus laevis:* analysis of mesoderm inducing activity. *Mech. Dev.* **33**, 191–199.

Krieg, P. A., Sakaguchi, D. S., and Kintner, C. R. (1989). Primary structure and developmental expression of a large cytoplasmic domain form of *Xenopus laevis* neural cell adhesion molecule (*N-CAM*). *Nucleic Acids Res* **17**, 10321–10335.

LaFlamme, S. E., and Dawid, I. B. (1990). XK endo B is preferentially expressed in several induced embryonic tissues during the development of *Xenopus laevis*. *Differentiation* **43**, 1–9.

Lai, E., Prezioso, V. R., Tao, W. F., Chen, W. S., and Darnell, J. J. (1991). Hepatocyte nuclear factor 3 alpha belongs to a gene family in mammals that is homologous to the *Drosophila* homeotic gene *fork head*. *Genes Dev.* **5**, 416–27.

Lamb, T. M., Knecht, A. K., Smith, W. C., Stuchel, S. E., Economides, A. N., Stahl, N., Yancopolous, G. D., and Harland, R. M. (1993). Neural induction by the secreted polypeptide noggin. *Science* **262**, 713–718.

Lehmann, F. (1926). Entwicklungsstorungen an der Medullaranlage von Triton, erzeugt durch Unterlagerungsdefekte. *Wilhelm Roux's Arch.* **122**, *Entwicklungsmech. Org.* 243–283.

Lindsley, D., and Zimm, G. (1992). "The Genome of *Drosophila melanogaster*." Academic Press, San Diego.

Machemer, H. (1932). Experimentelle Untersuchung uber die Induktionsleistunger der oberen Unmundlippe in alteren Urodelenkeim. *Wilhelm Roux's Arch. Entwicklungsmech. Org.* **126**, 391–456.

Mangold, O. (1933). Uber die Induktionsfahigkeit der verschiedenen Bezirke der Neurula von Urodelen. *Naturwissenschaften* **21**, 761–766.

Mangold, O., and Spemann, H. (1927). Uber Induktion von Medullarplatte durch Medullarplatte im jungeren. *Wilhelm Roux's Arch. Entwicklungsmech. Org.* **111**, 341–422.

Mansour, S., Goddard, J., and Cappecchi, M. (1993). Mice homozygous for a targeted disruption of the proto-oncogene *int-2* have developmental defects in the tail and inner ear. *Development* **117**, 13–28.

McGinnis, W., and Krumlauf, R. (1992). Homeobox genes and axial patterning. *Cell* **68**, 283–302.

McGrew, L., Otte, A., and Moon, R. (1992). Analysis of *Xwnt-4* in embryos of *Xenopus laevis*: a *wnt* family member expressed in the brain and floor plate. *Development* **115**, 463–473.

McMahon, A. P., and Bradley, A. (1990). The *Wnt-1* (*int-1*) proto-oncogene is required for development of a large region of the mouse brain. *Cell* **62**, 1073–1085.

Moon, R. T., and Christian, J. L. (1992). Competence modifiers synergize with growth factors during mesoderm induction and patterning in *Xenopus*. *Cell* **71**, 709–712.

Murre, C., McCaw, P. S., and Baltimore, D. (1989). A new DNA binding and dimerization motif in immunoglobulin enhancer binding, *daughterless, MyoD*, and *myc* proteins. *Cell* **56**, 777–783.

Nakamura, O., and Toivonen, S. (1978). "Organizer: A Milestone of Half-Century." Elsevier/North-Holland, New York.

Nakamura, T., Takio, K., Eto, Y., Shibai, H., Titani, K., and Sugino, H. (1990). Activin-binding protein from rat ovary is *follistatin*. *Science* **247**, 836–838.

Niehrs, C., and De Robertis, E. (1991). Ectopic expression of a homeobox gene changes cell fate in *Xenopus* embryos in a position-specific manner. *Embo J.* **10**, 3621–3629.

Niehrs, C., Keller, R., Cho, K. W., and De Robertis, E. M. (1993). The homeobox gene *goosecoid* controls cell migration in *Xenopus* embryos. *Cell* **72**, 491–503.F

Nieuwkoop, P. D. (1969). The formation of mesoderm in Urodelean amphibians. I. Induction by the endoderm. *Wilhem Roux's Arch. Entwicklungsmech. Org.* **162**, 341–373.

Nieuwkoop, P. D. (1973). The "organization center" of the amphibian embryo: its origin, spatial organization and morphogenetic action. *Adv. Morphog.* **10**, 1–39.

Noordermeer, J., Meijlinf, F., Verrijzer, P., Rijsewijk, F., and O. D. (1989). Isolation of a

Xenopus homolog of *int-1/wingless* and expression during neurula stages of early development. *Nucleic Acid Res.* **17**, 11–18.

Okada, Y., and Takaya, H. (1942a). Experimental investigation of the regional differences in the inductive capacity of the organizer. *Proc. Imp. Acad. (Tokyo)* **18**, 505–513.

Okada, Y., and Takaya, H. (1942b). Further studies upon the regional differences of the inductive capacity of the organizer. *Proc. Imp. Acad. (Tokyo)* **18**, 514–519.

Oliver, G., Wright, C. V. E., Hardwicke, J., and De Robertis, E. M. (1988). Differential antero-posterior expression of two proteins encoded by a homeobox gene in *Xenopus* and mouse embryos. *EMBO J.* **7**, 3199–3209.

Otte, A. P., and Moon, R. (1992). Protein kinase C isozymes have distinct roles in neural induction and competence in *Xenopus*. *Cell* **68**, 1021–1029.

Otte, A. P., Koster, C. H., Snoek, G. T., and Durston, A. J. (1988). Protein kinase C mediates neural induction in *Xenopus laevis*. *Nature (London)* **334**, 618–620.

Otte, A. P., van, R. P., Heideveld, M., van, D. R., and Durston, A. J. (1989). Neural induction is mediated by cross-talk between the protein kinase C and cyclic AMP pathways. *Cell* **58**, 641–648.

Otte, A. P., Kramer, I. M., and Durston, A. J. (1991). Protein kinase C and regulation of the local competence of *Xenopus* ectoderm. *Science* **251**, 570–573.

Papalopulu, N., and Kintner, C. (1992). Induction and patterning of the neural plate. *Semin. Neurosci.* **4**, 295–306.

Papalopulu, N., and Kintner, C. (1993). *Xenopus distal-less* related homeobox genes are expressed in the developing forebrain and are induced by planar signals. *Development* **117**, 961–975.

Papalopulu, N., Lovell, B. R., and Krumlauf, R. (1991). The expression of murine *Hox-2* genes is dependent on the differentiation pathway and displays a collinear sensitivity to retinoic acid in F9 cells and *Xenopus* embryos. *Nucleic Acids Res.* **19**, 5497–5506.

Perrin, S. F. (1992). *rel, NFKB,* and the *Brachyury* T gene. *Biochim. Biophys. Acta* **1171**, 129–131.

Pflugfelder, G. O., Roth, H., and Poeck, B. (1992). A homology domain shared between *Drosophila* optomotor-blind and mouse *Brachyury* is involved in DNA binding. *Biochem. Biophys. Res. Commun.* **186**, 918–925.

Rashbass, P., Cooke, L. A., Herrmann, B. G., and Beddington, R. S. (1991). A cell autonomous function of *Brachyury* in T/T embryonic stem cell chimaeras. *Nature (London)* **353**, 348–351.

Rashbass, J., Taylor, M., and Gurdon, J. (1992). The DNA-binding protein E12 cooperates with *XMyoD* in the activation of muscle specific gene expression in *Xenopus* embryos. *EMBO J.* **11**, 2981–2990.

Represa, J., Leon, Y., Miner, C., and Giraldez, F. (1991). The *int-2* protooncogene is responsible for induction of the inner ear. *Nature (London)* **353**, 561–563.

Ribera, A. B. (1990). A potassium channel gene is expressed at neural induction. *Neuron* **5**, 691–701.

Ruiz i Altaba, A. (1992). Planar and vertical signals in the induction and patterning of the *Xenopus* nervous system. *Development* **116**, 67–80.

Ruiz i Altaba, A., and Jessell, T. M. (1991). Retinoic acid modifies the pattern of cell differentiation in the central nervous system of neurula stage *Xenopus* embryos. *Development* **112**, 945–958.

Ruiz i Altaba, A., and Jessell, T. (1992). *Pintallavis,* a gene expressed in the organizer and midline cells of frog embryos: involvement in the development of the neural axis. *Development* **116**, 81–93.

Ruiz i Altaba, A., and Melton, D. A. (1989a). Bimodal and graded expression of the *Xenopus* homeobox gene *Xhox3* during embryonic development. *Development* **106**, 173–183.

Ruiz i Altaba, A., and Melton, D. A. (1989b). Involvement of the *Xenopus* homeobox gene *Xhox3* in pattern formation along the anterior–posterior axis. *Cell* **57**, 317–326.

Saha, M. S., and Grainger, R. M. (1992). A labile period in the determination of the anterior–posterior axis during early neural development in *Xenopus*. *Neuron* **8**, 1003–1014.

Saha, M. S., Michel, R. B., Gulding, K. M., and Grainger, R. M. (1993). A *Xenopus* homeobox gene defines dorsal–ventral domains in the developing brain. *Development* **118**, 193–202.

Sargent, M. G., and Bennett, M. F. (1990). Identification in *Xenopus* of a structural homologue of the *Drosophila* gene *snail*. *Development* **109**, 967–973.

Saxen, L., and Toivonen, S. (1961). The two-gradient hypothesis in primary induction. The combined effect of two types of inductors mixed in different ratios. *J. Embryol. Exp. Morphol.* **9**, 514–533.

Saxen, L., and Toivonen, S. (1962). "Primary Embryonic Induction." Academic Press, London.

Schechtman, A. (1938). Competence for neural plate formation in Hyla and the so-called nervous layer of the ectoderm. *Proc. Soc. Exp. Biol. Med.* **38**, 430–433.

Schulte, M. S., Ho, R. K., Herrmann, B. G., and Nusslein, V. C. (1992). The protein product of the zebrafish homologue of the mouse T gene is expressed in nuclei of the germ ring and the notochord of the early embryo. *Development* **116**, 1021–1032.

Servetnick, M., and Grainger, R. M. (1991). Homeogenetic neural induction in *Xenopus*. *Dev. Biol.* **147**, 73–82.

Sham, M. H., Vesque, C., Nonchev, S., Marshall, H., Frain, M., Gupta, R. D., Whiting, J., Wilkinson, D., Charnay, P., and Krumlauf, R. (1993). The zinc finger gene *Krox20* regulates *HoxB2* (*Hox2.8*) during hindbrain segmentation. *Cell* **72**, 183–196.

Sharpe, C. R. (1988). Developmental expression of a neurofilament-M and two vimentin-like genes in *Xenopus laevis*. *Development* **103**, 269–277.

Sharpe, C. R., and Gurdon, J. B. (1990). The induction of anterior and posterior neural genes in *Xenopus laevis*. *Development* **109**, 765–774.

Sharpe, C. R., Fritz, A., De Robertis, E. M., and Gurdon, J. B. (1987). A homoeobox-containing marker of posterior neural differentiation shows the importance of pre-determination in neural induction. *Cell* **50**, 749–758.

Sharpe, C. R., Pluck, A., and Gurdon, J. B. (1989). *XIF3*, a *Xenopus* peripherin gene, requires an inductive signal for enhanced expression in anterior neural tissue. *Development* **107**, 701–714.

Shih, J., and Keller, R. (1992). Cell motility driving mediolateral intercalation in explants of *Xenopus laevis*. *Development* **116**, 901–914.

Simonneau, L., Broders, F., and Thiery, J. P. (1992). N-cadherin transcripts in *Xenopus laevis* from early tailbud to tadpole. *Dev. Dyn.* **194**, 247–260.

Sive, H. L. (1993). The frog *prince-ss:* a molecular formula for dorsoventral patterning in *Xenopus*. *Genes Dev.* **7**, 1–12.

Sive, H. L., and Cheng, P. F. (1991). Retinoic acid perturbs the expression of *Xhox.lab* genes and alters mesodermal determination in *Xenopus laevis*. *Genes Dev.* **5**, 1321–1332.

Sive, H. L., Draper, B. W., Harland, R. M., and Weintraub, H. (1990). Identification of a retinoic acid-sensitive period during primary axis formation in *Xenopus laevis*. *Genes Dev.* **4**, 932–942.

Slack, J. M. (1990). Growth factors as inducing agents in early *Xenopus* development. *J. Cell Sci., Suppl.* **13**, 119–130.

Slack, J. M., and Tannahill, D. (1992). Mechanism of anteroposterior axis specification in vertebrates. Lessons from the amphibians. *Development* 114, 285–302.

Slack, J. M., Isaacs, H. V., and Darlington, B. G. (1988). Inductive effects of fibroblast growth factor and lithium ion on *Xenopus* blastula ectoderm. *Development* 103, 581–590.

Smith, J. C., and Watt, F. M. (1985). Biochemical specificity of *Xenopus* notochord. *Differentiation* 29, 109–115.

Smith, J. C., Price, B. M., Van Nimmen, K., and Huylebroeck, D. (1990). Identification of a potent *Xenopus* mesoderm-inducing factor as a homologue of activin A. *Nature (London)* 345, 729–731.

Smith, J. C., Price, B. M., Green, J. B., Weigel, D., and Herrmann, B. G. (1991). Expression of a *Xenopus* homolog of *Brachyury* (T) is a immediate-early response to mesoderm induction. *Cell* 67, 79–87.

Smith, W., and Harland, R. (1991). Injected *Xwnt-8* RNA acts early in *Xenopus* embryos to promote formation of a vegetal dorsalizing center. *Cell* 67, 753–765.

Smith, W., and Harland, R. (1992). Expression cloning of noggin, a new dorsalizing factor localized in the Spemann organizer in *Xenopus* embryos. *Cell* 70, 829–840.

Smith, W., Knecht, A., Wu, M., and Harland, R. (1993). Secreted noggin protein mimics the Spemann organizer in dorsalizing *Xenopus* mesoderm. *Nature (London)* 361, 547–549.

Sokol, S., and Melton, D. A. (1991). Pre-existent pattern in *Xenopus* animal pole cells revealed by induction with activin. *Nature (London)* 351, 409–411.

Sokol, S., Christian, J., Moon, R., and Melton, D. (1991). Injected *wnt* RNA induces a complete body axis in *Xenopus* embryos. *Cell* 67, 741–752.

Spemann, H., and Mangold, H. (1924). Über Induktion von Embryonalanlagen durch Implantation artfremder Organisatoren. *Wilhelm Roux's Arch. Entwicklungsmech. Org.* 100, 599–638.

Spitzer, N. C. (1991). A developmental handshake: neuronal control of ionic currents and their control of neuronal differentiation. *J. Neurobiol.* 22, 659–673.

Steinbeisser, H., De Robertis, E. M., Ku, M., Kessler, D. S., and Melton, D. A. (1993). *Xenopus* axis formation: Induction of *goosecoid* by injected *Xwnt-8* and activin mRNAs. *Development* 118, 499–507.

Stewart, R., and Gerhart, J. (1990). The anterior extent of dorsal development of the *Xenopus* embryonic axis depends on the quantity of organizer in the late blastula. *Development* 109, 363–372.

Su, M. W., Suzuki, H. R., Solursh, M., and Ramirez, F. (1991). Progressively restricted expression of a new homeobox-containing gene during *Xenopus laevis* embryogenesis. *Development* 111, 1179–1187.

Taira, M., Jamrich, M., Good, P., and Dawid, I. (1992). The *LIM* domain-containing homeobox gene *Xlim-1* is expressed specifically in the organizer region of *Xenopus* gastrula embryos. *Genes Dev.* 6, 356–366.

Tannahill, D., Isaacs, H., Close, M., Peters, G., and Slack, M. (1992). Develomental expression of the *Xenopus int-2* (FGF-3) gene: activation by mesodermal and neural induction. *Development* 115, 695–702.

Tashiro, K., Yamada, R., Asano, M., Hashimoto, M., Muramatsu, M., and Shiokawa, K. (1991). Expression of mRNA for activin-binding protein (*follistatin*) during early embryonic development of *Xenopus laevis*. *Biochem. Biophys. Res. Commun.* 174, 1022–1027.

Thisse, B., Stoetzel, C., Gorostiza, T. C., and Perrin, S. F. (1988). Sequence of the twist gene and nuclear localization of its protein in endomesodermal cells of early *Drosophila* embryos. *Embo J.* 7, 2175–2183.

Thomsen, G., Woolf, T., Whitman, M., Sokol, S., Vaughan, J., Vale, W., and Melton, D. A. (1990). Activins are expressed early in *Xenopus* embryogenesis and can induce axial mesoderm and anterior structures. *Cell* **63**, 485–493.

Treacy, M. N., and Rosenfeld, M. G. (1992). Expression of a family of *POU*-domain protein regulatory genes during development of the central nervous system. *Annu. Rev. Neurosci.* **15**, 139–165.

Umbhauer, M., Riou, J. F., Spring, J., Smith, J. C., and Boucaut, J. C. (1992). Expression of tenascin mRNA in mesoderm during *Xenopus laevis* embryogenesis: the potential role of mesoderm patterning in tenascin regionalization. *Development* **116**, 147–157.

Van Doren, M., Ellis, H. M., and Posakony, J. W. (1991). The *Drosophila extramacrochaetae* protein antagonizes sequence-specific DNA binding by *daughterless/achaete–scute* protein complexes. *Development* **113**, 245–255.

Villares, R., and Cabrera, C. V. (1987). The *achaete–scute* gene complex of *D. melanogaster:* conserved domains in a subset of genes required for neurogenesis and their homology to *myc. Cell* **50**, 415–424.

von Dassow, G., Schmidt, J. E., and Kimelman, D. (1993). Induction of the *Xenopus* organizer: expression and regulation of *Xnot*, a novel FGF and activin-regulated homeo box gene. *Genes Dev.* **7**, 355–366.

Walther, C., and Gruss, P. (1991). *Pax-6,* a murine paired box gene, is expressed in the developing CNS. *Development* **113**, 1435–1449.

Way, J. C., and Chalfie, M. (1988). *mec-3,* a homeobox-containing gene that specifies differentiation of the touch receptor neurons in *C. elegans. Cell* **54**, 5–16.

Weigel, D., Jurgens, G., Kuttner, F., Seifert, E., and Jackle, H. (1989). The homeotic gene *fork head* encodes a nuclear protein and is expressed in the terminal regions of the *Drosophila* embryo. *Cell* **57**, 645–658.

Wilkinson, D. G., Bhatt, S., Chavrier, P., Bravo, R., and Charnay, P. (1989). Segment-specific expression of a zinc-finger gene in the developing nervous system of the mouse. *Nature (London)* **337**, 461–464.

Wilkinson, D. G., Bhatt, S., and Herrmann, B. G. (1990). Expression pattern of the mouse T gene and its role in mesoderm formation. *Nature (London)* **343**, 657–659.

Wolda, S. L., and Moon, R. T. (1992). Cloning and developmental expression in *Xenopus laevis* of seven additional members of the *Wnt* family. *Oncogene* **7**, 1941–1947.

Wolda, S. L., Moody, C. J., and Moon, R. T. (1993). Overlapping expression of *Xwnt-3A* and *Xwnt-1* in neural tissue of *Xenopus laevis* embryos. *Dev. Biol.* **155**, 46–57.

Wright, C. V., Cho, K. W. Y., Hardwicke, J., Collins, R. H., and De Robertis, E. M. (1989). Interference with function of a homeobox gene in *Xenopus* embryos produces malformations of the anterior spinal cord. *Cell* **59**, 81–93.

Wright, C. V., Morita, E. A., Wilkin, D. J., and De Robertis, E. M. (1990). The *Xenopus XlHBox 6* homeo protein, a marker of posterior neural induction is expressed in proliferating neurons. *Development* **109**, 225–234.

Zaraisky, A. G., Lukyanov, S. A., Vasiliev, O. L., Smirnov, Y. V., Belyavsky, A. V., and Kazanskaya, O. V. (1992). A novel homeobox gene expressed in the anterior neural plate of the *Xenopus* embryo. *Dev. Biol.* **152**, 373–382.

Zimmerman, K., Shih, J., Bars, J., Collazo, A., and Anderson, D. (1993). *XASH-3,* a novel *Xenopus achaete–scute* homolog, provides an early marker of planar neural induction and position along the medio-lateral axis of the neural plate. *Development* **119**, 221–232.

CELL CYCLE GENES OF Drosophila

Cayetano Gonzalez, Luke Alphey, and David Glover

Department of Anatomy and Physiology, Medical Sciences Institute, University of Dundee, Dundee DD1 4HN, Scotland

I. Introduction

The genetic analysis of cell division in *Drosophila* has been the subject of several recent reviews (Ripoll *et al.*, 1987, 1992; Glover 1989, 1990, 1991; and Gatti and Goldberg 1991). All of these articles discuss to a greater or lesser extent the utility of genetic analysis in dissecting cell division, and the choice of *Drosophila* as a model system. They also introduce those aspects of *Drosophila* development which are more relevant for genetic analysis of cell division in this organism. Since our understanding of these matters has not changed substantially, we will not deal with these topics at any length in this chapter. Likewise, since the characterization of many of the genes mentioned in previous re-

ADVANCES IN GENETICS, Vol. 31

views has not yet gone beyond the stage of identification and preliminary study of the mutant phenotype, we will not discuss such genes any further.

Fortunately, there are areas in which our knowledge of how cell division works has improved, some at an incredible pace, and it is these topics that we have chosen to review. The advances that have taken place are due largely to a combination of extensive genetic analysis and thorough cytological characterization of the mutant phenotypes, facilitated by recent technical refinements and extensive molecular analysis. Thanks to this multidisciplinary approach, gene products with known sequences and biochemical activities are being added to the more fanciful names that have been assigned to cell cycle genes. That is not the only development. Phenotypic and molecular characterization have in some instances complemented each other. We will see, for example, how the realization that some gene products contain highly conserved motifs known to perform particular biochemical activities has prompted a review and reinterpretation of our understanding of the phenotypes resulting from mutations in such genes. Likewise, detailed phenotypic analysis of mutations in genes whose products lack sequence motifs corresponding to known biochemical functions is providing the grounds for the assignment of biological function to these enigmatic polypeptides.

We have arranged this chapter around those issues which have either undergone more significant advances in recent years, or have simply been neglected in previous reviews. While this will inevitably result in some areas being excluded altogether, we hope that this arrangement will complement previous papers and help to provide new insights into different aspects of the process of cell division.

II. Cyclin-Dependent Kinases in *Drosophila*

A. CYCLIN-DEPENDENT KINASES AND EUKARYOTIC CELL CYCLE REGULATION

Recent years have seen the identification of a conserved regulatory network that controls entry into mitosis. The genetic paradigm for the regulation of entry into mitosis was established in the fission yeast, *Schizosaccharomyces pombe,* where a 34-kDa protein kinase, p34[cdc2], encoded by *cdc2* mediates the G2–M transition. Activation of the p34[cdc2] kinase is in part controlled by its association with a cyclin subunit; in part by phosphorylation on threonine 167 (Gould *et al.*, 1991); and also by the dephosphorylation of a key tyrosine, Y15, principally by p80[cdc25] (Russell and Nurse, 1986; Moreno *et al.*, 1990) but also by the *pyp3* gene product (Millar *et al.*, 1992). Antagonistic control is provided by the *wee1* mitotic inhibitor (Russell and Nurse, 1987) and also by *mik1* (Lundgren *et al.*, 1991), both of which encode protein kinases that phosphorylate tyrosine 15. In addition to its G2–M role, the *cdc2* gene product is also required at the other major cell cycle control point, START or G1-S. The cloning of a human *cdc2* homolog by complementation (Lee and Nurse, 1987) demonstrated that a mammalian *cdc2* gene can functionally replace the yeast gene at both these control points. p34^{cdc2} kinase is also a component of M-phase promoting factor (MPF) (Gautier *et al.*, 1988), first defined by its ability to cause frog oocytes to enter their meiotic divisions. The mitotic cyclins A and B are the other components of MPF. These were originally identified as proteins that accumulate during the earliest cell divisions of marine invertebrates to be degraded at each metaphase-anaphase transition (Evans *et al.*, 1983).

The budding yeast homolog of *cdc2* is *CDC28*. Several sets of cyclins have been identified that interact with the kinase encoded by *CDC28*. A set of functionally overlapping G1 cyclins (CLN1, 2, 3, HCS26, ORF D) have been identified that are required for progression through START (Richardson *et al.*, 1989). Potential G1 cyclins have also been identified in higher eukaryotes by their ability to rescue a *CLN1 CLN2 CLN3* triple mutant strain. These have been termed cyclins C, D, and E (Koff *et al.*, 1991; Lew *et al.*, 1991; Matsushime *et al.*, 1991; Motokura *et al.*, 1991; Xiong *et al.*, 1991). These cyclins form complexes with various members of a family of cyclin-dependent protein kinases (cdks), which are all related to p34^{cdc2}. Cyclin A can complex with such a kinase, cdk2, and has a role in S-phase, whereas when complexed to p34^{cdc2}, it is thought to function in mitosis (Pines and Hunter, 1991a; Pagano *et al.*, 1992; Rosenblatt *et al.*, 1992; Girard *et al.*, 1991; Zindy *et al.*, 1992). The possibility of particular cyclins complexing with various cdk enzymes gives a large number of possible cdk/cyclin combinations, with potentially different and possibly overlapping roles (for reviews see Norbury and Nurse, 1992; Lew and Reed, 1993).

B. The *CDC2* Gene Family in *Drosophila*

Two homologs of *cdc2* have so far been identified in *Drosophila* of which *Dmcdc2* can complement both *cdc2* and *CDC28* mutations, whereas *Dmcdc2c* cannot (Jimenez *et al.*, 1990; Lehner and O'Farrell, 1990a). The enzymes encoded by these genes show about 50–60% amino acid identity to the p34^{cdc2} proteins of other organisms. The phenotypes of *Dmcdc2* mutants indicate a cell cycle requirement for this gene (see later discussion). *Dmcdc2c* does not obviously correspond to particular members of the vertebrate cdk family of kinases, and its sequence is only slightly more related to *cdk2* than to *cdc2*. It is a *bona fide* cyclin-dependent kinase, however, insomuch that it has proved possible to isolate genes encoding interacting cyclins (see later discussion). In the absence of mutations in the gene, its function is currently unclear. Nevertheless, both genes are expressed with a developmental profile characteristic of genes required for the cell cycle in *Drosophila*, with high levels of transcripts being produced in oogenesis to provide a maternal contribution to the embryo, in proliferating tissues in the embryo and larva, and in the proliferating cells of the gonads.

To date, mutants have only been described at the *Dmcdc2* locus, which has been defined as one of several lethal complementation groups within chromosome division 31 (Clegg *et al.*, 1993; Stern *et al.*, 1993). All but one of the ethyl methanesulfonate-induced *Dmcdc2* alleles are missense mutations at residues completely conserved in the p34^{cdc2} kinases of fission yeast, budding yeast, fruit flies, and humans. The exception is a missense mutation in a nonconserved residue. This allele is clearly hypomorphic. Animals hemizygous for strong alleles die at the larval-pupal boundary with missing or rudimentary imaginal discs and small brains, a phenotype previously recognized as being characteristic of mutations in genes essential for cell proliferation (Szabad and Bryant, 1982; Gatti and Baker, 1989). This is generally indicative of a maternal supply of wild-type protein to the embryo that can perdure throughout larval development (Carmena *et al.*, 1991), which consists primarily of cell growth and polytenization. The cells in most larval tissues do not undergo proliferation, but do undertake endoreplication cycles to produce polytene chromosomes. These cycles do not appear to be affected in *Dmcdc2* mutants, suggesting that p34^{cdc2} is actually not required for this process, although it is possible that the maternal contribution of p34^{cdc2} may persist longer or be required at a lower level for endoreduplication. The maternal contribution of *Dmcdc2* has been directly demonstrated in embryos derived from females carrying a temperature-sensitive allele of *Dmcdc2*. At the restrictive temperature, these embryos show abnormal divisions in the syncytial stage of embryogenesis with uneven nuclear density and "disrupted chromatin." The role of *Dmcdc2* in embryonic development has been examined by crossing females heterozygous for two hypomorphic alleles of *Dmcdc2* with males heterozygous for a deficiency of the region. The combined maternal and zygotic contribution of p34^{cdc2} in the resulting *cdc2*^{216P}/*Df(2L)J27* embryos permits them to develop past cellularization in division cycle 14 and to die later in embryogenesis. The earlier epidermal divisions appear normal, and cell cycle arrest is seen primarily in the cells of the peripheral nervous system, which appear to arrest in G2 with abnormally high levels of cyclin A (Stern *et al.*, 1993). These authors also showed that overexpression of *Dmcdc2c* could not rescue the observed *Dmcdc2* mutant phenotype, and that there is no enhancement of the *Dmcdc2* mutant phenotype in animals lacking one copy of the *Dmcdc2c* gene.

C. THE CYCLINS

1. Mitotic Cyclins

a. Mitotic Behavior of Cyclins A and B during Development. The *Drosophila* genes for cyclins A and B were first isolated by screening libraries using degenerate oligonucleotide probes (Lehner and O'Farrell, 1989; Whitfield *et al.*, 1989). Antibodies were raised against their gene products, thus permitting a study of their behavior during the cell cycle.

The *Drosophila* embryo is a syncytium for the first 13 mitotic cycles. The nuclei are initially found in the interior of the embryo, and migrate out to the cortex during division cycles 8—10 (Foe and Alberts, 1985). The chromosomes become oriented with their centromeres on the apical (surface) side of the cortical nuclei. Nuclear envelope breakdown begins on the apical side of the nucleus and then spreads basally. The spindle is also asymmetrically positioned on the apical side, and it is within this region that chromosome congression begins. The centromeric regions of the chromosomes are the first to align on the metaphase plate and are followed by the chromosome arms (Hiraoka *et al.*, 1990). Maternally contributed cyclin A and B proteins are at first uniformly distributed throughout the syncytial embryo. Cyclin A then comes to occupy a cortical layer extending to a depth of about 40 μ throughout the cytoplasm that surrounds the nuclei at blastoderm. Cyclin B, on the other hand, is found within a layer of approximately 10 μ, predominantly on the apical side of the cortical nuclei. Within the apical layer of cortical cytoplasm, cyclin B is most abundant in a cloud that surrounds the centrosome within the region occupied by the astral microtubules (Maldonado-Codina and Glover, 1992a). A similar pericentrosomal localization is also seen in yeast (Alfa *et al.*, 1990), starfish (Ookata *et al.*, 1992), and mammalian cells (Pines and Hunter, 1991b; Bailey *et al.*, 1992). This distribution would be consistent with a role for cyclin B in targeting p34[cdc2] to potential substrates, the nuclear lamins and astral microtubules. Cyclin A, however, appears to be predominantly cytoplasmic, with only extremely weak, punctate staining in the nucleus. In early prophase, the cytoplasmic staining persists but becomes more punctate. By late prophase, the cyclin A starts to show association with the chromatin. This association is very clear by metaphase and persists to telophase. The antigen then redistributes and is predominantly cytoplasmic during the next interphase.

Both cyclins A and B fail to be completely degraded at the metaphase–anaphase transition in the syncytium. This is not to say that localized degradation of cyclins does not occur. Indeed, some cyclical degradation of the cyclins can be detected by Western blotting experiments once the cycles begin to lengthen slightly during the syncytial blastoderm stages (Edgar et al., 1994; Fenton and Glover, 1993). Nevertheless, no cyclical degradation is detectable in earlier cycles. The extent of degradation appears to become more extensive at each blastoderm cycle. Cyclin levels may in part regulate the length of these syncytial blastoderm cycles because in embryos that receive only half the dose of either cyclin A or B from their mothers, the timing of the syncytial blastoderm cycle is extended (Edgar et al., 1993).

The individual nuclei of the syncytium become incorporated into cells in cycle 14, which differs from the previous cycles in having an extended G2 phase. Mitoses no longer occur synchronously throughout the embryo, but instead in a set of 27 spatially and temporally regulated domains (Foe, 1989). The complete degradation of both proteins now takes place at the metaphase–anaphase transition, cyclin A degradation preceding that of cyclin B by about 1–2 min (Whitfield et al., 1990; Lehner and O'Farrell, 1990b). The points at which the two cyclins are degraded can be clearly separated by blocking the mitotic cycle with the microtubule destabilizing drug colchicine. Larval neuroblasts treated in this way accumulate and degrade cyclin A normally, whereas cyclin B protein accumulates to a very high level (Whitfield et al., 1990). This implies that the cue for cyclin B degradation is dependent on the correct functioning of the spindle and suggests that cyclin B degradation is not necessarily the signal for the commencement of anaphase.

Other workers using stable cyclin derivatives in Xenopus egg extracts (Murray et al., 1989) or budding yeast (Ghiara et al., 1991) have concluded that the metaphase–anaphase transition depends on cyclin B breakdown. However, Surana et al. (1993) have recently shown that the destruction of the CDC28/CLB2 mitotic kinase is not required for the metaphase–anaphase transition. It should well be possible to dissect the events of the metaphase–anaphase transition by using mutations that affect mitosis at this point. Mutations in asp, for example, arrest larval neuroblasts in a metaphase-like state with both bipolar and hemispindles (Gonzalez et al., 1991). These arrested cells have high levels of cyclin B and low levels of cyclin A (C. Gonzalez and D. M. Glover, unpublished observations).

Two genes have been described that lead to a delay in the progression through metaphase in newly cellularized embryos. One of these, *fizzy*, was identified because development of the embryonic nervous system is severely disrupted (Dawson *et al.*, 1993). A high mitotic index with a striking predominance of metaphase figures is seen in *fizzy* embryos in cycle 15. *fizzy* appears to act earlier in the cycle than microtubule depolymerizing drugs insomuch that both A- and B-type cyclins are present at high levels in *fizzy* mutant cells (A. Philp and I. Dawson, personal communication). Mutations in the gene *three rows* have a similar phenotype, but do not result in cell cycle arrest, but rather a metaphase-like delay in mitosis 15 in which chromosomes are aligned on a normal-looking spindle with both cyclins A and B having been degraded (Philp *et al.*, 1993; D'Andrea *et al.*, 1994). The *thr* mutation prevents any anaphase chromatid movement, but mutant cells eventually enter interphase as the chromosomes decondense in the midregion of the spindle, and this is later followed by DNA replication. Since the regulation of the segregation of chromatids into daughter cells at anaphase is crucial in cell division, the failure of the cell cycle to arrest at this point in *thr* embryos is unexpected. The *thr* protein contains a small region of similarity with the fission yeast gene *nuc2*. Mutations in this gene block chromosome segregation and lead to metaphase arrest in fission yeast. However, since other characteristic features of *nuc2* are not present in *thr*, the significance of this remains unclear.

b. Genetic Studies on Cyclins A and B. Lehner and O'Farrell (1989) have described recessive embryonic lethal mutations in the *cyclin A* gene, thus demonstrating that it has essential functions that cannot be supplied by any of the other cyclins. They state that the maternal supply of cyclin A RNA is sufficient to permit the first 15 rounds of mitotic divisions and S-phase 16, leading to arrest in G2. However, the patterns of divisions in the cellularized embryo are complex, and have yet to be accurately mapped from cycle 15 onward. This is in contrast to the mitotic domains of cycle 14, the spatial and temporal patterns of which were accurately described by Foe (1989). The rate of accumulation of cyclin A protein in wild-type, cycle 14 embryos appears to be uniform, even though divisions no longer occur synchronously throughout the embryo, suggesting that cyclin A is unlikely to be rate limiting. This is further supported by the normal progression of cycle 14 divisions in the *cyclin A* mutant embryos, despite the low levels of cyclin A protein relative to the wild type (Lehner and O'Farrell, 1989).

Mutations in the *cyclin B* gene have yet to be isolated, although

Knoblich and Lehner (1993) have generated a deficiency, Df(2R)59AB, of the interval in which the gene lies. Embryos homozygous for this deficiency die late in their development, although it is not clear whether this is a result of lack of cyclin B or some other essential gene. These embryos are described as being able to complete mitosis 16 normally, although a detailed examination of these divisions has yet to be presented. Knoblich and Lehner (1993) also describe subtle abnormalities of mitotic spindles, and some elevation of the mitotic index of cyclin B-deficient embryos, which can be corrected by the expression of wild-type cyclin B from a heat shock promoter. The exact nature of the mitotic delay is unclear. Embryos mutant for both cyclin A and cyclin B functions arrest before mitosis 15, that is to say, at an earlier cycle than mutations in either cyclin gene alone, thus suggesting a synergistic effect for the lack of these two cyclins.

In vertebrate cells, cyclin A has been clearly demonstrated to have a role in DNA replication, consistent with its localization to the nucleus. The incorporation of bromodeoxyuridine (BrdU) into S-phase cells in cycle 16 of cyclin A mutant embryos, and the apparent G2 block in cycle 15 embryos in the double mutant, argues against an S-phase role for cyclin A in Drosophila, although a role in the completion of S-phase cannot be ruled out. Moreover, in the wild-type embryo, cyclin A is degraded after mitosis in cycles 14–16, to be followed immediately by the S-phase. Of course it is possible either that sufficient cyclin A persists following mitosis to permit entry into S, or that a cyclin A-dependent kinase acts before anaphase of one cycle to allow immediate entry into the S-phase of the next cycle at this developmental stage.

c. Maternal Cyclin B Transcripts and Pole Cell Development. Both the cyclin A and cyclin B genes are abundantly transcribed in the nurse cells during oogenesis (Dalby and Glover, 1992). Two forms of cyclin B transcript are synthesized in the developing egg chamber that differ as a result of a splicing event in the 3' nontranslated region. The spliced form of the transcript is found in the developing oocyte at earlier stages of oogenesis, whereas the unspliced form is synthesized in the nurse cells. The transcripts synthesized in the nurse cells are deposited in the developing oocyte. Cyclin A transcripts are uniformly distributed throughout the early syncytial embryo, whereas cyclin B transcripts are in addition concentrated at the posterior pole of the embryo. This concentration begins at stages 13 and 14 of oogenesis onward (Dalby and Glover, 1992). The maternal RNA becomes incorporated into the pole cells and persists at high levels until at least stage 8 of embryogenesis (Whitfield et al., 1989; Lehner and O'Farrell, 1990b; Raff et al., 1990). That the pole cell cyclin B RNA is maternal in origin is known from several lines of evidence: tagged cyclin B RNAs are found in these cells only if inherited from the mother (Dalby and Glover, 1993); the RNA is still found in embryos homozygous for a deficiency that removes the entire cyclin B gene (Knoblich and Lehner, 1993); and injection of the RNA polymerase II inhibitor, α-amanatin, does not prevent the accumulation and localization of the RNA (Raff *et al.*, 1990).

The accumulation of cyclin B maternal RNA at the posterior pole is disrupted in the eggs of females homozygous for mutations in *cappuccino, spire, staufen,* and *oskar* (Raff *et al.*, 1990). These gene functions are also required for the formation of polar granules, electron-dense structures which are putative germ-line determinants (Counce, 1963; Mahowald, 1962, 1968). The posterior pole accumulation seems to be a late event of oogenesis, and the transcripts appear to be interacting with some component of the polar granules.

While maternal cyclin A transcripts, like many other maternal RNAs, are found in the cortical cytoplasm of the syncytial embryo when the nuclei migrate to the surface, cyclin B transcripts have a much tighter perinuclear distribution and become concentrated on the apical (surface) side of the nuclei. Raff *et al.*, (1990) found that this association is disrupted by microtubule-destabilizing drugs and therefore suggested that either nascent cyclin B polypeptides on polysomes may be associating with microtubules and dragging the RNA with them, or else that the cyclin B RNA is itself targeted to microtubules. Dalby and Glover (1992) showed that a sequence element within the 3′ untranslated sequence is required both for the posterior accumulation of the RNA and also for the perinuclear localization in the somatic part of the embryo, and that this sequence could be used to localize foreign mRNAs to these regions in germ-line transformation experiments. The region required for posterior retention of cyclin B mRNA was more tightly defined by experiments in which biotinylated RNAs tagged with segments of this sequence were injected into early syncytial embryos (Dalby and Glover, 1993). Injected RNAs having 3′ sequence elements become stabilized within the pole cells. The minimum sequence requirement for this localization and stabilization has been defined as lying within a 181-nucleotide region. It is possible that the mechanisms that localize maternal RNAs in the perinuclear and posterior polar regions could share a common intermediary, since a microtubular transport system is thought to localize other RNAs to the posterior pole. Moreover, the polar granules are also known to associate with microtubules, and cluster around centrioles in a number of *Drosophila* species (Rabinowitz, 1941; Counce, 1963; Mahowald, 1968). Pole cell formation itself appears to be triggered by an interaction between the pole plasm and a centrosome-associated factor (Raff and

Glover, 1989). It is not inconceivable that such a factor could also be involved in the transport of posterior pole components from the nurse cells.

Once the pole cells have formed, they undergo two divisions that are completed shortly before the onset of gastrulation. The pole cells then migrate from the posterior pole of the cellularized embryo to become incorporated into the gonads of the stage 14 embryo, over a period of about 9 hr. During this time the pole cells cease dividing, to resume in the last 2–3 hr of embryogenesis (Sonnenblick, 1950). Dalby and Glover, (1993) have found that the pole cell cyclin B transcripts contain a control element that represses the translation of this RNA until late stage 14 of embryogenesis, when the divisions resume. The region required for this translational repression contains sequence motifs similar to the *nanos*-response element that mediates the *nanos*-dependent repression of *bicoid* and *hunchback* translation (Wharton and Struhl, 1991), although the functional significance of this is not yet clear. It is speculated that repression of translation of maternal cyclin B RNA results in cyclin B becoming rate limiting for mitosis until the gonads form. Translation of cyclin B could be the trigger for the resumption of cell division in this lineage. Zygotic transcription of cyclin B in the pole cells is not seen until the first larval instar (Dalby and Glover, 1993).

2. Other Cyclins

Three approaches have been used to identify other cyclin genes in *Drosophila*. Two groups have independently identified a *Drosophila* gene able to complement mutant *Saccharomyces cerevisiae* that are defective for *CLN1, 2,* and *3* (Lahue *et al.*, 1991; Leopold and O'Farrell, 1991). This new cyclin is closely related to a human cyclin identified by a similar approach, and is given the name cyclin C. *Drosophila cyclin C* maps to cytological position 88E on chromosome 3R. A single transcript of about 1.2 kb is found in Kc tissue culture cells and at all developmental stages examined (Lahue *et al.*, 1991). Its tissue distribution during development is broadly consistent with a role in the cell cycle. However, since no mutants in *cyclin C* have yet been identified, its role is as enigmatic as those of cyclin C-like molecules in vertebrates.

Another successful approach has been to use the interaction trap, in which the gene for a protein of interest, in this case *Dmcdc2* or *Dmcdc2c* is expressed as a LexA fusion in a yeast strain containing LexA binding sites upstream of a selectable marker gene, *LEU2*. A cDNA library is then expressed in this strain with the cDNA encoded proteins fused to a transcription activation domain such that those interacting with the LexA fusion will activate transcription of *LEU2*. Using this system, Finley and Brent (personal communication) have identified four unique cDNAs encoding proteins that interact with *Dmcdc2;* they are termed CDIs for cdc2 interactors. Of these, one (CDI2) is a homolog of the *S. pombe* gene *suc1*, and another (CDI3) is a homolog of the vertebrate cyclin D. These two genes plus the gene for one other protein interacting with *Dmcdc2* were also picked out in a search for interactors with *Dmcdc2c*. This second screen identified three further genes, one of which (CDI5) identifies a new type of cyclin sharing some identity with cyclins A and B, but missing amino acids that would place it in either of these classes. Both CDI3 and CDI5 are able to complement an *S. cerevisiae cln1 cln2 cln3* strain, indicating that they are *bona fide* cyclins. The proteins identified by the remaining CDI genes are of unknown function, but could conceivably be substrates for cyclin-associated kinases.

Finally, a *Drosophila* cyclin E gene has been identified through its homology with its vertebrate counterpart (Richardson *et al.*, 1993). By analogy with the vertebrate kinases, it seems likely that this cyclin associates with a kinase having a role in S-phase. Maternal transcripts are supplied to the syncytial embryo and in cell cycles 14–16 of the cellularized embryo, zygotic transcripts appear to be constitutively expressed. Concomitant with the introduction of a G1 phase into the cells of the central and peripheral nervous system in cycle 17, transcription of the gene appears to become cell cycle regulated around the time of the G1–S transition. A thorough understanding of the functions of the *Drosophila* cyclins C, D, and E awaits the finding of mutations in their genes.

D. CONTROL OF P34^{cdc2} ACTIVITY

1. Mitotic Regulation in the Syncytial Embryo

The rapid mitotic cycles of the syncytium are unusual in several respects. They lack G1 and G2 phases, and do not have effective feedback controls to ensure that S-phase is completed before initiation of M-phase. Thus if an inhibitor of DNA synthesis such as aphidicolin is injected into the syncytium, repeated cycles of centrosome replication, nuclear envelope breakdown-reassembly, and chromatin condensation-decondensation occur (Raff and Glover, 1988).

What controls these unusual cycles? As discussed above, there seems to be only limited degradation of the two mitotic cyclins. Edgar and colleagues have studied the phosphorylation state of p34^{cdc2} and found it to oscillate in these syncytial cycles in a manner indicative of activation by phosphorylation on threonine 161. This could well regulate the association of the kinase subunit with the cyclins, and may well be the principal way in which p34^{cdc2} is regulated in the syncytium.

The enzyme appears to be dephosphorylated on tyrosine 15 throughout the syncytial cycles and the *Drosophila cdc25* homolog *string (stg)* is present and apparently constitutively active at this stage (Edgar *et al.*, 1993).

2. G2–M Regulation in the Cellularized Embryo: string

The embryo is supplied with a dowry of maternal RNA for *stg*. However, this RNA is apparently not translated until the first mitotic division is under way. $p34^{cdc2}$ is then dephosphorylated on tyrosine 15 until the advent of the extended G2 phase in cycle 14, whereupon it becomes tyrosine phosphorylated (Edgar *et al.*, 1993). The entry into mitosis appears then to come under the control of *stg* from cycle 14 onward (Edgar and O'Farrell, 1989). Transcription of *stg* occurs in a spatially and temporally regulated manner that anticipates mitosis by about 30 min. Recessive embryonic lethal alleles of *stg* have a rather nonspecific reduction in the number of differentiated cuticular structures (Jürgens *et al.*, 1984). This is due to a reduction in the number of cells in later embryos, relative to the wild type. Edgar and O'Farrell (1989) identified a P element induced allele of *stg*, *l(3)neo62*, which enabled them to clone the gene. Subsequently, complementation experiments were used to clone *stg* cDNAs able to rescue a *cdc25^{ts}* mutation in fission yeast (Jimenez *et al.*, 1990). Bacterially expressed *stg* gene product has been shown to be able to activate $p34^{cdc2}$ kinase from *Xenopus* extracts (Kumagai and Dunphy, 1991; Gautier *et al*, 1991). The amino acid sequence of *stg* and all other *cdc25* homologies shows conservation in the C-terminal domain of the protein which has some limited similarity with a tyrosine-serine phosphatase from vaccinia virus (Guan *et al.*, 1991; Moreno and Nurse, 1991), and mutation of these conserved residues results in the loss of activity of either the *Drosophila* or fission yeast protein (Dunphy and Kumagai, 1991; Gautier *et al.*, 1991).

Edgar and O'Farrell (1990) used a heat-shock inducible *stg* transgene to investigate the effect of ectopic expression of *string* during cycle 14. This induced mitosis throughout these embryos, followed by immediate entry into S-phase. They concluded that the timing of cell division at this developmental stage is controlled by *stg*. Together, the expression pattern, the phenotype of *string* mutants, and the effects of ectopic expression of *string* under the heat shock promoter all indicate that *stg* does behave as a mitotic inducer, as expected from its fission yeast counterpart. However, a degree of caution should be exercised in interpreting experiments that use the heat shock promoter to study cell cycle regulation because Maldonado-Codina *et al.*, (1992b) have shown that a heat shock pulse administered in the S-phase of cycle 14 prevents progression through the cell cycle. Furthermore, heat shock in G2 can delay mitosis, leading to a synchronization of divisions that would otherwise occur at different times.

In contrast to the effect of a heat shock pulse in transgenic embryos that contain a copy of *stg* driven by a heat shock promoter, cyclin degradation appears to be delayed in wild-type embryos subjected to a heat pulse. It therefore seems likely that the phenotypic consequences of heat shock on the *hs-stg* embryo result from a combined effect of heat shock *per se* and ectopic *string* expression. The introduction of tyrosine 15 phosphorylation in cycle 14 strongly suggests that the checkpoint controls that block mitosis in the presence of unreplicated DNA might be mediated through *string* as they are through *cdc25* in fission yeast (Enoch and Nurse, 1990). Edgar and O'Farrell's experiments showing that heat-shock induced expression of *stg* does not overcome a replication block would on first sight appear to suggest an alternative mechanism in *Drosophila*. However, the interpretation of their experiment must be clouded by the severe effects of heat shock on *Drosophila* embryos in S-phase.

3. G2–M Regulation in Meiosis: twine

twine was identified as the second *cdc25* homolog in *Drosophila* by complementation of *cdc25–22* in *S. pombe* (Jimenez *et al.*, 1990; Alphey *et al.*, 1992). The sequence conservation between *cdc25* genes from different species also allowed Courtot *et al.* (1992) to isolate it by using a polymerase chain reaction (PCR) method. *twine* expression has been detected only in the adult germ line.

a. Male Meiosis. The apex of the testes contains a population of dividing stem cells that produce the gonial precursor cells. *string* but not *twine* transcripts are seen in these cells. The precursor cells undergo four rounds of mitotic division to produce a cyst of 16 cells which remain linked by cytoplasmic bridges or ring canals. These cells grow over a period of about 90 hr during which time *twine* (but not *string*) is expressed and following which the two meiotic divisions occur to generate 64 cells within the cyst. These cells then elongate and differentiate into mature sperm. A single mutant allele of *twine* has been identified that appears to be a null. Spermatogenesis appears normal up to the end of the growing stages in *twine* males but the cysts do not go through meiosis (Alphey *et al.*, 1992). The synthesis of *twine* transcripts in the growing stage of primary spermatocytes suggests a role in regulating the G2–M transition of meiosis. The *twine*[HB5] allele has a missense mutation that changes a conserved proline residue in the tyrosine phosphatase domain of the protein to a leucine (Courtot *et al.*, 1992). It would therefore seem that p34^{cdc2} kinase regulation is not correctly activated during meiosis in *twine*^{HB5} mutants. However, although a meiotic spindle never forms, chromosome condensation does occur and moreover it is accompanied by breakdown of the nuclear envelope (White-Cooper *et al.*, 1993). This could be explained if enzymes other than p34^{cdc2} could mediate some aspects of the G2–M transition for the entry into male meiosis, or if some forms of the p34^{cdc2} complex might not be regulated by the phosphorylation and dephosphorylation of tyrosine 15. In extracts of activated *Xenopus* eggs, p34^{cdc2} complexed to cyclin A is not subject to inhibitory phosphorylation of tyrosine 15, in contrast to the p34^{cdc2} / cyclin B complex (Clarke *et al.*, 1992; Devault *et al.*, 1992). If this were the case in the *Drosophila* spermatocyte, *twine* function would only be required to activate the p34^{cdc2} / cyclin B complex, which may be specifically needed to modify microtubule behavior for spindle formation.

b. Female Meiosis. Oogenesis in *Drosophila* begins with four mitotic divisions of a precursor cell in the germarium of the ovary to produce a cyst of 16 cells interconnected by cytoplasmic bridges. As the egg chamber matures, one cell becomes the oocyte while the other cells develop into highly polyploid nurse cells. *twine* and *string* are expressed in the nurse cells together with many other maternally provided mRNAs that are required for early embryonic development. Prophase of meiosis is initiated in the presumptive oocyte nucleus in the germarium and continues during oogenesis. The meiotic spindle is not formed until stage 13 of oogenesis. It forms as a short bipolar structure which then lengthens before metaphase (Theurkauf and Hawley, 1992; Hatsumi and Endow, 1992). In mature stage 14 oocytes, meiosis is arrested in metaphase I, with the exchange chromosomes in a bundle at the metaphase plate and the nonexchange fourth chromosomes displaced from the plate toward the poles. The meiotic divisions are completed only after activation of the oocyte on entry into the oviduct.

Meiosis does not arrest at metaphase I in *twine* females, but continues abnormally, resulting in gross nondisjunction (White-Cooper *et al.*, 1993). The phenotype suggests that *twine* function is required to maintain metaphase arrest by keeping p34^{cdc2} dephosphorylated at tyrosine 15 and thereby active. Meiotic recombination occurs only in females in *Drosophila* and this could explain the differing requirement for *twine* (and p34^{cdc2}) meiotic function between the two sexes. The major microtubule nucleating activity in female meiosis is provided by paired centromeres of the major chromosomes rather than the centrosomes

(Theurkauf and Hawley, 1992) (see later discussion). Such a different mechanism of spindle formation could well be under different regulation and so not blocked by the *twine* mutation as in male meiosis. Repeated abnormal meiotic divisions are seen in *twine* oocytes that appear to resemble reductional division rather than equational division. Repeated attempts at reductional division would explain the dramatic non-disjunction that occurs during *twine* meiosis.

Arrest at metaphase I in female meiosis is normally also dependent upon recombination having taken place to produce chiasmate bivalents (McKim *et al.*, 1993). Thus in mutants that prevent recombination, the meiotic arrest at metaphase I does not occur. However, the absence of any significant zygotic lethality indicates that meiosis is otherwise normal and relies entirely upon the mechanisms for segregating nonexchange chromosomes. Thus the failure to arrest in *twine* mutants differs profoundly from the effects of mutations preventing meiotic recombination. It has been suggested that the formation of chiasmata leads to the establishment of mechanical tension at the metaphase plate that signals a meiotic block (McKim *et al.*, 1993). The gross abnormalities in meiosis in *twine* females suggest that its function is likely to be a prerequisite for the block imposed through the mechanism that senses the presence of chiasmata.

III. Other Mitotic Kinases

It is clear that other kinases, as yet poorly characterized in *Drosophila*, are required to phosphorylate p34^{cdc2} kinase and the components of its regulatory network. Moreover, several genes encoding other protein kinases have been identified as part of the machinery that regulates progression through other cell cycle stages in the fungi. We therefore expect that many *Drosophila* cell cycle regulatory genes will encode protein kinases. To date, however, few such genes have been identified. One such enzyme is the product of the gene *polo*, first identified through a female sterile mutant allele. Embryos derived from mothers homozygous for the *polo¹* mutation undergo highly abnormal mitoses that show disorganized arrays of condensed chromatin associated with disorganized microtubules. Homozygous *polo* flies reach adulthood, and can produce some sperm and eggs in spite of showing defective mitoses during larval development, as well as defective meiotic divisions (Sunkel and Glover, 1988; Llamazares *et al.*, 1991). The defects include a high mitotic index in larval neuroblasts, which display both monopolar spindles and spindles with broad poles, and nondisjunction on multipolar spindles in male meiosis. *polo* encodes a 577-amino-acid protein that has an N-terminal kinase domain and a 300-residue C-terminal domain (Llamazares *et al.*, 1991). Extracts of individual syncytial embryos show a peak of polo kinase activity against the *in vitro* substrate casein during late anaphase-telophase (Fenton and Glover, 1993).

The *polo* kinase is highly conserved. Its homolog from budding yeast, CDC5 (Kitada *et al.*, 1993), and mouse, Plk (polo-like kinase) (Clay *et al.*, 1993) and Snk (serum-induced kinase) (Simmons *et al.*, 1992) show respective identities of 52, 65, and 52% in the kinase domain, and 8, 43, and 35% in the C-terminal domain, the latter homologies lying in three distinct blocks. The murine genes are expressed only in proliferating cells, but whether they have a role in mitosis has not yet been established. In mutants of the budding yeast homolog, *CDC5*, nuclear division is arrested at a late stage and the spindle is elongated (Byers and Goetsch, 1974). Furthermore, when *cdc5* (ts) mutants are shifted to restrictive temperature after the first meiotic division is mitiated, two spindles form which do not elongate, and the four poles encapsulate into two diploid spores (Schild and Byers. 1980). A number of circumstantial observations suggest that *polo/CDC5* may regulate microtubule behavior. *cdc5* mutants show an unusual interaction with methyl benzimidazole-2-yl-carbamate, a drug that binds tubulin and depolymerizes microtubules (Wood and Hartwell, 1982); and there is a strong interaction of *polo* with *asp* mutations that appears to affect microtubule stability (C. Gonzalez, C. E. Sunkel, and D. Glover, unpublished observations). The mitotic stage at which *polo* kinase appears maximally active would be consistent with a role in orchestrating the changes in microtubule organization that have to occur late in anaphase and in telophase. The availability of mutations in the genes and of antibodies to their proteins now offers the means of exploring these potential roles for the enzyme.

IV. Serine–Threonine Protein Phosphatases

A. BACKGROUND

The family of serine-threonine phosphatases has been subdivided into four major types, PP1, PP2A, PP2B, and PP2C on the basis of their substrate specificities, cation requirements, and their sensitivity to certain activators and inhibitors. (Cohen *et al.*, 1990). Conflicting biochemical studies have suggested various mitotic roles for both PP1 and PP2A in regulating the p34[cdc2] mitotic kinase. cdc25 protein undergoes an increase in activity prior to mitosis assayed in *Xenopus* extracts, resulting from its phosphorylation (Kumagai and Dunphy, 1992). This cdc25 activation is opposed by a phosphatase, which from its sensitivity to okadaic acid appears to be like type 2A. These experiments might help clarify a number of earlier experiments showing that treatment of *Xenopus* extracts with okadaic acid accelerates the G2–M transition (Felix *et al.*, 1990; Jessus *et al.*, 1991; Lee *et al.*, 1991; Solomon *et al.*, 1990; Picard *et al.*, 1991). However, Maller's laboratory has attributed this function to PP1. Lee *et al.* (1991) have purified an activity, INH, which inhibits the activation of maturation-promoting factor in amphibian oocytes, and showed it to be a member of the PP2A family. This activity would directly dephosphorylate the purified p34^{cdc2}/cyclin complex, leading these workers to suggest that PP2A acts directly to dephosphorylate p34^{cdc2}, presumably at threonine 161, and thereby participates in regulation at G2. Lorca *et al.* (1992) conclude from its insensitivity to low concentrations of okadaic acid that dephosphorylation of threonine-161 at the metaphase–anaphase transition is in part due to PP1. Nevertheless, Inhibitor 1, which is specific for PP1, is not sufficient to prevent dephosphorylation, suggesting the involvement of another phosphatase.

The mitotic cycle requires many proteins to become phosphorylated and dephosphorylated, and kinases other than p34^{cdc2} have been implicated in these processes. It seems likely therefore that protein phosphatases must also have important roles other than the regulation of p34^{cdc2}. Both PP1 and PP2A have been independently suggested to be the enzymes that dephosphorylate p34^{cdc2} phosphorylation sites in histone H1. Kinoshita *et al.* (1991) showed that histone H1 dephosphorylation was sensitive to a PP1 inhibitor and was reduced in extracts prepared from the *dis2–11* mutant. Sola *et al.* (1991), on the other hand, find that PP2A is the most active histone H1 phosphatase, and that the trimeric form is the most effective form of the enzyme.

Studies using drugs such as okadaic acid to classify types of phosphatase activity are limited by the difficulty in knowing the spectrum of response of the large number of PP1 and PP2A-related genes that are now being identified (Cohen *et al.*, 1990). Several of these genes encode enzymes intermediate in primary structure to PP1 and PP2A, and they may be impossible to classify by the criterion of sensitivity to inhibitors such as okadaic acid. *Drosophila* has genes for four isoforms of PP1, and one for PP2A. In addition, three genes have been identified that encode PPs closely resembling mammalian PP2C, and so far seven other genes have been identified whose proteins are intermediate in primary sequence to any of the archetypal family members (Chen *et al.*, 1992). Mutations have only been identified in a small number of this gene family, and it might be expected that several of these enzymes could have mitotic roles.

B. Protein Phosphatase 1

Mutations affecting PP1 have been characterized in *Aspergillus* (Doonan and Morris, 1989); fission yeast (Ohkura et al., 1989), and *Drosophila* (Axton et al., 1990; Baksa et al., 1993), all of which lead to defective mitosis. The activity of p34^{cdc2} kinase remains high in the late mitotic stages of the fission yeast mutant, *dis2–11* (Kinoshita et al., 1990), which is consistent with PP1 having a role in inactivating p34^{cdc2}. Mutations in the major PP2A catalytic subunit gene lead to a wee phenotype (Kinoshita et al., 1990). This could accommodate either the data of Kumagai and Dunphy (1992), indicating that PP2A has a role in regulating cdc25 activation, or the suggestion of Lee et al. (1991) that PP2A can act directly upon p34^{cdc2} itself. Null alleles of the gene for the major PP1 isoform at 87B in *Drosophila* lead to an elevated mitotic index and an extensive degree of aberrant chromosome condensation. This suggests a delay in the transit through the mitotic state, and is seen in *Drosophila* cells following extensive mitotic arrest either in mutants or following treatment with drugs such as colchicine. Spindle organization is also affected by the loss of PP1 activity, leading to the apparent aggregation of spindle microtubules. Cytokinesis might also be affected, since some cells are observed with tetrapolar spindle structures.

C. Protein Phosphatase 2A: The 55-kDa Regulatory Subunit

The *Drosophila* genes for the catalytic subunit of PP2A (Orgad et al., 1990; Mayer-Jaekel et al., 1992), its PR65 regulatory subunit (Mayer-Jaekel et al., 1992), and its 55-kDa subunit have all been cloned. However, mutations have so far only been described in the gene for the 55-kDa subunit which corresponds to a locus independently named *aar* (*abnormal anaphase resolution*) (Gomes et al., 1993; Mayer-Jaekel et al., 1993) and *twins* (Uemura et al., 1993). The name *twins* reflects the ability of certain alleles to lead to duplication of pattern within the wing imaginal disc. The extent to which this aspect of the phenotype reflects defects in cell cycle progression is not clear, but those alleles showing this phenotype all display mitotic abnormalities characteristic of the *aar* alleles.

The mutants show an elevated index of both metaphase and anaphase figures in homozygous larval neuroblasts (Gomes et al., 1993), although in individuals hemizygous for the *aar¹* allele, the mitotic index is not as high, but the proportion of figures in metaphase is increased. This suggests that mutation in this gene also delays progression through mitosis. A second allele, aar^2, is viable when heterozygous with aar^1 and does not show a phenotype in larval neuroblasts, but females are sterile and embryos derived from them show a mitotic phenotype. In some cells of *aar* mutants, bipolar spindles do not form, but the most characteristic feature of the mutant phenotype is abnormal anaphases showing either bridging chromosomes or lagging chromatids. The bridging chromatin could represent dicentric chromosomes that have arisen through some abnormal recombination event, or sister chromatids that have failed to separate. Connections between sister chromatids could be the result of either protein or DNA interactions. The latter could arise through defects in DNA replication *per se,* or by some failure to correctly resolve the topology of an intermediate complex of sister chromatid DNA molecules. The second type of anaphase defect points to a role of the PR55 subunit in the segregation of chromatids. At least some lagging third chromosomes are intact chromatids in which pericentromeric satellite sequences and both telomeres are present. This suggests that sister chromatid separation is taking place correctly, and that the mutation is affecting the ability of chromatids either to attach to the spindle or to move upon it.

It will be of future interest to see whether PP2A regulation can be placed into an anaphase control network. The phenotype of the PR55 mutant is similar to that resulting from a mutation in *lodestar,* a *Drosophila* gene encoding a DEAH-box protein that may also have a role in controlling DNA topology (Girdham and Glover, 1991). Mutations in two other *Drosophila* genes, *zw10* (Gatti and Baker, 1989; Williams *et al.,* 1992) and *rough deal* (*rod*) (Karess and Glover, 1989) have also been described in which chromatids are observed to lag at the metaphase plate when others have already migrated to the poles. *zw10* encodes a protein that associates with the kinetochore regions of the chromatids at the metaphase–anaphase transition (Williams *et al.,* 1992). The *rod* gene has not yet been cloned.

V. Cytoskeletal Proteins and Cell Division

A. MICROTUBULES AND THEIR ASSOCIATED PROTEINS

1. Tubulin Gene Family

Microtubules are the primary structural component of the mitotic and meiotic spindles, where they play a major role in chromosome segregation. Microtubules are made of tubulin dimers containing one α-tubulin and one β-tubulin molecule. The molecular identification and characterization of the genes encoding the *Drosophila* tubulins was carried out in the early 1980s, facilitated by the high conservation of these genes. Using interspecific probes, Sanchez *et al.* (1980) identified the genes that encode the α- and β-subunits of tubulin in *Drosophila melanogaster*. These proteins are encoded by two multigene families, each including four members (however, see Section V,A,3). Their names, given after their location in the polytene chromosomes, are as follows: αTub67C, αTub84B, αTub84D, αTub85E, βTub56D, βTub60D, βTub85D, and βTub97EF. *In situ* localization with specific probes made it clear that the transcripts of some of these genes are regulated in a stage- or tissue-specific manner and in some instances even in specific cells within a tissue (see review in Fyrberg and Goldstein, 1990). One α-tubulin gene (84D) and one β-tubulin gene (56D) make up most of the tubulin present in *Drosophila* cells.

The presence of multigene tubulin families is not unique to *Drosophila*, but a common feature of many other eukaryotes. Two hypotheses have traditionally been considered to explain such a situation (see review in Cleveland, 1987). The first maintains that diversity is due to the fact that individual tubulin genes perform unique roles. The second hypothesis claims that the different polypeptides are the products of duplicated genes evolved to posses different regulatory sequences. The genetic analysis of tubulins in *Drosophila* has shed some light on this problem. Matthews and Kaufman (1987) studied the developmental effects of mutations in the gene encoding α-tubulin 84B, which accounts for the majority of α-tubulin in all tissues and at all developmental stages. All six mutant alleles studied produced at least partially stable 84B protein, although genetic analysis revealed that two of them were severe hypomorphs. These two produced a semi-dominant maternal effect polyphasic lethality plus a predominantly recessive zygotic lethality. Phenotypic defects associated with the six mutant alleles include disrupted embryos, cuticular abnormalities in pharate adults, and a short life span in the escapers. Similar conclusions can be drawn from the results of the genetic analysis of βTub85D carried out by Kemphues *et al.* (1979, 1980, 1982, 1983) and Fuller *et al.* (1987, 1988). Mutations in this gene, which accounts for most of the β tubulin in *Drosophila* testes, disrupt a variety of microtubular arrays, including those in the meiotic spindle, the axoneme, and the cytoplasm of developing spermatids. The wide variety of microtubule-related functions which are altered by mutation in these two genes clearly indicates that tubulin isoforms are multifunctional, and argues against the notion of one gene product for each microtubule class.

The third tubulin gene from *Drosophila* that has been studied by genetic analysis is βTub60D (Kimble *et al.*, 1990). The developmental pattern of expression of this gene indicates that its primary role is the assembly of microtubules involved in changes of cell shape or tissue organization, suggesting that the functionality of this isoform is very restricted. This hypothesis is substantiated by the observations made by Hoyle and Raff (1990) on the effect of ectopic expression of βTub60D under the control of the promoter of the testis-specific β-tubulin. These authors showed that in the presence of endogenous testes, specific β-tubulin, βTub60D, does not interfere with meiotic spindles or cytoskeletal microtubules, but alters the morphology of the doublet microtubules of the axoneme. The genetic analysis carried out by Kimble *et al.* (1990) has shown that βTub60D is essential for viability and fertility.

2. Microtubule-Associated Proteins

a. Biochemical Approaches. When microtubules are purified from cells, they are found to be associated with a series of proteins that, often in the absence of any other information, have been named microtubule associated proteins (MAPs). The first *Drosophila* MAP to be studied in any detail so far is Dmap 205 kDa, thought to be the *Drosophila melanogaster* homolog to mammalian MAP4. The gene that encodes this polypeptide was cloned by Goldstein *et al.* (1986), who also showed that Dmap 205 kDa colocalizes with mitotic spindle and cytoplasmic microtubules. The primary sequence and microtubule binding domain were further analyzed by Irminger-Finger *et al.* (1990). Using truncated versions of the protein, these authors defined a region containing 232 amino acids which is required for the microtubule binding activity. The protein also contains a potential MPF phosphorylation site. However, mutants in this gene appear perfectly normal, which suggests a redundant gene function (Pereira *et al.*, 1992).

A systematic search for *Drosophila* MAPs was carried out by Kellogg *et al.* (1989). These authors identified more than fifty proteins from *Drosophila* embryos which are able to bind to microtubule affinity columns. As judged by immunofluorescence, most of these proteins localize to one or more different components of the mitotic machinery, including the spindle, the centrosome, kinetochores, etc. We will come back to some of these in more detail later on in Section V,A,3.

b. The Genetic Approach. Screens for proteins that interact with tubulin have been carried out to look for second-site mutations that fail to complement mutations in tubulin. A detailed discussion of the rationale behind this approach can be found in Raff and Fuller (1984) and Hays *et al.*, (1989). In summary, it is expected that in the absence of half the normal dose of a tubulin isoform, the presence of a mutant protein able to bind to the remaining tubulin, but incapable of a specific function, might result in a mutant phenotype. The efficiency of such an approach is illustrated by the work of Stearns and Botstein (1988) in the budding yeast *Saccharomyces cerevisiae*. These authors carried out a screen for unlinked mutations that fail to complement mutations in the single β-tubulin encoding gene present in this organism (*TUB2*) and recovered a conditional lethal allele of the major α-tubulin coding gene (*TUB1*). By further applying this method, and searching for extragenic mutations that fail to complement *TUB1*, they were able to recover new mutations in *TUB2* and *TUB3* (the minor α-tubulin encoding gene). They also showed that noncomplementation between *TUB1* and *TUB2* is allele specific, suggesting that, as expected, the phenotype is due to interaction at the protein level. These studies indicate that unlinked noncomplementation can provide a fruitful method to identify mutations in interacting gene products.

Similar screens have been undertaken to find second-site mutations that fail to complement a mutation in the testes specific β-tubulin (Fuller, 1986). These mutations belong to a series of complementation groups designated *nc* (*non complementors*) followed by a number. Until now only two of them have been characterized in some detail: *nc2*, renamed *haywire* (*hay*) (see later discussion); and *nc4*, now called *whirligig* (*wrl*). The mutant allele *wrl^nc4* has been shown by Green *et al.* (1990) to fail to complement mutations not only in *β2-tubulin*, but also in the *α1-tubulin* and *hay^nc2* loci, an observation that they interpreted as an indication that the *wrl* product physically interacts with microtubule proteins. Deficiencies for the *wrl* locus and possible loss-of-function alleles obtained as revertants of the *nc* phenotype are dominant male sterile. Interestingly, this haploinsufficiency can be rescued by mutation in *β2-tubulin*, the *hay^nc2* allele, or a deficiency for *α1-tubulin*. This was interpreted as a suggestion that it is not the absolute amount of *wrl* product, but its level relative to other microtubule components that is important for male fertility. The molecular nature of *wrl* is not yet known but it seems clear from the genetic analysis carried by Green *et al.* (1990) that it is specific to spermatogenesis because mutations in this gene seem to affect only the postmeiotic stages where abnormal flagellar axonemes are observed.

The gene *exuperantia* (*exu*) was originally identified by Schupbach and Wieschaus (1986) as a maternal effect gene and later shown to be required for proper localization of *bicoid (bcd)* RNA during oogenesis (Berleth *et al.*, 1988), a process that requires functional microtubules. More recently, Hazelrigg *et al.* (1990) found that the gene is also required in the male germ line. Genetic analysis showed that six of the seven extant alleles of *exu* are male sterile, and cytological analysis confirmed that the sterility is associated with a variety of abnormal phenotypes during spermatogenesis. The first defects that are detectable appear during meiosis when considerable chromosome nondisjunction takes place. After meiosis, both early spermatid nuclei and mitochondrial derivatives are variable in size and do not keep the one-to-one association found in wild-type testes. At this stage, the phenotype is indistinguishable from that shown by any other mutant that leads to a high level of meiotic chromosome nondisjunction, such as *asp* (Ripoll *et al.*, 1985) or *polo* (Sunkel and Glover, 1988). Spermatid nuclei do not fully condense and no functional motile sperm is formed in *exu* males. All these processes are known to be mediated by microtubules (Tokuyasu, 1974a–c, 1975).

Hazelrigg *et al.* (1990) put forward two hypotheses to interpret these results. According to the first, exu is a cytoskeletal element. Thus processes which are dependent upon the cytoskeleton are altered in *exu* mutants, including the localization of *bcd*. The second hypothesis is that the *exu* protein localizes different RNAs by binding directly to them. This interpretation, which would readily explain the effect on *bcd* RNA, could also account for the effect on spermiogenesis if, as the authors propose, precise localization of RNA during the growing stage is essential for the complex processes of cell differentiation that lay ahead. Further evidence suggesting a role for the exu product in binding *bcd* RNA to microtubules has been shown by Pokrywka and Stephenson (1991). The cloning and sequencing of the *exu* gene (McDonald *et al.*, 1991) has not revealed any significant homologies to other known polypeptides.

3. Microtubule Organizing Centers

A genetic approach to microtubule function in *Aspergillus nidulans* has recently identified a protein that may play a key role in organizing microtubules. Looking for suppressors of a conditional lethal β-tubulin mutation, Weil *et al.* (1986) identified the gene *mipA* (*microtubule interacting protein A*). The sequence of this gene was found later on to be related, but significantly different from those of the α- and β-tubulin isoforms (Oakley and Oakley, 1989) and was accordingly called γ-tubulin. This γ-tubulin gene had escaped previous molecular screens. Unlike the α and β isoforms, γ-tubulin is not used to build up microtubules. Instead, it is mainly restricted to the microtubule organizing centers (MTOCs) in all the organisms in which it has been examined, from yeasts to humans. γ-Tubulin has been proposed to act as the link between an MTOC and the microtubules that it organizes, and is thought to be acting as a seed from which the α-β-tubulin dimers polymerize to form microtubules. γ-Tubulin has become one of the best characterized components of the centrosome and might provide the ideal hook with which to fish out other components of this elusive organelle. Functional assays performed by Joshi et al. (1992) suggest that it is required for cell cycle-dependent microtubule nucleation, an essential role that would be in agreement with the high degree of evolutionary conservation of the γ-tubulin genes shown by Zheng et al. (1991). A Drosophila gene encoding g-tubulin has been reported (Zheng et al., 1991).

Other centrosome associated proteins have already been identified. Several years ago, Frasch and colleagues (1986), screening libraries of monoclonal antibodies raised against nuclear fractions of cultured Drosophila cells and embryos, identified one monoclonal antibody, Bx63, which recognizes a protein that is nuclear during interphase and associated with the centrosome during mitosis. The gene encoding this protein was cloned using antibodies to screen an expression library (Whitfield et al., 1988). Since then, the same protein has been identified as one of several centrosomal associated antigens within a collection of Drosophila MAPs purified by affinity chromatography to a microtubule preparation (Kellogg and Alberts, 1992). The Bx63 antigen corresponds to the protein known as DMAP-190 by Kellogg and colleagues. When injected into syncytial Drosophila embryos, different fragments of the protein have been demonstrated to undergo cyclical association with centrosomes and nucleus in a cell cycle dependent manner (W. G. F. Whitfield; personal communication). Antibodies against this protein have been used successfully by Kellogg and Alberts (1992) to immunopurify a complex of Drosophila proteins, some of which behave as centrosomal MAPs. It is assumed that these proteins are retained in the column via an interaction with the Bx63 protein. Two of these have been characterized and correspond to DMAP60 (Kellogg and Alberts, 1992) and γ-tubulin (Raff et al., 1993). This observation suggests that these three proteins are components of a complex involved in the interaction between centrosomes and microtubules, of which very little is known at this moment.

If little is known about the manner whereby the centrosome nucleates microtubules, less is known about the mechanisms that mediate and control its replication. Some answers to these puzzles may arise from studies of mutants in which centrosome behavior is impaired. Mutations at the loci merry-go-round (Gonzalez et al., 1988), and aurora (M. Leibowitz et al., unpublished observations) appear to permit centrosome replication but prevent centrosome separation. This leads to monopolar structures in which the duplicated centrosome nucleates an aster-like arrangement of microtubules to which are attached a circular array of chromosomes (Glover *et al.*, 1993; M. Leibowitz *et al.*, unpublished observations). Similar arrays of monoastral half-spindles and associated chromosomes have been described in cells mutant for *urchin*, a member of the *bimC* subfamily of proteins (Wilson *et al.*, 1992). In *urchin* mutant cells, each monaster contains multiple centrosomes, suggesting that the wild-type product of this gene is required for centrosome segregation or to maintain the separation of spindle poles. Monopolar half-spindles can also be induced by drug treatment. Callaini and colleagues (1989) have shown that diazepam induces abnormal mitosis in the early *Drosophila* embryo. The polyploid and aneuploid figures they observe appear to result from spindle fusion, inhibition of centrosome movement, and the formation of monopolar spindles with circular mitotic figures. Similar mitotic figures are also seen in mammalian cell lines treated with diazepam (see references cited in Callaini *et al.*, 1989).

The ability to nucleate microtubules is not restricted to the centrosome. Indeed, terminally differentiated cells, including those of the wing blade of pharate *Drosophila* adults, that lack a classical centrosome have been shown to contain large arrays of microtubules that run all the way across the cytoplasm and seem to be anchored to the cell membrane (Mogensen and Tucker, 1987; Mogensen *et al.*, 1989). Chromosomes have also been shown to be able to organize microtubule growth. This centrosome-independent polymerization of microtubules has been observed during meiosis in *Drosophila melanogaster* oocytes (Theurkauf and Hawley, 1992). *Drosophila* oocytes do not contain centrioles because they are eliminated during oogenesis (Carpenter, 1975). Theurkauf and Hawley (1992) have shown that microtubule assembly during prometaphase actually takes place on the chromatin, giving rise to an anastral meiotic spindle. This situation is reminiscent of that found in mice where centrosomes are also absent. Interestingly, in this system it has been shown that the cytoplasm of metaphase oocytes contains an activity that enables sperm chromatin to induce microtubule assembly (Harrouk and Clarke, 1993).

4. The Kinesin Family of Motor Proteins

The kinesin family is a series of mechanochemically active proteins which are able to move along microtubules and transport other proteins or organelles with them (Allen *et al.*, 1985; Vale *et al.*, 1985; see reviews in Goldstein, 1991; Endow, 1991, 1993, Endow and Titus,

1992). In *Drosophila*, as in other organisms, the kinesin family has many members (Endow and Hatsumi, 1991; Stewart *et al.*, 1991), of which so far only three have been characterized in any detail, namely, the *kinesin heavy chain (khc)*, protein and the products of two genes that were initially identified by mutation analysis to be required for chromosome stability, *no distributive disjunction (nod)* and *non claret disjunctional (ncd)* (Carpenter, 1973; Endow *et al.*, 1990; McDonald *et al.*, 1990).

a. kinease heavy chain (khc). The *Drosophila* homolog of kinesin, the founder of the family found in a variety of organisms and cell types, was first isolated by Saxton *et al.* (1988) and Yang *et al.* (1988). Microtubule binding analyses have shown that khc has an N-terminal 50-kDa globular motor domain that contains an ATP binding consensus sequence, together with a 50- to 60-kDa domain with the characteristic features of an α-helix. The C-terminal domain is thought to be responsible for the interaction with other molecules, vesicles, and organelles. The function of khc has been examined *in vivo* by genetic analysis and immunolocalization (Saxton *et al.*, 1991). These authors concluded that mitosis can proceed in the absence of khc function and that khc is not a major component of the mitotic apparatus.

b. nonclaret disjunctional (ncd). ncd was originally known as ca^{nd}, an allele of the eye color locus *claret (ca)* (Lewis and Gencarrella, 1952). It is now known that this mutation is a deficiency that uncovers both the *ca* and *ncd* loci, which are tightly linked (Yamamoto *et al.*, 1989). *ncd* females show nondisjunction and loss of chromosomes in meiosis, in addition to a maternal effect in which chromosome loss occurs in the early embryonic mitoses (Lewis and Gencarrella, 1952). All chromosome pairs are affected by the meiotic nondisjunction produced by *ncd* while the embryonic chromosome loss affects almost exclusively maternally supplied chromosomes. Meiotic nondisjunction affects almost all the nonexchange chromosomes, but a significant percentage of those that have undergone exchange are also affected. Cytological analysis of *ncd* females reveals diffuse spindles with abnormal poles and nuclei of varying number and size (Wald, 1936; Kimble and Church, 1983; Hatsumi and Endow, 1992). The bundling of microtubules emanating from the chromosomal nucleation points appears to require the activity of the ncd protein (Walker *et al.*, 1990; Hatsumi and Endow, 1992; Sequeira *et al.*, 1989). In *ncd* mutants, this bundling is not complete, leading to spindles with broad poles that are often distorted around the metaphase plate. Later in development, branched spindles that extend to free chromosomes can be observed during the early cleavage divisions of embryos from *ncd* females (Hatsumi and Endow, 1992). *ncd* was shown by Endow *et al.* (1990) and McDonald and Goldstein (1990) to encode a kinesin homolog. Thus it became the first of the kinesin proteins to be identified that is required for meiotic chromosome segregation. The protein has been shown *in vitro* to be a microtubule motor protein that, somewhat unexpectedly, translocates toward the minus ends of microtubules, opposite to the direction of kinesin movement (Walker *et al.*, 1990; McDonald *et al.*, 1990). Furthermore, the protein is also capable of rotating and bundling microtubules (Walker *et al.*, 1990; McDonald *et al.*, 1990).

   c. *no distributive disjunction (nod)*. Females mutant for *nod* (Carpenter, 1973; Zhang and Hawley; 1990) exhibit increased frequencies of nondisjunction of those chromosomes that have failed to recombine during meiosis I. None of the extant alleles of *nod* causes lethality, decreases the frequency of exchange, or impairs the segregation of the exchange chromosomes. Mutation in *nod* also results in misbehavior of maternal chromosomes in the first zygotic division of embryos derived from *nod/nod* females although at a very low rate. nod is also a member of the kinesin family (Zhang *et al.*, 1990). Cytological analyses revealed that the defects caused by mutation in *nod* include abnormal positioning of nonexchange chromosomes away from the spindle (Hatsumi and Endow, 1992; Theurkauf and Hawley, 1992). *nod* exorts a force on these chromosomes in the direction of the metaphase plate, and is counteracted by a poleward-directed force that allows nonexchange chromosomes to move toward the poles in a size-dependent manner (Zhang and Hawley, 1990). In this way, the tiny fourth chromosome becomes positioned between the poles and the equator. Mutation in *nod* leads to the dissociation of nonexchange chromosomes from the spindle or their premature movement to the pole. This observation has been interpreted to mean that nod acts like chiasmata holding chromosomes to the metaphase plate. It is therefore predicted that if nod is a microtubule motor, its direction of translocation will be the opposite to that shown by ncd.

   While loss-of-function alleles of *nod* have been shown not to affect mitosis, one unusual allele of *nod* (nod^{DTW}) does result in alterations of chromosome transmission in the somatic cycle, in addition to affecting meiosis. Originally called $l(1)TW^{6cs}$, the most conspicuous phenotype of this cold-sensitive, recessive, lethal mutation is the formation of anaphase bridges in mitosis and consequential chromosome breakage (Gatti and Baker, 1989). Rasooly *et al.* (1991) were able to show that $l(1)TW^{6c}$ is a dominant antimorphic allele of *nod*, and they re-named it nod^{DTW}. Sequencing reveals a single amino acid change in the putative ATP binding domain of the *nod* kinesin-like protein. It is proposed that the defective protein interferes with proper chromosome movement, perhaps by binding irreversibly to microtubules. This interpretation is based on the knowledge that hydrolysis of ATP is not required for kinesin to bind to microtubules, but is needed for its release from microtubules once it is bound (Vale *et al.*, 1985). nod^{DTW} might therefore bind to microtubules and by so doing might inactivate them, thus "poisoning" the mitotic apparatus.

The close functional relation between the *ncd* and *nod* products suggested by the similar phenotypes of mutations in these loci and their amino acid sequence is further substantiated by the observation that despite their recessive nature, the double heterozygote shows frequencies of chromosome nondisjunction higher than those shown by any of the single heterozygous combination (Knowles and Hawley, 1991). These authors believe that this effect is due to dosage sensitivity of essential components of the meiotic apparatus rather than to physical interaction between these two products.

The implications of the realization that both ncd and nod are members of the kinesin family for the interpretation of chromosome segregation during female meiosis in *Drosophila* have been discussed in an excellent review by Carpenter (1991). A recent review on progress in the functional characterization of newly identified microtubule motors has been published by Sawin and Endow (1993).

B. ACTIN FILAMENT NETWORKS

1. The Actin Gene Family

Actin is a major contractile protein that plays essential roles in many cellular processes, from morphogenesis to cell motility, and cytokinesis. With the exception of yeasts, all eukaryotes studied contain different forms of actin encoded by highly conserved multigene families.

The *Drosophila* actin genes were identified and cloned by Fyrberg *et al.* (1980) and Tobin *et al.* (1980) using homologous genes from chicken and *Dictyostelium* as probes. *In situ* hybridization localized these sequences in the polytene chromosomes at positions 5C, 42A, 57A, 79B, 87E, and 88F. The expression patterns of the members of this gene family suggest different functional roles. *Act5C* and *Act42A* appear to encode cytoplasmic actins; *Act79B* encodes muscle actin in adult legs and thorax; *Act57B* and *Act87E* seem to be expressed in intersegmental muscles in larvae and adults; and *Act88F* appears to encode flight muscle-specific actin (Zulauf *et al.*, 1981). Mutagenesis of the *Act88F* locus has generated many mutants that exhibit a range of flight defects but are otherwise normal, supporting the conclusion that the product of the actin locus at 88F functions specifically in flight muscles. The screen for mutant alleles of the A*ct87E* carried out by Manseau *et al.* (1988) failed to identify any mutations. Since the region is near saturation for loci that can be mutated to lethal or visible phenotypes, this result suggests that either the loss of function of *Act87E* results in a phenotype other than recessive lethality, or its function can be performed by one or more of the other actin genes.

2. Actin-Binding Proteins

By using F-actin affinity chromatography columns, Miller *et al.* (1989) identified more than forty proteins from *Drosophila* embryos which are able to bind to actin. Some of them colocalize with actin throughout the cell cycle. Others remain in the furrow region after actin has disappeared. Antigens of a third group are not present in the cap structures above the blastoderm nuclei; their location is restricted to the cap border or to the leading edge of the membrane furrows. Taken together, these observations suggest that the group of actin interacting proteins is highly complex. The cloning and genetic analysis of these antigens, and others that can be isolated in the future, will provide the basis for the dissection of the actin network.

3. Other Genes Involved in Actin-Mediated Processes

Genes other than actin that are involved in actin-mediated processes such as pseudocleavage furrow formation, cellularization, cytokinesis, and ventral closure have now been identified. These genes are expected to encode products that are members of the actin filament network and will be instrumental in its dissection.

a. Cellularization. By cellularization we refer to the formation of pole and somatic cells after the first syncytial divisions. The formation of the pole cells was mentioned in Section II,C. Little is known about the control of this process other than that it requires an interaction between the polar granules and centrosomes (Raff and Glover, 1989). The cellularization of the remaining somatic nuclei occurs during the interphase of cycle 14. It takes place in two discernible phases: a slow phase during which new membrane is added and a fast phase during which membrane from the surface is pulled down into the cleavage furrows. The distribution of actin and tubulin arrays in precellularization embryos has been reported by Karr and Alberts (1986) and Warn et al. (1984). They showed that although syncytial blastoderm nuclei are not separated by cell membranes, each nucleus has its own individualized region of cytoplasm with associated arrays of actin filaments and microtubules. The actin filaments are arranged in domains above each nucleus, closely associated with the plasma membrane. This structure shows dramatic reorganization between mitosis and interphase. Myosin colocalizes with plasma membrane-associated actin during cellular blastoderm formation.

Although most of the functions required for cellularization are expected to be contributed maternally, it has been known for a long time that some zygotic activity is required for cellularization because it can be disrupted by inhibitors of RNA synthesis (Zalokar and Erk, 1976). Wieschaus and Sweeton (1988) have carried out a detailed analysis to identify the X-linked zygotic genes required for this process and Marril et al. (1988) have extended this study to those located on the autosomes. These studies, based on the production of embryos deficient for large autosomal regions, showed that just a few genes were required zygotically for cellularization. They map to 6F1–2, 26B-F, 40C, 99D-F, and 100A-C.

A few maternal effect genes involved in this process have already been reported. The analysis of the phenotypic effects of mutation in one of them, sponge, has yielded new insights into different aspects of the formation and function of actin structures in the developing Drosophila embryo (Postner et al., 1992). Embryos laid by homozygous sponge mothers lack the dome-shaped actin caps which are present above interphase nuclei in the wild type. The maternal protein is essential for the formation of actin caps and metaphase furrows, but is not required for the formation of the actin arrays that mediate cellularization. Despite the lack of actin caps, the actin binding protein 13D2 (Miller et al., 1989) forms normal structures in sponge-derived embryos. Thus the 13D2 protein seems to be required upstream of actin assembly as far as the formation of the caps is concerned. Interestingly, Postner et al. (1992) also observed that sponge-derived embryos display a variety of mitotic defects, including mitotic spindles nucleated between centrosomes from different nuclei; chromosomes from different nuclei segregating to a common pole; and a high frequency of polyploid and aneuploid cells. Similar alterations are found in embryos mutant for daughterless-abolike (Sullivan et al., 1990) and embryos derived from three loci, grapes, scrambled, and nuclear fall-out, which have been shown to be required for pseudocleavage furrow formation and cellularization (W. Sullivan et al., personal communication). The onset of the phenotypic alterations produced by mutation in these loci does not take place until the nuclei have migrated to the surface of the embryo. After nuclear migration, the embryos derived from females homozygous for any of these loci exhibit extensive nuclear division abnormalities. The same alterations have also been reported following treatment with the actin filament depolymerizing drug, cytocholasin (Zalokar and Erk, 1976; Edgar *et al.*, 1987). Taken together, these observations show that, even prior to cellularization, the actin network is essential for maintaining the identity of each nucleus and its mitotic machinery.

b. Cytokinesis. Cytokinesis is the process whereby the plasma membrane "pinches in" at the midpoint of the decaying mitotic apparatus to produce the two daughter cells. This process does not continue to completion in all cell types. We have already seen how cysts are produced during gametogenesis in both sexes in which the cells remain connected by canals (Section II,D,3,a). Cytokinesis is largely mediated by the contractile ring, a structure that forms beneath the cell membrane during anaphase and whose contraction produces the cleavage furrow that will eventually split the cell in two. The contractile ring contains both actin and the cytoplasmic isoform of myosin.

A genetic analysis has identified the first nonactin component of the mechanism that drives cytokinesis in *Drosophila*. It corresponds to the regulatory light chain of nonmuscle myosin which is encoded by *spaghetti squash* (*sqh;* Karess *et al.*, 1991). Both standard microscopic analysis of fixed tissue and real time observation of mitosis in the ganglia of mutant larvae (R. E. Karess, personal communication.) indicate that cytokinesis is disrupted in individuals mutant for *sqh*. The brains of mutant larvae can reach sizes 50 to 100 × the volume of a wild-type sibling and contain many highly polyploid cells with a wide range of chromosome numbers extending beyond octoploidy. Polyploid cells in *sqh* mutants fall into two categories, depending on whether they carry several individual nuclei within them or a single large one. The same is true for other systems in which cytokinesis has been inhibited (Hatzfeld and Buttin, 1975; Kiehart *et al.*, 1982). It is thought that multinucleate cells are the result of exponential accumulation of nuclei in cytokinesis-defective cell cycles and that single large nuclei are formed at a later stage by fusion. Although the possible analogy between actin-myosin-based muscle movement and cytokinesis is tantalizing, Karess *et al.* (1991) point out that the contractile ring differs markedly from the muscle sarcomeres in that it is a transient, cell cycle-regulated structure whose rapid assembly and disassembly must be coordinated with other cellular events.

The gene encoding the *Drosophila* cytoplasmic, myosin heavy chain itself has also been identified recently (Young et al., 1993). It corresponds to zipper, a gene that was known to be required for the completion of dorsal closure. The fact that mutation in sqh does not show a zipper phenotype is probably due to the combined action of the perdurance of the maternal myosin and the leakiness of the sqh mutation. The possible mitotic effect of mutation in zipper has not yet been addressed.

Another gene that has been reported to be required for cytokinesis is pebble. Mutation in pebble results in embryos that contain fewer and larger cells than wild type embryos (Hime and Saint, 1992; Lehner, 1992). This results from mitoses 14, 15, and 16 taking place without cytokinesis. During mitosis, cells containing duplicated mitotic figures can be observed, leading to an increase in the average cell size as cells become multinucleate. Unlike nuf, grapes, and scrambled, pebble seems to be specific for cytokinesis and does not seem to be required for pseudocleavage furrow formation or cellularization. The molecular nature of these genes is not known yet.

C. NUCLEAR ENVELOPE

1. Behavior of the Nuclear Envelope in the Mitotic Cycle

The behavior of the nuclear envelope throughout the cell cycle varies among organisms. In vertebrates it undergoes a total breakdown at the onset of mitosis, whereas in yeasts it remains almost intact throughout mitosis. In Drosophila, the situation is somewhat intermediate. Detailed electron microscopic analysis of mitosis during the syncytial blastoderm stage has shown that the nuclear envelope undergoes complete breakdown only at the spindle poles. Large holes are left behind where the nuclear pores disassemble during mitosis, but elsewhere the nuclear envelope remains largely intact (Stafstron and Staehelin, 1984). This residual envelope has been termed the "spindle envelope." It is associated with a second layer of membranous cisternae which is built up just outside the original membranes. Harel et al. (1989) have further studied the partial breakdown of the nuclear envelope during mitosis in Drosophila embryos using immunofluorescence microscopy to follow the behavior of three nuclear envelope components, lamin, otefin (a 53-kDa lamina component), and gp 188 (a putative pore complex component). They showed that the "spindle envelope" in Drosophila embryos contains lamin and otefin, but no gp188. Interestingly, the maintenance of the spindle envelope is dependent upon microtubule integrity. When fixation is carried out under conditions that depolymerize microtubules, lamin and otefin become delocalized during mitosis, but if taxol (a potent microtubule stabilizer) is added before fixation, the envelope also becomes stabilized. More recent work has shown that the behavior of the nuclear envelope in somatic tissues (third instar larval brains) is comparable to that reported here (A. T. C. Carpenter, personal communication; Gonzalez, 1994).

Recent evidence confirms that the nuclear envelope plays an active role in mitosis in *D. melanogaster* (Hiraoka *et al.*, 1989). Using three-dimensional *in vivo* time-lapse microscopy, these authors followed the chromosomes from telophase until they entered the next mitosis. These studies revealed that certain chromosomal regions distinct from centromeres and telomeres serve as foci for the decondensation and condensation of diploid chromosomes. Chromosome regions that serve as attachment sites to the nuclear envelope in polytene chromosomes have also been described by Mathog and Sedat (1989), but as yet there is no evidence of the correlation between these and those seen by Hiraoka *et al.* in diploid nuclei.

2. *The Lamins*

The nuclear envelope is made up of two lipid bilayers which join at the nuclear pore complexes. Underneath the inner membrane is the nuclear lamina whose main component is lamin, an intermediate filament-like protein. There are three forms of nuclear lamin protein in *Drosophila* (Smith *et al.*, 1987). Lamin Dm0 (76-kDa) is processed into Dm1 (74-kDa) which is then incorporated into the nuclear envelope. A third form, Dm2, is also seen as a result of phosphorylation on one of two sites that leads to a mobility shift to an apparent molecular weight of 76-kDa. Gruenbaum *et al.* (1988) have reported that the *Drosophila* nuclear lamin precursor Dm0 is translated from either of two developmentally regulated mRNA species apparently encoded by a single gene at 25F. The cDNA encoding the *Drosophila* lamins was identified by screening an expression library. Conceptual translation shows that the polypeptide encoded by this gene is highly homologous to human lamins A and C. Recently, Bossie and Sanders (1993) have identified a novel intermediate filament protein called pG-IF which shows high homology to the lamin C subtype and has a predicted molecular weight of about 70-kDa. The gene encoding this protein maps to 51A. There are no known mutants at either loci.

3. *fs(1)Ya*

Females homozygous for mutations in *fs(1)Ya* (Lin and Wolfner, 1989) lay eggs that fail to develop. The earliest phenotype of the embryos laid by these females is developmental arrest from the very first nuclear divisions, thus suggesting that the product of *fs(1)Ya* is involved in initiating the first embryonic mitosis. The embryos that complete this stage are arrested in later stages of development, suggesting that the gene might also be required for subsequent embryonic mitoses. *fs(1)Ya* encodes a 2.35 kb maternal transcript which produces a 91.3-kDa protein that is hydrophilic and contains putative MPF phosphorylation sites as well as potential nuclear localization signals (Lin and Wolfner, 1991). *fs(1)Ya* also contains the sequence UUUUUAU, a meiotic maturation-specific polyadenylation signal that is found upstream of the canonical polyadenylation signal in certain *Xenopus* and mouse maternal mRNAs and is responsible for their translational activation (Lin and Wolfner, 1991, and references cited therein). The *fs(1)Ya* protein is translated during oocyte maturation independent of fertilization and persists throughout embryogenesis. Immunohistological studies have shown it to localize to the nuclear envelope in a cell cycle-dependent manner that differs from that shown by lamin. It disperses from anaphase to telophase and is localized to the nuclear envelope during interphase up to the next metaphase. Thus the *fs(1)Ya* gene encodes a cell cycle-dependent nuclear envelope component required for embryonic mitoses.

VI. The Chromosome Cycle: Replication, Repair, and Segregation

A. DNA REPLICATION

1. Genes for the Replication Enzymes

Replication of the entire genome is achieved within 4 min in each of the rapid division cycles of the syncytial *Drosophila* embryo. This highly efficient replication is achieved in part because of the density of replication origins along the DNA at approximately 9 kb intervals (Blumenthal *et al.*, 1973). Furthermore, the embryo has large maternal stockpiles of the replication proteins. At least three major DNA polymerases are utilized for eukaryotic DNA replication, of which DNA-polymerase α has been purified from embryo extracts and is the best characterized. This enzyme has four subunits; the largest 180-kDa subunit has polymerase activity; two subunits of 50 and 60 kDa constitute primase activity; and a 73-kDa subunit appears to mask the exonuclease activity of the 180-kDa subunit (Cotterill *et al.*, 1987a–c). The genes for the 180- and 73-kDa subunits have been cloned (Hirose *et al.*, 1991; Melov *et al.*, 1992; Cotterill *et al.*, 1992), but mutations have not yet been identified in these genes.

The gene for *proliferating cell nuclear antigen* (PCNA) has also been cloned (Yamaguchi *et al.*, 1990). PCNA is a highly conserved protein now known to be an auxiliary factor of DNA polymerases δ and ε. It shows a cyclical behavior revealed by immunostaining during the cell cycle, leading it to be originally termed "cyclin," a term that was subsequently abandoned. Unlike the original competitors for this nomenclature, cyclins A and B, PCNA does not undergo degradation each cell cycle. Indeed its cyclical behavior is not recognized by all antibodies. In S-phase nuclei, however, the protein localizes to sites of ongoing DNA synthesis. It is postulated to act as a processivity factor for DNA polymerase δ and to coordinate leading and lagging strand synthesis at the replication fork. Several studies have also suggested a role for PCNA in DNA repair. Recent work by Henderson and colleagues (1994) has demonstrated that the *Drosophila PCNA* gene corresponds to *mutagen sensitive 209 (mus209)*. *mus209* is one of a series of *mutagen sensitive* mutants that are hypersensitive to the killing effects of DNA-damaging agents (see review in Boyd *et al.*, 1987). Several of these genes are likely to have overlapping roles in DNA repair, synthesis, and recombination, and at least three loci, *mus101, mus105*, and *mus 109*, are known to be required for cell cycle progression (Gatti, 1979; Gatti and Baker, 1989; Baker *et al.*, 1982). Henderson and co-workers show that maternal expression of PCNA is required for zygotic development, which is consistent with the requirement for DNA synthesis in the early embryo; and that the later zygotic requirement for PCNA is correlated with periods of cell proliferation. Intriguingly, certain *mus209* alleles behave as recessive suppressors of position effect variegation, the process by which euchromatic genes can be inactivated following transposition into a heterochromatic environment. This suggests either that PCNA might have an additional role in regulating chromatin structure directly, or that the organization of DNA or chromatin at the euchromatin-heterochromatin boundary is affected by DNA replication and repair processes.

2. Mutants Showing Uncontrolled Replication in Syncytial Embryos

Mutations have been described at three loci, *gnu (giant nuclei), plu (plutonium)*, and *pgu (pan gu)*, that result in the repeated initiation of DNA replication in the nuclei of either unfertilized or fertilized eggs to produce giant nuclei (Freeman *et al.*, 1986; Freeman and Glover, 1987; Shamanski and Orr-Weaver, 1991). Unfertilized *gnu* eggs also undertake extensive DNA replication, leading to the suggestion that *gnu* regulates the onset of DNA replication following the completion of female meiosis. In fertilized eggs derived from females homozygous for mutations at these loci, up to five giant nuclei can develop. These correspond to the replicated products of the three polar bodies and the male and female pronuclei. In one weak allele of *pgu*, however, 50% of the embryos have between 6 and 16 nuclei. This indicates that a limited number of mitoses have taken place and suggests that *pgu* at least has an additional role in regulating the entry into S-phase in the cleavage divisions as well as at fertilization. If, however, females homozygous for this weak allele of *pgu* are also deficient for one wild-type copy of either *plu* or *gnu*, then these limited mitoses cannot take place. This indicates that the three genes control a common regulatory function.

Interestingly, double mutants between *gnu* and *fs(1)Ya* fail to develop giant nuclei and show the *fs(1)Ya* phenotype (Lin and Wolfner, 1991). This is interpreted as suggesting either that the action of *gnu* is mediated through *fs(1)Ya* or that the two genes work in different pathways with *fs(1)Ya* establishing a precondition for the action of *gnu*. Of the three genes, only *plu* has been cloned (T. Orr-L. Weaver, personal communication), but the sequence of the encoded protein offers few clues as to its molecular function.

B. DNA REPAIR

Several of the *mutagen-sensitive loci* discussed above are likely to have a role in both DNA recombination and repair. Furthermore, several loci have been described that have a role in mediating DNA recombination during female meiosis (Hawley, 1988). A discussion of these genes is outside the scope of this review. Two genes have been recently described that encode members of the helicase superfamily that may have mitotic roles, although in both cases their true function remains enigmatic.

1. haywire (hay)

The gene *hay* was identified in the previously described screen for mutations that fail to complement certain β-tubulin mutations (Regan and Fuller, 1988). Consequently the wild-type product of this gene was originally expected to be a molecule that would associate primarily with microtubules. Males homozygous for mutations in *hay* have abnormal meiotic spindles. Some cells show clustered chromosomes typical of metaphase I, and have four asters as if they were undergoing meiosis II after failure of both chromosome segregation and cytokinesis. The chromosomes seem to remain arrested in a metaphase I-like stage as unresolved bivalents, even in postmeiotic stages. Thus, during the onion stage, highly condensed chromatin is often found in very small nuclei which are associated with much larger Nebenkerns.

In their genetic analysis of this locus, Regan and Fuller (1990) showed that two classes of mutations revert the failure of hay^{nc2} to complement mutations in tubulin genes: those viable over a deficiency and with less severe phenotypic consequences than hay^{nc2} itself, and those that behave as recessive lethals. All the revertants behave as intragenic mutations belonging to a single complementation group. Viable alleles show a variety of defects in spermatogenesis which include defective flagellar elongation, abnormal nuclear shaping, alterations of the meiotic spindle, and failure of homologs to separate during meiosis I. These results were then interpreted as an indication that the product of *hay* itself is involved in spermatogenesis, which is consistent with the expectation that the *hay* product is a structural component of microtubules in *Drosophila* males.

The finding, following the molecular cloning of *hay,* that this gene encodes a protein member of the helicase superfamily, with 66% identity to the product of the human gene *ERCC3* was therefore unexpected (Mounkes *et al.,* 1992). The *ERCC3* gene is involved in DNA repair and is associated with the diseases xeroderma pigmentosum and Cockayne's syndrome. Human patients with these conditions often show CNS disorders and hypogonadism in addition to extreme sensitivity to ultraviolet irradiation. These results prompted a reexamination of the phenotypes associated with mutations in *hay,* and the finding of increased UV sensitivity, motor defects, and reduced life span. These observations, which are in accordance with the high degree of amino-acid identity between *hay* and *ERCC3,* raise the question of how *hay* mutations can fail to complement β-tubulin mutations. Mounkes *et al.* (1992) argue that since the levels of β-tubulin are at a threshold in the β-tubulin mutant used, a reduction in gene expression consequential to either failure to repair DNA lesions or impaired transcription could lead to the mutant phenotype. The latter seems more likely since ERCC3 has recently been shown to encode the basic transcription factor BTF2 (Schaefer *et al.,* 1993). These authors also suggest the possibility that the *hay* protein might be multifunctional, but a precise explanation remains to be found.

2. *lodestar*

lodestar (84D13–14) was originally identified through maternal effect mutations that result in chromatin bridges at anaphase. Immunocytological analyses have shown that the *lodestar* protein undergoes a cell cycle-dependent relocalization in mitosis in embryos. It is found in the cytoplasm during interphase but rapidly enters the nucleus at prophase. It then remains restricted to the area of the spindle envelope until telophase, when it is found almost exclusively within each newly formed nucleus. The sequence of the protein encoded by *lodestar* indicates that it is another member of the superhelicase family (Girdham and Glover, 1991). It has strong homology to the human gene *ERCC6*, and the yeast genes *RAD 5*, *RAD16*, and *RAD54*, genes that are required for DNA repair, together with another group of genes that appear to have a role in transcriptional regulation (Troelstra *et al.*, 1993). Repair defects in a *Drosophila* function could result in errors in DNA replication or in the resolution of concatenated sister chromatids and thus help explain the types of mitotic defect observed.

C. Chromosome Pairing and Segregation:
The Centromeric Function

The pairing between sister chromatids is mediated by the centromere. This also contains the kinetochore that mediates the interaction between the chromosomes and the spindle microtubules. Our picture about the organization of the *Drosophila* centromere is still very sketchy. In mammalian systems, different kinds of proteins have been found at the contact points between sister chromatids and along the chromosome region that organizes the kinetochore (Brinkley *et al.*, 1992). CLIP antigens are present along contact points within and outside the centromere (Rattner *et al.*, 1988). INCENP antigens, on the other hand, are confined to the inner centromeric region (where they accumulate following colchicine treatment) until anaphase. Both CLIP and INCENP antigens lose their association with the chromatids during anaphase. Some INCENP antigens have been described as remaining at the metaphase plate as the chromosomes migrate to the poles (Cooke *et al.*, 1987) and to relocate to the spindle where they are thought to perform cytoskeletal roles (Earnshaw and Cooke, 1991). Recent reviews on chromosome structure and function during mitosis have been published by Earnshaw (1988, 1991), Earnshaw and Bernat (1991), Earnshaw and Tomkiel (1992), and Rattner (1992).

The *Drosophila* homologs of these antigens are not known yet. Tudor *et al.* (1992) have shown that conceptual translation of *pogo*, a transposable element of *Drosophila melanogaster,* results in a protein which shares 51% similarity and 21% homology with the major human centromere antigen CENP-B (Earnshaw *et al.*, 1987). Nevertheless, as Tudor *et al.* (1992) point out, the functional implication of such homology is doubtful since *pogo* elements are restricted to *D. melanogaster* and are not found in other *Drosophila* species. The only centromeric antigen that has been identified thus far in *Drosophila* is S5–39 (Kellogg *et al.*, 1989). The antibody specific for this antigen recognizes a 59-kDa protein that can be isolated by its ability to bind microtubules and which is localized to the kinetochore region during metaphase–anaphase. A strong S5–39 signal can also be seen over the centrosomal region from prophase to telophase.

The functional DNA component of the centromeres must lie within the heterochromatic regions of the chromosomes, the molecular characterization of which has been impeded in the past by the repetitive nature of most of its DNA. Thus, although many genetic functions— including vital loci—have been identified within the heterochromatin in *Drosophila* (see review in Gatti and Pimpinelli, 1992), their molecular nature remains largely unknown. A way forward lies in the characterization of minichromosome derivatives that retain their ability to segregate correctly (Abad *et al.*, 1992; Carmena *et al.*, 1993). One such minichromosome, *Thelma*, reported by Wynes and Henikoff (1992), is peculiar in that it carries very little centromeric heterochromatin. Although stable enough to be kept, the transmission of the *Thelma* chromosome is not perfect and missegregation occurs both in mitosis and meiosis. This feature has allowed these authors to make the remarkable observation that centromere function in this chromosome is affected by well-known modifiers of position effect variegation. Moreover, these effects take place in the opposite way in which they modify variegation of euchromatic genes, that is suppressors of variegation enhance the instability of *Thelma* while enhancers of variegation reduce it. In this respect, the *Thelma* centromere function behaves in the same way as some heterochromatic genes, such as *light* (see Gatti and Pimpinelli, 1992, for a review on functional elements in heterochromatin).

Whether it takes place between homologous chromosomes during meiosis I or between sister chromatids during mitosis and meiosis II, pairing is essential to ensure the equitable segregation of the genetic material. Here we review recent findings on chromosome pairing. We have not included the process of pairing in female meiosis I, which has been extensively reviewed elsewhere (see, e.g., Hawley, 1988).

1. Pairing of Homologs in Male Meiosis

The pairing between chromosome homologs during male meiosis in *Drosophila* was postulated by Cooper (1964) to occur at specific pairing sites. That simple homology between the X and Y chromosomes is not sufficient is shown by the fact that naturally occurring sequences that are common to both the X and the Y chromosome cannot promote pairing between rDNA-deficient sex chromosomes (Lindsley and Sandler, 1957). Although the close link between the rDNA and the colochore responsible for the achiasmatic pairing between the X and Y chromosome has been known for a long time, the pairing sequences been characterized at the molecular level only recently (McKee and Karpen, 1990). By producing transgenic flies, these authors were able to observe that a single rRNA gene, only about 12 kb in length, inserted in a heterochromatin deficient X chromosome is able to drive the pairing and disjunction from the Y chromosome. McKee and Karpen also observed that autosomal rDNA insertions do not interfere with X-Y pairing or disjunction of these two chromosomes. They argue this effect is probably due to a dose-dependent nature of the process of pairing, so that more than one dose of the rDNA gene is required to make the affected autosome compete efficiently for the colochores of the X and Y chromosomes.

Later on, McKee et al. (1992) were able to map more precisely the sequences within one of these inserted X-linked rRNA genes by generating deficiencies via P element-mediated destabilization of the insert. They showed that a rDNA fragment containing only a few copies of the 240 bp repeats from the intergenic spacer is able to stimulate X-Y disjunction as effectively as a complete rRNA gene. The precise mechanism of pairing is as yet unknown, but based on these results, McKee et al. (1992) have suggested a model for achiasmatic pairing which involves the formation of a nonrecombinogenic heteroduplex.

Unlike the pairing between sex chromosomes, the DNA sequences involved in pairing between autosomes are not yet known. The distribution of these sites on chromosome 2 has been recently studied by McKee et al. (1993) by analyzing the meiotic pairing and segregation of a series of 2-Y transpositions. This work has shown that the shortest euchromatic regions tested were still able to promote pairing, as seen by the appearance of quadrivalents involving the two sex and second chromosome bivalents. Moreover, the frequency of such events was proportional to the length of the duplicated region. McKee et al. (1993) concluded that autosomal pairing in *Drosophila melanogaster* males is based on general homology, despite the lack of homologous recombination. They also confirmed that heterochromatin does not play a role in this process.

2. Sister Chromatid Pairing and Separation

Pairing at the centromeric region has been shown to be resistant to colcemid and to provide the last point of attachment between the sister chromatids prior to the onset of anaphase. In *Drosophila*, sister chromatids remain in close contact along their heterochromatic regions well after euchromatic regions have separated (Kaufmann, 1934; Gonzalez *et al.*, 1991). This tight association is lost only at the onset of anaphase. The mechanisms that drive this process are not known. At this moment we can only hypothesize that it involves both structural proteins and specific target DNA sequences, coordinated by a control mechanism that ensures the correct timing is achieved. None of these have yet been identified.

It has often been postulated that concatenation of DNA after DNA replication could ensure proper binding between sister chromatids (Murray and Szostak, 1985). However, Koshland and Hartwell (1987) have shown that there is no extensive interlocking of plasmid yeast prior to anaphase in mitosis. If DNA concatenation does play a role in holding together sister chromatids, it is reasonable to assume that proteins able to resolve these structures might be essential during the metaphase–anaphase transition. The work of Shamu and Murray (1992) seems to indicate that this is the case at least in *Xenopus* from their finding that topoisomerase II is required at the onset of anaphase. DNA topoisomerase II is also essential for chromosome segregation in *S. pombe* (Uemura *et al.*, 1987). The *Drosophila* topoisomerase II gene was identified several years ago by Berrios *et al.* (1985), but we are not aware of the existence of any mutants at this locus. Nevertheless, the results of a functional study of topoisomerase II in *Drosophila* carried out by Buchenau *et al.* (1993) seem to substantiate the hypothesis that this enzyme is required for the onset of anaphase in *Drosophila*. These authors treated living embryos with antibodies against topoisomerase II or with specific inhibitors of its enzymatic activity. They observed that these treatments hindered the segregation of chromatin at anaphase. Furthermore, increased concentrations of the drug also resulted in defects in chromosome condensation and inability of the chromosomes to collect at the metaphase plate. The observation that topoisomerase II is a component of the mitotic chromosomes scaffold (Earnshaw *et al.*, 1985; Earnshaw and Heck, 1985) and that its expression is coupled with cell proliferation (Heck and Earnshaw, 1986) adds further support to this hypothesis.

Genetic analysis in *Drosophila melanogaster* has identified several genes that are good candidates for part of the mechanism that ensures sister chromatid pairing. Here we mention *mei-S322* and *ord,* which act in meiosis, and *pasc*, which functions in mitosis. Other genes that have an involvement in chromatid separation were discussed earlier in this chapter. These include mutations in genes that result in lagging chromatids during anaphase (*aar, rod,* and *zw10;* Section IV,C). Other mutations such as *thr* totally prevent chromatid separation and may represent functions that are a prerequisite for this process (see Section II,C,1,a).

a. *mei-S332.* Unlike most meiotic mutants which are sex specific, *mei-S332* and *ord* affect meiosis in both sexes. In fact they provided some of the first evidence that the processes governing chromosome behavior are similar in males and females. *mei-S332* was isolated from a natural population of *Drosophila* melanogaster as a mutant that produced a high frequency of meiotic nondisjunction and affected all chromosomes independently of meiotic exchange (Sandler *et al.*, 1968). The first detailed meiotic analysis of *mei-S332* was carried out by Davis (1971), who showed that the primary effect of mutation in this gene is the precocious separation of sister centromeres which leads to frequent equational nondisjunction. He also observed low levels of reductional nondisjunction and chromosome loss. Davis concluded that the wild-type function of *mei-S332* regulates a component of the centrosome or the process of sister chromatid separation.

More recently, Kerrebrock *et al.* (1992) have confirmed these observations and showed that the product of this gene promotes sister chromatid cohesion in meiosis following kinetochore differentiation before anaphase II in both sexes. The timing of the requirement for the *mei-S332* gene product coincides with the transformation of the single hemispherical kinetochore into two distinct sister kinetochores (Goldstein, 1981). All alleles obtained by Kerrebrock *et al.* (1992) are phenotypically very similar to the original *mei-S332.* This includes full viability even over a deficiency for the locus, and so it is concluded that the gene cannot be essential for mitosis. These observations differ from those of Baker *et al.* (1978), who detected increased levels of mitotic recombination or nondisjunction in homozygous *mei-S332* mutants. A possible interpretation that reconciles these results is that the *mei-S332* function contributes to, but is not essential for mitosis. Alternatively, other unknown mutations present in the *mei-S332* chromosome used by Baker *et al.* (1978) might have caused the mitotic instability reported by these authors.

b. *ord, orientation disruptor.* The first mutation in this gene was reported by Mason (1976), who showed that it produces nondisjunction in both sexes and lowers the frequency of recombination in females. Both reductional and equational exceptional offspring are produced. Genetic and cytological analysis of *ord* led Goldstein (1980) to conclude that this gene is required for sister chromatid cohesion throughout meiosis. This conclusion was substantiated by Lin and Church (1982), who also observed that mutation in *ord* causes a high level of nondisjunction during spermatogonial mitotic divisions. The mitotic instability affects the larger second and third chromosomes more frequently than the sex chromosomes or the fourth chromosome.

A detailed genetic analysis of the *ord* locus has recently been carried out by Miyazaki and Orr-Weaver (1992), which has rendered the isolation of five new mutant alleles of this locus. The have shown that precocious separation of sister chromatids in individuals mutant for *ord* occurs in prometaphase I, earlier that in *mei-S332* mutants. As a result, homozygous *ord* individuals show significant levels of nondisjunction during the first meiotic division. The work of Miyazaki and Orr-Weaver has also confirmed that some *ord* alleles seem to impair the mitotic divisions that take place in the germ line of both males and females, the second and third chromosomes being more affected than the X and the fourth. This is in contrast with the full viability of all extant alleles, even when in trans to a deficiency, which seems to suggest that the gene is not essential for mitosis. Additional evidence that the *ord* function may play a role in somatic cell division was presented by Baker *et al.* (1978).

c. pasc, parallel sister chromatids. This gene, first identified by Perrimon *et al.* (1985) and named *C204*, has recently been renamed by M. L. Goldberg and M. Gatti (personal communication). Homozygous *pasc* larvae show mitotic cells in which sister chromatids are aligned side by side with no visible contact between them. Although the original alleles were reported as late larval lethals with small discs, it is now known that the gene function is also required during embryogenesis because germ-line clones produce eggs that arrest development during preblastoderm. An intriguing aspect of the *pasc* pheno-type is the absence of mitotic nondisjunction. The fact that chromosomes in cells showing the *pasc* phenotype are able to segregate properly throws doubt on the requirement for sister chromatid cohesion until the metaphase–anaphase transition.

VII. Concluding Remarks

In this chapter we have concentrated on describing what is known about genes whose function is required for the cell cycle per se. The decision as to whether a cell should or should not proliferate is influenced by a large number of factors. Genetic analysis has identified a set of genes whose function prevents hyperplastic overgrowth of imaginal tissues (Jacob et al., 1987; Bryant et al., 1988; Klambt et al., 1989; Loffler et al., 1990; Szabad et al., 1991; Woods and Bryant, 1991). Several overgrowth mutations display a tumorous phenotype (Gatef, 1978; Gatef and Mechler, 1989; Bryant and Schmidt, 1990; Mahoney et al., 1991). Some of these mutations are thought to interfere with cell proliferation by disrupting the mechanisms for cell–cell communication. Another group of genes has been identified that may regulate cell proliferation through signal transduction pathways. The Drosophila homolog of human C-raf is one such example. Mutations at this locus, D-raf, result in larval lethality with small or absent imaginal tissues (Nishida et al., 1988). Dsor1, a dominant suppressor of D-raf, encodes a protein kinase with striking similarity to the MAP kinase activator (Tsuda et al., 1993), placing raf kinase upstream of the MAP kinase activator in Drosophila.

A most striking case of cell cycle control during development is provided by the tip cell that controls mitosis during development of the Malpighian tubule in D. melanogaster (Skaer, 1989). The tip cell does not incorporate BUdR or divide during the stage at which other cells in the tubule are proliferating. Ablation experiments have shown that tubules without a tip cell do not undertake cell proliferation. Although other possibilities exist, Skaer (1989) favors a model in which the tip cell plays an active secretory role in regulating the cell cycle. Regulation of cell division by tip cells is not unique to Drosophila (see references cited in Skaer, 1989). Another clear example of the developmental control of cell cycle progression is found in the optic lobe of the larva, where the lamina precursor cells require contact with axons growing out from the developing eye in order to undergo their final round of division (Selleck et al., 1992).

Although it is outside the scope of this chapter to discuss these areas fully, we anticipate that it will not be long before direct links are made between the mechanisms that regulate cell proliferation and cell cycle regulatory processes per se. An exciting time lies ahead in relating the regulation of cell division to development.

ACKNOWLEDGMENTS

We are grateful to Daryl Henderson for his comments on this manuscript. Our review is not comprehensive and we apologize to colleagues whose work may not have been cited. We thank the editors for their patience as we grew weary of our task. The work in our laboratory is supported by the Cancer Research Campaign.

REFERENCES

Abad, J. P., Carmena, M., Baars, S., Saunders, R. D. C., Glover, D. M., Ludena, P., Sentis, C., Tyler-Smith, C., and Villasante, A. (1992). Dodeca-satellite: a conserved GC-rich satellite from the centromeric heterochromatin of *Drosophila melanogaster. Proc. Natl. Acad. Sci. U.S.A.* **89**, 4663–4667.

Alfa, C. E., Ducommun, B., Beach, D., and Hyams, J. S. (1990). Distinct nuclear and spindle pole populations of cyclin-cdc2 in fission yeast. *Nature (London)* **347**, 680–682.

Allen, R. D., Weiss, D. G., Hayden, J. H., Brown, D. T., Fujiwake, H., and Simpson, M. (1985). Gliding movement of and bidirectional transport along single native microtubules from squid axoplasm: evidence for an active role of microtubules in cytoplasmic transport. *J. Cell Biol.* **100**, 1736–1752.

Alphey, L., Jimenez, J., White-Cooper, H., Dawson, I., Nurse, P., and Glover, D. (1992). *twine*, a *cdc25* homolog that functions in the male and female germline of *Drosophila. Cell* **69**, 977–988

Axton, J. M., Dombradi, V., Cohen, P. T. W., and Glover, D. M. (1990). One of the protein phosphatase 1 isozymes in *Drosophila* is essential for mitosis. *Cell* **63**, 33–46.

Bailey, E., Pines, J., Hunter, T., and Bornens, M. (1992). Cytoplasmic accumulation of cyclin B1 in human cells: association with a detergent resistant compartment and with the centrosome. *J. Cell Sci.* **101**, 529–545.

Baker, B. S., Carpenter, A. T. C., and Ripoll, P. (1978). The utilization during mitotic cell division of loci controlling meiotic recombination and disjunction in *Drosophila melanogaster. Genetics* **90**, 531–578.

Baker, B. S., Smith, D. A., and Gatti, M. (1982). Region-specific effects on chromosome integrity of mutations at essential loci in *Drosophila melanogaster. Proc. Natl. Acad. Sci. U.S.A.* **79**, 1205–1209.

Baksa, K., Morawietz, H., Dombradi, V., Axton, J. M., Taubert, H., Szabo, G., Torok, I., Udvardy, A., Gyurkovics, H., Glover, D. M., Reuter, G., and Gausz, J. (1993). Mutations in the protein phosphatase 1 gene at 87B can differentially affect suppression of position-effect variegation and mitosis in *Drosophila melanogaster. Genetics* **135**, 117–125.

Berleth, T., Burri, M., Thoma, G., Bopp, D., Richstein, S., Frigerio, G., Noll, M., and Nusslein-Volhard, C. (1988). The role of localization of *bicoid* RNA in organising the anterior pattern of the *Drosophila* embryo. *EMBO J.* **7**, 1749–1756.

Berrios, M., Osheroff, N., and Fisher, P. A. (1985). *In situ* localization of DNA topoisomerase II, a major polypeptide component of the *Drosophila* nuclear matrix fraction. *Proc. Natl. Acad. Sci. U.S.A.* **82**, 4142–4146.

Blumenthal, A. B., Kriegstein, H. J., and Hogness, D. S. (1973). The units of DNA replication in *Drosophila melanogaster* chromosomes. *Cold Spring Harbor Symp. Quant. Biol.* **38**, 205–223.

Bossie, C. A., and Sanders, M. M. (1993). A cDNA from *Drosophila melanogaster* encodes a lamin C-like intermediate filament protein. *J. Cell Sci.* **104**, 1263–1272.

Boyd, J. B., Mason, J. M., Yamamoto, A. H., Brodberg, R. K., Banga, S. S., and Sakaguchi, K. (1987). A genetic and molecular analysis of DNA repair in *Drosophila. J. Cell Sci., Suppl.* **6**, 39–60.

Brinkley, B. R., Ouspenski, I., and Zinkowski, R. (1992). Structure and molecular characterisation of the centromere-kinetochore complex. *TICB* **2**, 15–21.

Bryant, P. J., and Schmidt, O. (1990). The genetic control of cell proliferation in *Drosophila* imaginal discs. *J. Cell Sci., Suppl.* **13**, 169–189.

Bryant, P. J., Huettner, B., Held, L. I., Jr., Ryerse, J., and Szidonya, J. (1988). Mutations at the *fat* locus interfere with cell proliferation control and epithelial morphogenesis in *Drosophila. Dev. Biol.* **129**, 541–554.

Buchenau, P., Saumweber, H., and Arndt-Jovin, D. (1993). Consequences of topoisomerase II inhibition in early embryogenesis of *Drosophila* revealed by *in vivo* confocal laser scanning microscopy. *J. Cell Sci.* **104**, 1175–1185.

Byers, B., and Goetsch, L. (1974). Duplication of spindle plaques and integration of the yeast cell cycle. *Cold Spring Harbor Symp. Quant. Biol.* **38**, 123–131.

Callaini, G., Dallai, R., and Riparbelli, M. G. (1989). Diazepam induces abnormal mitosis in the early *Drosophila* embryo. *Biol. of the Cell* **67**, 313–320.

Carmena, M., Gonzalez, C., Casal, J., and Ripoll, P. (1991). Dosage-dependence of maternal contribution to somatic cell division in *Drosophila. Development* **113**, 1357–1364.

Carmena, M., Abad, J. P., Villasante, A., and Gonzalez, C. (1993). The *Drosophila melanogaster* dodecasatellite sequence is closely linked to the centromere and can form connections between sister chromatids during mitosis. *J. Cell Sci.* **105**, 41–50.

Carpenter, A. T. C. (1973). A mutant defective in distributive disjunction in *Drosophila melanogaster. Genetics* **73**, 393–428.

Carpenter, A. T. C. (1975). Electron microscopy of meiosis in *Drosophila melanogaster* females. I Structure, arrangement and temporal change of the synaptonemal complex in wild-type. *Chromosoma* **51**, 157–182.

Carpenter, A. T. C. (1991). Distributive segregation: motors in the polar wind? *Cell* **64**, 885–890.

Chen, M. X., Chen, Y. H., and Cohen, P. T. W. (1992). Polymerase chain reactions using *Saccharomyces, Drosophila,* and human DNA predict a large family of protein serine/threonine protein phosphatases. *FEBS Lett.* **306**, 54–58.

Clarke, P. R., Leiss, D., Pagano, M., and Karsenti, E. (1992). Cyclin A and cyclin B dependent protein kinases are regulated by different mechanisms in *Xenopus* egg extracts. *EMBO J.* **11**, 1751–1761.

Clay, F., McEwen, S. J., Bertoncello, I., Wilks, A. F., and Dunn, A. (1993). Identification and cloning of a protein kinase-encoding mouse gene, *plk,* related to the *polo* gene of *Drosophila. Proc. Natl. Acad. Sci. U.S.A.* **90**, 4882–4886.

Clegg, N. J., Whitehead, I. P., Brock, J. K., Sinclair, D. A., Mottus, R., Stromotich, G., Harrington, M. J., and Grigliatti, T. A. (1993). A cytogenetic analysis of chromosomal region 31 of *Drosophila melanogaster. Genetics* **134**, 221–230.

Cleveland, D. W. (1987). The multitubulin hypothesis revisited: what have we learned? *J. Cell Biol.* **104**, 381–383.

Cohen, P. T. W., Brevis, N. D., Hughes, V., and Mann, D. J. (1990). Protein serine/threonine phosphatases; an expanding family. *FEBS Lett.* **268**, 355–359.

Cooke, C. A., Heck, M. M. S., and Earnshaw, W. C. (1987). The inner centromere protein (INCENP) antigens: movement from inner centromere to midbody during mitosis. *J. Cell Biol.* **105**, 2053–2067.

Cooper, K. W. (1964). Meiotic conjunctive elements not involving chiasmata. *Proc. Natl. Acad. Sci. U.S.A.* **52**, 1248–1255.

Cotterill, S., Chui, G., and Lehman, I. R. (1987a). DNA polymerase-primase from embryos of *Drosophila melanogaster*: the polymerase subunit. *J. Biol. Chem.* **262**, 16,100–16,104.

Cotterill, S., Chui, G., and Lehman, I. R. (1987b). DNA polymerase-primase from embryos of *Drosophila melanogaster*: DNA primase subunits. *J. Biol. Chem.* **262**, 16,105–16,108.

Cotterill, S., Reyland, M., Lehman, I. R., and Loeb, L. (1987c). A cryptic proofreading 3' to 5' exonuclease activity associated with the polymerase subunit of the DNA polymerase-primase from *Drosophila melanogaster* embryos. *Proc. Natl. Acad. Sci. U.S.A.* **84**, 5635–5639.

Cotterill, S., Lehman, I. R., and McLachlan, P. (1992). Cloning of the gene for the 73 kD subunit of the DNA polymerase a primase of *Drosophila melanogaster. Nucleic Acids Res.* **20**, 4325–4330.

Counce, S. J. (1963). Developmental morphology of polar granules in *Drosophila* including observations on pole cell behaviour during embryogenesis. *J. Morphol.* **112**, 129–145.

Courtot, C., Fankhauser, C., Simanis, V., and Lehner, C. F. (1992). The *Drosophila cdc25* homologue *twine* is required for meiosis. *Development* **116**, 405–416.

Dalby, B., and Glover, D.M. (1992). 3' non translated sequences in the maternal cyclin B transcript direct posterior pole localisation in oogenesis and peri-nuclear localisation in embryogenesis. *Development* **115**, 989–997.

Dalby, B., and Glover, D. (1993). Discrete sequence elements control the posterior pole accumulation and translational repression of maternal cyclin B RNA in *Drosophila. EMBO J.* **12**, 1219–1227.

D'Andrea, R. J., Stratmann, R., Lehner, C. F., John, V. P., and Saint, R. (1994). The three rows gene of *Drosophila melanogaster* encodes a novel protein that is required for chromosome disjunction during mitosis. *Mol. Biol. Cell* **4**, 1161–1174.

Davis, B.R. (1971). Genetic analysis of a meiotic mutant resulting in precocious sister-centromere separation in *Drosophila melanogaster. Mol. Gen. Genet.* **113**, 251–272.

Dawson, I. A., Roth, S., Akam, M., and Artavanis-Tsakonas, S. (1993). Mutations of the *fizzy* locus cause metaphase arrest in *Drosophila melanogaster* embryos. *Development* **117**, 359–376.

Devault, A., Fesquet, D., Cavadore, J.-C., Garrigues, A.-M., Labbé, J.-C. Lorca, T., Picard, A., Philippe, M., and Dorée, M. (1992). Cyclin A potentiates maturation-promoting factor activation in the early *Xenopus* embryo via inhibition of the tyrosine kinase that phosphorylates CDC2. *J. Cell Biol.* **118**, 1109–1120.

Doonan, J. H., and Morris, N. R. (1989). The *bimG* gene of *Aspergilus nidulans*, required for completion of anaphase, encodes a homolog of mammalian phosphoprotein phosphatase 1. *Cell* **57**, 987–996.

Dunphy, W. G., and Kumagai, A. (1991). The cdc25 protein contains an intrinsic phosphatase activity. *Cell* **67**, 189–196.

Earnshaw, W. C. (1988). Mitotic chromosome structure. *BioEssays* **9**, 147–150.

Earnshaw, W. C. (1991). When is a centromere not a kinetochore? *J. Cell Sci.* **99**, 1–4.

Earnshaw, W. C., and Bernat, R. L. (1991). Chromosomal passengers: toward an integrated view of mitosis. *Chromosoma* **100**, 139–146.

Earnshaw, W. C., and Cooke, C. A. (1991). Analysis of the distribution of the INCENPs throughout mitosis reveals the existence of a pathway of structural changes in the chromosomes during metaphase and early events in cleavage furrow formation *J. Cell Sci.* **98**, 443–461.

Earnshaw, W. C., and Heck, M. M. S. (1985). Localization of topoisomerase II in mitotic chromosomes. *J. Cell Biol.* **100**, 1716–1725.

Earnshaw, W. C., and Tomkiel, J. E. (1992). Centromere and kinetochore structure. *Curr. Opinion Cell Biol.* **4**, 86–93.

Earnshaw, W. C., Halligan, B., Cooke, C. A., Heck, M. M. S., and Liu, L. F. (1985). Topoisomerase II is a structural component of mitotic chromosome scaffolds. *J. Cell Biol.* **100**, 1706–1715.

Earnshaw, W. C., Sullivan, K. F., Machlin, P. S., Cooke, C. A., Kaiser, D. A., Pollard, T. D., Rothfield, N. F., and Cleveland, D. W. (1987). Molecular cloning of cDNA for *CENP-B*, the major human centromere auto-antigen. *J. Cell Biol.* **104**, 817–829.

Edgar, B. A. and O'Farrell, P. H. (1989). Genetic control of cell division patterns in the *Drosophila* embryo. *Cell* **57**, 177–187.

Edgar, B. A., and O'Farrell, P. H. (1990). The three postblastoderm cell cycles of *Drosophila* embryogenesis are regulated in G2 by *string*. *Cell* **62**, 469–480.

Edgar, B. A., Odel, G. M., and Schubiger, G. (1987). Cytoarchitecture and the patterning of *fushi tarazu* expression in the *Drosophila* blastoderm. *Genes Dev.* **1**, 1126–1132.

Edgar, B. A., Sprenger, F., Duronio, R. J., Leopold, P., and O'Farrell, P. (1994). Distinct molecular mechanisms time mitosis at four successive stages of *Drosophila* embryogenesis. Submitted.

Endow, S. A. (1991). The emerging kinesin family of microtubule motor proteins. *TIBS* **16**, 221–225.

Endow, S. A. (1993). Chromosome distribution, molecular motors and the *claret* protein. *TIG* **9**, 52–55.

Endow, S. A., and Hatsumi, M. (1991). A multimember kinesin gene family in *Drosophila*. *Proc. Natl. Acad. Sci. U.S.A.* **88**, 4424–4427.

Endow, S. A., and Titus, M. A. (1992). Genetic approaches to molecular motors. *Annu. Rev. Cell Biol.* **8**, 29–66.

Endow, S.A., Henikoff, S., and Soler-Niedziela, L. (1990). Mediation of meiotic and early mitotic chromosome segregation in *Drosophila* by a protein related to kinesin. *Nature (London)* **345**, 81–83.

Enoch, T., and Nurse, P. (1990). Mutations of fission yeast cell cycle control genes abolishes dependence of mitosis on DNA replication. *Cell* **60**, 665–673.

Evans, T. E., Rosenthal, E. T., Youngbloom, J., Distal, D., and Hunt, T. (1983). Cyclin: a protein specified by maternal RNA in sea urchin eggs that is destroyed at each cell division. *Cell* **33**, 389–396.

Felix, A. M., Cohen, P., and Karsenti, E. (1990). cdc2 H1 kinase is negatively regulated by a type 2A phosphatase in the early embryonic cell cycles of *Xenopus*; evidence from the effects of okadaic acid. *EMBO J.* **11**, 675–683.

Fenton, B., and Glover, D. M. (1993). A conserved mitotic kinase active at late anaphase–telophase in syncytial *Drosophila* embryos. *Nature (London)* **363**, 637–639.

Foe, V. E. (1989). Mitotic domains reveal early commitment of cells in *Drosophila* embryos. *Development* **107**, 1–22.

Foe, V. E., and Alberts, B. M. (1985). Reversible chromosome condensation induced in *Drosophila* embryos by anoxia: visualisation of interphase nuclear organisation. *J. Cell Biol.* **100**, 1623–1636.

Frasch, M., Glover, D. M., and Saumweber, H. (1986). Nuclear antigens follow different pathways into daughter nuclei during mitosis in early *Drosophila* embryos. *J. Cell Sci.* **82**, 155–172.

Freeman, M., and Glover, D. M. (1987). The *gnu* mutation of *Drosophila* causes inappropiate DNA synthesis in unfertilised and fertilised eggs. *Genes Dev.* **1**, 924–930.

Freeman, M., Nusslein-Volhard, C., and Glover, D. M. (1986). The dissociation of nuclear and centrosomal division in *gnu*, a mutation causing giant nuclei in *Drosophila*. *Cell* **46**, 457–468.

Fuller, M. T. (1986). Genetic analysis of spermatogenesis in *Drosophila*: the role of the testis specific beta tubulin and interacting genes in cellular morphogenesis. In "Gametogenesis and the Early Embryo." (J. G. Gall, ed.), pp. 19–41. Alan R. Liss. New York.

Fuller, M. T., Caulton, J. H., Hutchens, J. A., Kaufman, T. C., and Raff, E. C. (1987). Genetic analysis of microtubule structure: a beta-tubulin mutation causes the formation of aberrant microtubules *in vivo* and *in vitro*. *J. Cell Biol.* **104**, 385–394.

Fuller, M. T., Caulton, J. H., Hutchens, J. A., Kaufman, T. C., and Raff, E. C. (1988). Mutations that encode partially functional beta2 tubulin have different effects on structurally different microtubule arrays. *J. Cell Biol.* **107,** 141–152.

Fyrberg, E., and Goldstein, L. S. B. (1990). The *Drosophila* cytoskeleton. *Annu. Rev. Cell Biol.* **6,** 559–596.

Fyrberg, E. A., Kindle, K. L., Davidson, N., and Sodja, A. (1980). The actin genes of *Drosophila*: a dispersed multigene family. *Cell* **19,** 365–378.

Gateff, E. (1978). Malignant and benign neoplasms of *Drosophila. In* "The Genetics and Biology of *Drosophila*," (M. Ashburner and T. R. F. Wright, eds.), Vol. 2b, pp. 187–275. Academic Press, London.

Gateff, E., and Mechler, B. M. (1989). Tumor-suppressor genes of *Drosophila melanogaster. Crit. Rev. Oncol.* **1,** 221–245.

Gatti, M. (1979). Genetic control of chromosome breakage and rejoining in *Drosophila melanogaster*: spontaneous chromosome aberrations in X-linked mutants defective in DNA metabolism. *Proc. Natl. Acad. Sci. U.S.A.* **76,** 1377–1381.

Gatti, M., and Baker, B. S. (1989). Genes controlling essential cell-cycle functions in *Drosophila melanogaster. Genes Dev.* **3,** 438–453.

Gatti, M., and Goldberg, M. L. (1991). Mutations affecting cell division in *Drosophila. Methods Cell Biol.* **35,** 543–586.

Gatti, M., and Pimpinelli, S. (1992). Functional elements in *Drosophila melanogaster* heterochromatin. *Annu. Rev. Genet.* **26,** 239–275.

Gautier, J., Norbury, C., Lohka, M., Nurse, P., and Maller, J. (1988). Purified maturation-promoting factor contains the product of a *Xenopus* homolog of the fission yeast cell cycle control gene *cdc2* $^+$. *Cell* **54,** 433–439.

Gautier, J., Solomon, M. J., Booher, R. N., Bazan, J. F., and Kirschner, M. W. (1991). *cdc25* is a specific tyrosine phosphatase that directly activates p34^{cdc2}. *Cell* **67,** 197–211.

Ghiara, J. B., Richardson, H. E., Sugimoto, K., Henze, M., Lew, D. J., Wittenberg, C., and Reed, S. I. (1991). A cyclin B homolog in *S. cerevisiae*: chronic activation of the *CDC28* protein kinase by cyclin prevents exit from mitosis. *Cell* **65,** 163–174.

Girard, F., Strausfeld, U., Fernandez, A., and Lamb, N. J. C. (1991). Cyclin A is required for the onset of DNA replication in mammalian fibroblasts. *Cell* **67,** 1169–1179.

Girdham, C., and Glover, D. M. (1991). Chromosome tangling and breakage at anaphase result from mutations in *lodestar,* a *Drosophila* gene encoding a putative nucleoside triphosphate binding protein. *Genes Dev.* **5,** 1786–1799.

Glover, D. M. (1989). Mitosis in *Drosophila. J. Cell Sci.* **92,** 137–146.

Glover, D. M. (1990). Abbreviated and regulated cell cycles in *Drosophila. Curr. Opinion Cell Biol.* **2,** 258–261.

Glover, D. M. (1991). Mitosis in the *Drosophila* embryo—in and out of control. *Trends Genet.* **7,** 125–132.

Glover, D. M., Gonzalez, C., and Raff, J. R. (1993). The centrosome. *Sci. Am.* **268,** 62–68.

Goldstein, L. S. B. (1980). Mechanisms of chromosome orientation revealed by two meiotic mutants of *Drosophila melanogaster. Chromosoma* **78,** 79–111.

Goldstein, L. S. B. (1981). Kinetochore structure and its role in chromosome orientation during the first meiotic division in male D. *melanogaster. Cell* **25,** 591–602.

Goldstein, L. S. B. (1991). The kinesin superfamily: tails of functional redundancy. *TICB* **1,** 93–98.

Goldstein, L. S. B., Laymon, R. A., and McIntosh, J. R. (1986). A microtubule-associated protein in *Drosophila melanogaster*: identification, characterisation, and isolation of coding sequences. *J. Cell Biol.* **102,** 2076–2087.

Gomes, R., Karess, R. E., Ohkura, H., Glover, D. M., and Sunkel, C. E. (1993). *abnormal anaphase resolution (aar)*: a locus required for progression through mitosis in *Drosophila. J. Cell Sci.* **104,** 583–593.

Gonzalez, C. (1994). Submitted.

Gonzalez, C., Casal, J., and Ripoll, P. (1988). Functional monopolar spindles caused by mutation in *mgr,* a cell division gene of *Drosophila melanogaster. J. Cell Sci.* **89,** 39–47.

Gonzalez, C., Jimenez, J. C., Ripoll, P., and Sunkel, C. E. (1991). The spindle is required for the process of sister chromatid separation in *Drosophila* neuroblasts. *Exp. Cell Res.* **192,** 10–15.

Gould, K. L., Moreno, S., Owen, D. J., Sazer, S., and Nurse, P. (1991). Phosphorylation at Thr167 is required for *Schizosaccharomyces pombe* p34 cdc2 function. *EMBO J.* **11,** 3297–3309.

Green, L. L., Wolf, N., McDonald, K. L., and Fuller, M. T. (1990). Two types of genetic interaction implicate the *whirligig* gene of *Drosophila melanogaster* in microtubule organization in the flagellar axoneme. *Genetics* **126,** 961–973.

Gruenbaum, Y., Landesman, Y., Drees, B., Bare, J. W., Saumweber, H., Paddy, M. R., Sedat, J., Smith, D. E., Benton, B. E., and Fisher, P. A. (1988). *Drosophila* nuclear lamin precursor *Drosophila melanogaster* 0 is translated from either of two developmentally regulated mRNA species apparently encoded by a single gene. *J. Cell Biol.* **106,** 585–596.

Guan, K., Broyles, S. S., and Dixon, J. E. (1991). A tyr/ser protein phosphatase encoded by vaccinia virus. *Nature (London)* **350,** 359–362.

Harel, A., Zlotkin, E., Nainudel-Epszteyn, S., Feinstein, N., Fisher, P. A., and Gruenbaum, Y. (1989). Persistence of major nuclear envelope antigens in an envelope-like structure during mitosis in *Drosophila melanogaster* embryos. *J. Cell Sci.* **94,** 463–470.

Harrouk, W., and Clarke, H. J. (1993). Sperm chromatin acquires an activity that induces microtubule assembly during residence in the cytoplasm of metaphase oocytes of the mouse *Chromosoma* **102,** 279–286.

Hatsumi, M., and Endow, S. (1992). Mutants of the microtubule protein, *non claret disjunctional,* affect spindle structure and chromosome movement in meiosis and mitosis. *J Cell Sci.* **101,** 547–559.

Hatzfeld, J., and Buttin, G. (1975). Temperature sensitive cell cycle mutants: a chinese hamster cell line with a reversible block in cytokinesis. *Cell* **5,** 123–129.

Hawley, R. S. (1988). Exchange and chromosomal segregation in eukaryotes. *In* "Genetic Recombination." (R. Kucherlapati and G. R. Smith, eds.), pp. 497–527. Am. Soc. Microbiol., Washington, D.C.

Hays, T. S., Deuring, R., Robertson, B., Prout, M., and Fuller, M. T. (1989). Interacting proteins identified by genetic interactions: a missense mutation in alpha-tubulin fails to complement alleles of the testis-specific beta-tubulin gene of *Drosophila melanogaster. Mol. Cell. Biol.* **9,** 875–884.

Hazelrigg, T., Watkins, W. S., Marcey, D. T. U. C., Karow, M., and Lin, X. R. (1990). The *exuperantia* gene is required for *Drosophila* spermatogenesis as well as anterioposterior polarity of the developing oocyte and encodes overlapping sex specific transcripts. *Genetics* **126,** 607–617.

Heck, M. M. S., and Earnshaw, W. C. (1986). Topoisomerase II: a specific marker for cell proliferation. *J. Cell Biol.* **103,** 2569–2581.

Henderson, D. S., Banga, S. S., Grigliatti, T. A., and Boyd, J. B. (1994). Mutagen sensitivity and suppression of position-effect variegation result from mutations in *mus 209,* the *Drosophila* gene encoding PCNA. *EMBO J., in press.*

Hime, G., and Saint, R. (1992). Zygotic expression of the *pebble* locus is required for cytokinesis during the postblastoderm mitoses of *Drosophila*. *Development* **114**, 165–171.

Hiraoka, Y., Minden, J. S., Swedlow, J. R., Sedat, J. W., and Agard, D. A. (1989). Focal points for chromosome condensation and decondensation revealed by three-dimensional *in vivo* time-lapse microscopy. *Nature (London)* **342**, 293–296.

Hiraoka, Y., Agard, D. A., and Sadat, J. W. (1990). Temporal and spatial coordination of chromosome movement, spindle formation, and nuclear envelope breakdown during prometaphase in *Drosophila melanogaster* embryos. *J. Cell Biol.* **6**, 2815–2828.

Hirose, S., Yamaguchi, M., Nishida, Y., Matsutani, M., Miyazawa, H., Hanoaka, F., and Matsukage, A. (1991). Structure and expression during development of the *Drosophila melanogaster* gene for the DNA polymerase α. *Nucleic Acids Res.* **10**, 4991–4998.

Hoyle, H. D., and Raff, E. C. (1990). Two *Drosophila* beta tubulin isoforms are not functionally equivalent. *J. Cell Biol.* **111**, 1009–1026.

Irminger-Finger, I., Laymon, R. A., and Goldstein, L. S. B. (1990). Analysis of the primary sequence and microtubule binding domain of the *Drosophila* 205K MAP. *J. Cell Biol.* **111**, 2563–2572.

Jacob, L., Opper, B., Metzroth, B., Phannavong, B., and Mechler, B. M. (1987). Structure of the *l(2)gl* gene of *Drosophila* and delimitation of its tumor-suppressor domain. *Cell* **50**, 215–225.

Jessus, C., Rime, H., Haccard, O., VanLint, J., Goris, J., Merlevede, W., and Ozon, R. (1991). Tyrosine phosphorylation of p34[cdc2] and p42 during meiotic maturation of Xenopus oocyte. *Development* **111**, 813–820.

Jimenez, J., Alphey, L., Nurse, P., and Glover, D. M. (1990). Complementation of fission yeast *cdc2*[ts] and *cdc25*[ts] mutants identifies two cell cycle genes from *Drosophila*: a *cdc2* homologue and *string*. *EMBO J.* **9**, 3565–3571.

Joshi, H. C., Palacios, M. J., McNamara, L. R., and Cleveland, D. N. (1992). Gamma tubulin is a centrosomal protein required for cell cycle dependent microtubule nucleation. *Nature (London)* **356**, 80–83.

Jürgens, G., Wieschaus, E., Nusslein-Volhard, C., and Kluding, H. (1984). Mutations affecting the pattern of the larval cuticle in *Drosophila melanogaster*. II. Zygotic loci on the third chromosome. *Wilhelm Roux's Arch. Dev. Biol.* **193**, 283–295.

Karess, R. E., and Glover, D. M. (1989). *rough deal*: a gene required for proper mitotic segregation in *Drosophila*. *J. Cell Biol.* **109**, 2951–2961.

Karess, R. E., Chang, X., Edwards, K. A., Kuckarni, S., Aguilera, I., and Kiehart, D. P. (1991). The regulatory light chain of nonmuscle myosin is encoded by *spaghetti squash*, a gene required for cytokinesis in *Drosophila*. *Cell* **64**, 49–62.

Karr, T. L., and Alberts, B. M. (1986). Organisation of the cytoskeleton in early *Drosophila* embryos. *J. Cell Biol.* **102**, 1494–1509.

Kaufmann, B. F. (1934). Somatic mitoses of *Drosophila melanogaster*. *J. Morphol.* **56**, 125–155.

Kellogg, D. R., and Alberts, B. M. (1992). Purification of a multiprotein complex containing centrosomal proteins from the *Drosophila* embryo by chromatography with low affinity polyclonal antibodies. *Mol. Biol. Cell* **3**, 1–11.

Kellogg, D. R., Field, C. M., and Alberts, B. M. (1989). Identification of microtubule-associated proteins in the centrosome, spindle and kinetochore of the early *Drosophila* embryos. *J. Cell Biol.* **109**, 2977–2991.

Kemphues, K. J., Raff, R. A., Kaufman, T. C., and Raff, E. C. (1979). Mutation in a structural gene for a β-tubulin specific to testis in *Drosophila melanogaster*. *Proc. Natl. Acad. Sci. U.S.A.* **76**, 3991–3995.

Kemphues, K. J., Raff, E. C., Raff, R. A., and Kaufman, T. C. (1980). Mutation in a testis-specific β-tubulin in *Drosophila*: analysis of its effects on meiosis and map location of the gene. *Cell* **21**, 445–451.

Kemphues, K. J., Kaufman, T. C., Raff, R. A., and Raff, E. C. (1982). The testis-specific β-tubulin subunit in *Drosophila melanogaster* has multiple functions in spermatogenesis. *Cell* **31**, 655–670.

Kemphues, K. J., Raff, E. C., and Kaufman, T. C. (1983). Genetic analysis of β2t, the structural gene for the testis-specific beta-tubulin subunit in *Drosophila melanogaster*. *Genetics* **105**, 345–356.

Kerrebrock, A. W., Miyazaki, W. Y., Birnsby, D., and Orr-Weaver, T. L. (1992). The *Drosophila mei-S332* gene promotes sister chromatid cohesion in meiosis following kinetochore differentiation. *Genetics* **130**, 827–841.

Kiehart, D. P., Mabuch, I., and Inoue, S. (1982). Evidence that myosin does not contribute to force production in chromosome movement. *J. Cell Biol.* **94**, 165–178.

Kimble, M., and Church, K. (1983). Meiosis and early cleavage in *Drosophila melanogaster* eggs: effects of the *claret nondisjunctional* mutation. *J. Cell Sci.* **62**, 301–318.

Kimble, M., Dettman, R. W., and Raff, E. C. (1990). The beta-3-tubulin gene of *Drosophila melanogaster* is essential for viability and fertility. *Genetics* **126**, 991–1005.

Kinoshita, N., Ohkura, H., and Yanagida, M. (1990). Distinct, essential roles of type 1 and 2A protein phosphatases in the control of the fission yeast cell division cycle. *Cell* **63**, 405–415.

Kinoshita, N., Yamano, H., LeBouffant-Sladeczek, F., Kurooka, H., Ohkura, H., Stone, E. M., Takeuchi, M., Toda, T., Yoshida, T., and Yanagida, M. (1991). Sister-chromatid separation and protein dephosphorylation in mitosis. *Cold Spring Harbor Symp. Quant. Biol.* **56**, 621–628.

Kitada, K., Johnson, A. L., Johnston, L. H., and Sugino, A. (1993). A multicopy suppressor gene of the *Saccharomyces cerevisiae* G1 cell cycle mutant gene *DBF4* encodes a protein kinase and is identified as *CDC5*. *Mol. Cell Biol.* **13**, 4445–4457.

Klambt., C., Lutzelschwab, R., Muller, S., Rossa, R., Schmidt, O., and Totzke, F. (1989). The *Drosophila melanogaster 1(2)g1* gene encodes a protein homologous to the cadherin cell-adhesion molecule family. *Dev. Biol.* **133**, 425–436.

Knoblich, J. A., and Lehner, C. F. (1993). Synergistic action of *Drosophila* cyclins A and B during the G2-M transition. *EMBO J..* **12**, 65–74.

Knowles, B. A., and Hawley, R. S. (1991). Genetic analysis of microtubule motor proteins in *Drosophila*: a mutation at the *ncd* locus is a dominant enhancer of *nod. Proc. Natl. Acad. Sci. U.S.A.* **88**, 7165–7169.

Koff, A., Cross, F., Fisher, A., Schumacher, J., Leguellec, K., Phillipe, M., and Roberts, J. M. (1991). Human cyclin E, a new cyclin that interacts with two members of the CDC2 gene family. *Cell* **66**, 1217–1228.

Koshland, D., and Hartwell, L. (1987). The structure of sister minichromosome DNA before anaphase in *Saccharomyces cerevisiae*. *Science* **238**, 1713–1716.

Kumagai, A., and Dunphy, W. G. (1991). The *cdc25* protein controls tyrosine dephosphorylation of the *cdc2* protein in a cell-free system. *Cell* **64**, 903–914.

Kumagai, A., and Dunphy, W. G. (1992). Regulation of the *cdc25* protein during the cell cycle in *Xenopus* extracts. *Cell* **70**, 139–151.

Lahue, E. E., Smith, A. V., and Orr-Weaver, T. L. (1991). A novel cyclin gene from *Drosophila* complements CLN function in yeast. *Genes Dev.* **5**, 2166–2175.

Lee, M., and Nurse, P. (1987). Complementation used to clone a human homologue of the fission yeast cell cycle control gene *cdc2. Nature (London)* **327**, 31–35.

Lee, T. H., Solomon, M. J., Mumby, M. C., and Kirschner, M. W. (1991). *INH*, a negative regulator of MPF, is a form of protein phosphatase 2A. *Cell* **64**, 415–423.

Lehner, C. F. (1992). The *pebble* gene is required for cytokinesis in *Drosophila*. *J. Cell Sci.* **103**, 1021–1030.

Lehner, C. F., and O'Farrell, P. H. (1989). Expression and function of *Drosophila* cyclin A during embryonic cell cycle progression. *Cell* **56**, 957–968.

Lehner, C. H., and O'Farrell, P. H. (1990a). *Drosophila cdc2* homologs: a functional homolog is coexpressed with a cognate variant. *EMBO J.* **9**, 3573–3581.

Lehner, C. H., and O'Farrell, P. H. (1990b). The roles of *Drosophila* cyclins A and B in mitotic control. *Cell* **61**, 535–547.

Léopold, P., and O'Farrell, P. H. (1991). An evolutionarily conserved cyclin homolog from *Drosophila* rescues yeast deficient in G1 cyclins. *Cell* **66**, 1207–1216.

Lew, D. J., and Reed, S. I. (1993). Morphogenesis in the yeast cell cycle–regulation by *CDC28* and cyclins. *J. Cell Biol.* **120**, 1305–1320.

Lew, D. J., Dulic, V., and Reed, S. I. (1991). Isolation of three novel human cyclins by rescue of G1 cyclin (*cln*) function in yeast. *Cell* **66**, 1197–1206.

Lewis, E. B., and Gencarrella, W. (1952). *claret* and non-disjunction in *Drosophila melanogaster*. *Genetics* **37**, 600–601.

Lin, H. P. P., and Church, K. (1982). Meiosis in *Drosophila melanogaster*. III. The effect of *orientation disruptor* (*ord*) on gonial mitotic and the meiotic divisions in males. *Genetics* **102**, 751–770.

Lin, H., and Wolfner, M. F. (1989). Cloning and analysis of *fs(1)Ya*, a maternal effect gene required for the initiation of *Drosophila* embryogenesis. *Mol. Gen. Genet.* **215**, 257–265.

Lin, H., and Wolfner, M. F. (1991). The *Drosophila* maternal effect gene *fs(1)Ya* encodes a cell cycle-dependent nuclear envelope component required for embryonic mitosis. *Cell* **64**, 49–62.

Lindsley, D. L., and Sandler, L. (1957). The meiotic behavior of grossly deleted X chromosomes in *Drosophila melanogaster*. *Genetics* **43**, 547–563.

Llamazares, S., Moreira, M. A., Tavares, A., Girdham, C., Gonzalez, C., Karess, R. E., Glover, D. M., and Sunkel, C. E. (1991). *polo* encodes a protein kinase homologue required for mitosis in *Drosophila*. *Genes Dev.* **5**, 2153–2164.

Loffler, T., Wismar J., Sass, H., Miyamoto, T., Becker, G., Konrad, L., Blondeau, M., Protin, U., Kaiser, S., Graf, P., Haas, M., Schuler, G., Schmidt, J., Plannavang, B., Gundacker, D., and Gateff, E. (1990). Genetic and molecular analysis of 6 tumorsuppressor genes in *Drosophila melanogaster*. *Environ. Health Perspect.* **88**, 157–161.

Lorca, T., Labbe, J. C., Devault, A., Fesquet, D., Capony, J.-P., Cavadore, J.-C., Le-Bouffant, F., and Doree, M. (1992). Dephosphorylation of cdc2 on threonine 161 is required for cdc2 kinase inactivation and normal anaphase. *EMBO J.* **11**, 2381–2390.

Luca, F. C., Shibuya, E. K., Dohrmann, C. E., and Ruderman, J. V. (1991). Both cyclin A Δ60 and D97 are stable and arrest cells in M-phase but only cyclin B D97 turns on cyclin destruction. *EMBO J.* **10**, 4311–4320.

Lundgren, K., Walworth, N., Booher, B., Dembski, M., Kirschner, M., and Beach, D. (1991). *mik1* and *wee1* cooperate in the inhibitory tyrosine phosphorylation of *cdc2*. *Cell* **64**, 1111–1122.

Mahoney, P. A., Onofrechuk, P., Biessmann, H., Weber, U., Bryant, P. J., and Goodman, C. S. (1991). The *fat* tumor suppressor gene in *Drosophila* encodes a novel member of the cadherin gene superfamily. *Cell* **67**, 853–868.

Mahowald, A. P. (1962). Fine structure of pole cells and polar granules in *Drosophila melanogaster*. *J. Exp. Zool.* **151**, 201–205.

Mahowald, A. P. (1968). Polar granules of *Drosophila*. II. Ultrastructural changes during early embryogenesis. *J. Exp. Zool.* **167**, 237–262.

Maldonado-Codina, G., and Glover, D. M. (1992a). Cyclins A and B associate with chromatin and the polar regions of spindles, respectively, and do not undergo complete degradation at anaphase in syncytial Drosophila embryos. J. Cell Biol. 116, 967–976.

Maldonado-Codina, G., and Glover, D. M. (1992b). Heat shock can delay and thereby synchronise the cell cycle in cellularised Drosophila embryos. J. Cell Sci. 105, 711–720.

Manseau, L. J., Ganetzky, B., and Craig, E. A. (1988). Molecular and genetic characterisation of the Drosophila melanogaster 87E actin gene region. Genetics 119, 407–420.

Marril, P. T., Sweeton, D., and Wieschaus, E. (1988). Requirements for autosomal gene activity during precellular stages of Drosophila melanogaster. Development 104, 495–509.

Mason, J. M. (1976). orientation disruptor (ord): a recombination-defective and disjunction-defective meiotic mutant in Drosophila melanogaster. Genetics 84, 545–572.

Mathog, D., and Sedat, J. W. (1989). The 3-dimensional organisation of polytene nuclei in male Drosophila melanogaster with compound XY-chromosome or ring X-chromosome. Genetics 121, 293–311.

Matsushime, H., Roussel, M. F., Ashmun, R. A., and Sherr, C. J. (1991). Colony stimulating factor 1 regulates novel cyclins during the G1 phase of the cell cycle. Cell 65, 701–713.

Matthews, K. A., and Kaufman, T. C. (1987). Developmental consequences of mutations in the 84B alpha-tubulin of Drosophila melanogaster. Dev. Biol. 119, 100–114.

Mayer-Jaekel, R. E., Baumgartner, S., Bilbe, G., Ohkura, H., Glover, D. M., and Hemmings, B. A. (1992). Molecular cloning and developmental expression of the catalytic and 65-kDa regulatory subunits of protein phosphatase 2A in Drosophila. Mol. Biol. Cell 3, 287–298.

Mayer-Jaekel, R. E., Ohkura, H., Gomes, R., Sunkel, C. E., Baumgartner, S., Hemmings, B. A., and Glover, D. M. (1993). The 55kD regulatory subunit of Drosophila protein phosphatase 2A is required for anaphase. Cell. 72, 621–633.

McDonald, H. B., and Goldstein, L. S. B. (1990). Identification and characterisation of a gene encoding a kinesin-like protein in Drosophila. Cell 61, 991–1000.

McDonald, H. B., Stewart, R. J., and Goldstein, L. S. B. (1990). The kinesin-like ncd protein of Drosophila is a minus end-directed microtubule motor. Cell 63, 1159–1165.

McDonald, P. M., Luk, S. K., and Kilpatrick, M. (1991). The protein encoded by the exuperantia gene is concentrated at sites of bicoid mRNA accumulation in Drosophila nurse cells but not in oocytes or embryos. Genes Dev. 5, 2455–2466.

McKee, B. D., and Karpen, G. H. (1990). Drosophila ribosomal RNA gene function as an X-Y pairing site during male meiosis. Cell 61, 61–72.

McKee, B. D., Ledare, H., and Vrana, J. A. (1992). Evidence that intergenic spacer repeats of Drosophila melanogaster rRNA genes function as X-Y pairing sites in male meiosis, and a general model for achiasmatic pairing. Genetics 132, 529–544.

McKee, B. D., Lumsden, S. E., and Das, S. (1993). The distribution of male meiotic pairing sites on chromosome 2 of Drosophila melanogaster: meiotic pairing and segregation of 2-Y transpositions. Chromosoma 102, 180–194.

McKim, K. S., Jang, J. K., Therkauf, W. E., and Hawley, R. S. (1993). The mechanical basis of meiotic metaphase arrest. Nature (London) 362, 346–366.

Melov, S., Vaughan, H., and Cotterill, S. (1992). Molecular characterisation of the gene for the 180 kDa subunit of the DNA polymerase-primase of Drosophila melanogaster. J. Cell Sci. 102, 847–856.

Millar, J. B. A., Lenears, G., and Russell, P. (1992). *pyp3* PTPase acts as a mitotic inducer in fission yeast. *EMBO J.* **11**, 4933–4941.

Miller, K. G., Field, C. M., and Alberts, B. M. (1989). Actin binding proteins from *Drosophila* embryos: a complex network of interacting proteins detected by F-actin affinity chromatography. *J. Cell Biol.* **109**, 2963–2975.

Miyazaki, W. Y., and Orr-Weaver, T. L. (1992). Sister chromatid misbehaviour in *Drosophila ord* mutants. *Genetics* **342**, 1047–1061.

Mogensen, M. M., and Tucker, J. B. (1987). Evidence of microtubule nucleation at plasma membrane-associated sites in *Drosophila*. *J. Cell Sci.* **88**, 95–107.

Mogensen, M. M., Tucker, J. B., and Stebbing, S. H. (1989). Microtubule polarities indicate that nucleation and capture of microtubules occurs at the cell surfaces in *Drosophila*. *J. Cell Biol.* **108**, 1445–1452.

Moreno, S., and Nurse, P. (1991). Clues to the action of *cdc25* protein. *Nature (London)* **351**, 194.

Moreno, S., Nurse, P., and Russell, P. (1990). Regulation of mitosis by cyclic accumulation of p80 cdc25 mitotic inducer in fission yeast. *Nature (London)* **344**, 549–552.

Motokura, T., Bloom, T., Kim, H. G., Jüppner, H., Ruderman, J. V., Kronenberg, H. M., and Arnold, A. (1991). A novel cyclin encoded by a *bc11*-linked candidate oncogene. *Nature (London)* **350**, 512–515.

Mounkes, L. C., Jones, R. S., Liang, B., Gelbart, W., and Fuller, M. T. (1992). A *Drosophila* model for Xeroderma pigmentosum and Cockayne's syndrome: *haywire* encodes the fly homologue of *ERCC3*, a human excision repair gene. *Cell* **71**, 925–937.

Murray, A. W., and Szostak, J. W. (1985). Chromosome segregation in mitosis and meiosis *Annu. Rev. Cell Biol.* **1**, 289–315.

Murray, A. W., Solomon, M. J., and Kirschner, M. W. (1989). The role of cyclin synthesis and degradation in the control of maturation promoting factor activity. *Nature (London)* **339**, 280–286.

Nishida, Y., Hata, M., Ayaki, T., Ryo, H., Yamagata, M., Shimizo, K., and Nishizuka, Y. (1988). Proliferation of both somatic and germ cells is affected in the *Drosophila* mutants of *raf* proto-oncogene. *EMBO J.* **7**, 775–778.

Norbury, C., and Nurse, P. (1992). Animal cell cycles and their control. *Annu. Rev. Biochem.* **61**, 441–470.

Oakley, C. E., and Oakley, B. R. (1989). Identification of gamma-tubulin, a new member of the tubulin superfamily encoded by the *mipA* gene of *Aspergillus nidulans*. *Nature (London)* **338**, 662–664.

Ohkura, H., Kinoshita, N., Miyatani, S., Toda, T., and Yanagida, M. (1989). The fission yeast *dis2* + gene required for chromosome disjoining encodes one of two putative type 1 protein phosphatases. *Cell* **57**, 997–1007.

Ookata, K., Hisanaga, S., Okano, T., Tachibana, K., and Kishimoto, T. (1992). Relocation and distinct subcellular localisation of p34 cdc2-cyclin B complex at meiosis reinitiation in starfish oocytes. *EMBO J.* **11**, 1763–1772.

Orgad, S., Brewis, N. D., Alphey, L. S., Axton, J. M., Dudai, Y., and Cohen, P. T. W. (1990). The structure of protein phosphatase 2A is as highly conserved as that of protein phosphatase 1. *FEBS Lett.* **275**, 44–48.

Pagano, M., Pepperkok, R., Verde, F., Ansorge, W., and Draetta, G. (1992). Cyclin A is required at two points in the human cell cycle. *EMBO J.* **11**, 961–971.

Pereira, A., Doshen, J., Tanaka, E., and Goldstein, L. S. B. (1992). Genetic analysis of a *Drosophila* microtubule-associated protein. *J. Cell Biol.* **116**, 377–383.

Perrimon, N., Engstrom, L., and Mahowald, A. P. (1985). Developmental genetics of the 2C-D region of the *Drosophila* X chromosome. *Genetics* **111**, 23–41.

Philp, A. V., Axton, J. M., Saunders, R. D. C., and Glover, D. M. (1993). Mutations in the *Drosophila melanogaster* gene *three rows* permit aspects of mitosis to continue in the absence of chromatid segregation. *J. Cell Sci.* **106**, 87–98.

Picard, A., Labbe, J. C., Barakat, H., Cavadore, J.-C., and Doree, M. (1991). Okadaic acid mimics a nuclear component required for cyclin B-cdc2 kinase microinjection to drive starfish oocytes into M phase. *J. Cell Biol.* **115**, 337–344.

Pines, J., and Hunter, T. (1991a). Cyclin-dependent kinases: a new cell cycle motif? *Trends Cell Biol.* **1**, 117–121.

Pines, J., and Hunter, T. (1991b). Human cyclins A and B1 are differentially located in the cell and undergo cell-cycle dependent nuclear transport. *J. Cell Biol.* **115**, 1–17.

Pokrywka, N. J., and Stephenson, E. C. (1991). Microtubules mediate the localisation of *bicoid* RNA during *Drosophila* oogenesis. *Development* **113**, 55–66.

Postner, M. A., Miller, K. G., and Wieschaus, E. F. (1992). Maternal effect mutations of the *sponge* locus affect actin cytoskeletal rearrangements in *Drosophila melanogaster* embryos. *J. Cell Biol.* **119**, 1205–1218.

Rabinowitz, M. (1941). Studies on the cytology and early embryology of the egg of *Drosophila melanogaster*. *J. Morphol.* **69**, 1–49.

Raff, E. C. (1984). Genetics of microtubule systems. *J. Cell Biol.* **99**, 1–10.

Raff, E. C., and Fuller, M. T. (1984). Genetic analysis of microtubule function in *Drosophila*. *In* "Molecular Biology of the Cytoskeleton" (G. G. Borisy, D. W. Cleveland, and D. B. Murphy, eds.), pp. 293–304. Cold Spring Harbor Lab. Press, Cold Spring Harbor, New York.

Raff, J. W., and Glover, D. M. (1988). Nuclear and cytoplasmic mitotic cycles continue in *Drosophila* embryos in which DNA synthesis is inhibited with aphidicolin. *J. Cell Biol.* **107**, 2009–2019.

Raff, J. W., and Glover, D. M. (1989). Centrosomes, and not nuclei, initiate pole cell formation in *Drosophila* embryos. *Cell* **57**, 611–619.

Raff, J. W., Whitfield, W. G. F., and Glover, D. M. (1990). Two distinct mechanisms localise cyclin B transcripts in syncytial *Drosophila* embryos. *Development* **110**, 1249–1261.

Raff, J. W., Kellogg, D. R., and Alberts, B. M. (1993). *Drosophila* gamma tubulin is part of a complex containing two previously identified centrosomal MAPs. *J. Cell Biol.* **121**, 823–835.

Rasooly, R. S., New, C. M., Zhang, P., Hawley, R. S., and Baker, B. S. (1991). The *lethal(1)TW-6cs* mutation of *Drosophila melanogaster* is a dominant antimorphic allele of *nod* and is associated with a single base change in the putative ATP-binding domain. *Genetics* **129**, 409–422.

Rattner, J. B. (1992). Integrating chromosome structure with function. *Chromosoma* **101**, 259–264.

Rattner, J. B., Kingwell, B. G., and Fritzler, M. J. (1988). Detection of distinct structural domains within the primary constriction using autoantibodies. *Chromosoma* **96**, 360–367.

Regan, C. L., and Fuller, M. T. (1988). Interacting genes that affect microtubule function: the *nc2* allele of the *haywire* locus fails to complement mutations in the testis-specific beta-tubulin gene of *Drosophila*. *Genes Dev.* **2**, 82–92.

Regan, C. L., and Fuller, M. T. (1990). Interacting genes that affect microtubule function in *Drosphila melanogaster*: two classes of mutations revert the failure to complement between *hay*[nc2] and mutations in tubulin genes. *Genetics* **125**, 77–90.

Richardson, H. E., Wittenburg, C., Cross, F., and Reed, S. I. (1989). An essential G1 function for cyclin-like proteins in yeast. *Cell* **59**, 1127–1133.

Richardson, H. E., O'Keefe, L. V., Reed, S. I., and Saint, R. (1993). A *Drosophila* G1-specific cyclin E homolog exhibits different modes of expression durib embryogenesis. *Development* 119, 673–690.

Ripoll, P., Pimpinelli, S., Valdivia, M. M., and Avila, J. (1985). A cell division mutant of *Drosophila* with a functionally abnormal spindle. *Cell* 41, 907–912.

Ripoll, P., Casal, J., and Gonzalez, C. (1987). Mitosis in *Drosophila*. *BioEssays* 7, 204–210.

Ripoll, P., Carmena, M., and Molina, I. (1992). Genetic analysis of cell division in *Drosophila melanogaster*. *Curr. Top. Dev. Biol.* 27, 275–307.

Rosenblatt, J., Gu, Y., and Morgan, D. O. (1992). Human cyclin-dependent kinase 2 is activated during the S and G2 phases of the cell cycle and associates with cyclin A. *Proc. Natl. Acad. Sci. U.S.A.* 89, 2824–2828.

Russell, P., and Nurse, P. (1986). *cdc25*⁺ functions as an inducer in the mitotic control of fission yeast. *Cell* 45, 145–153.

Russell, P., and Nurse, P. (1987). Negative regulation of mitosis by *wee1* a gene encoding a protein kinase homolog. *Cell* 49, 559–567.

Sanchez, F., Natzle, J. E., Cleveland, D. W., Kirschner, M. W., and McCarthy, B. J. (1980). A dispersed multigene family encoding tubulin in *Drosophila melanogaster.Cell* 22, 845–854.

Sandler, L., Lindsley, D. L., Nicoletti, B., and Trippa, G. (1968). Mutants affecting meiosis in natural populations of *Drosophila melanogaster*. *Genetics* 60, 525–558.

Sawin, K. E., and Endow, S. A. (1993). Meiosis mitosis and microtubule motors. *BioEssays* 15, 399–407.

Saxton, W. M., Porter, M. E., Colin, S. A., Scholey, J. M., Raff, E. C., and McIntosh, J. R. (1988). *Drosophila* kinesin: characterisation of microtubule motility and ATPase. *Proc. Natl. Acad. Sci. U.S.A.* 85, 1109–1113.

Saxton, W. M., Hicks, J., Goldstein, L. S. B., and Raff, E. C. (1991). Kinesin heavy chain is essential for viability and neuromuscular functions in *Drosophila,* but mutants show no defects in mitosis. *Cell* 64, 1093–1102.

Schaefer, L., Roy, R., Humbert, S., Moncollin, V., Vermeulen, W., Hoeijmakers, J. H. J., Chambon, P., and Egly, J.-M. (1993). DNA repair helicase: a component of BTF2 (TFIIH) basic transcription factor. *Science* 260, 58–63.

Schild, D., and Byers, B. (1980). Diploid spore formation and other meiotic effects of 2 cell-division-cycle mutations of *Saccharomyces cerevisiae*. *Genetics* 96, 859–876

Schupbach, T., and Wieschaus, E. (1986). Maternal effect mutations altering anterior posterior pattern in the *Drosophila melanogaster* embryo. *Wilhem Roux's Arch Dev. Biol.* 195, 302–317.

Selleck, S. B., Gonzalez, C., Glover, D., and White, K. (1992). Regulation of the G1-S transition in post-embryonic neuronal precursors by axon ingrowth. *Nature (London)* 355, 253–255.

Sequeria, W., Nelson, C. R., and Szauter, P. (1989). Genetic analysis of the *claret* locus of *Drosophila melanogaster*. *Genetics* 123, 511–524.

Shamanski, F., and Orr-Weaver, T. L. (1991). The *Drosophila plutonium* and *pan gu* genes regulate entry into S phase at fertilisation. *Cell* 66, 1289–1300.

Shamu, C. E., and Murray, A. W. (1992). Sister chromatid separation in frog egg extracts requires DNA topoisomerase II activity during anaphase. *J. Cell. Biol.* 117, 921–934.

Simmons, D. L., Neel, B. G., Stevens, R., Evett, G., and Erikson, R. (1992). Identification of an early growth response gene encoding a novel putative protein kinase. *Mol. Cell Biol.* 12, 4164–4169.

Skaer, H. (1989). Cell division in Malpighian tubule development in *D. melanogaster* is regulated by a single tip cell. *Nature (London)* **342**, 566–569.

Smith, D. E., Gruenbaum, Y., Berrios, M., and Fisher, P.A. (1987). Biosynthesis and interconversion of *Drosophila* nuclear lamin isoforms during normal growth and in response to heat shock. *J. Cell Biol.* **99**, 20–28.

Sola, M. M., Langan, T., and Cohen, P. (1991). p34[cdc2] phosphorylation sites in histone H1 are dephosphorylated by protein phosphatase 2A. *Biochim. Biophys. Acta* **1094**, 211–216.

Solomon, M. J., Glotzner, M., Lee, T. H., Philippe, M., and Kirschner, M. W. (1990). Cyclin activation of p34[cdc2]. *Cell* **63**, 1013–1024.

Sonnenblick, B. P. (1950). The early embryology of *Drosophila melanogaster. In* "Biology of *Drosophila*" (M. Demerec, ed.), Chapter 2. Wiley, New York.

Stafstron, J. P., and Staehelin, L. A. (1984). Dynamics of the nuclear envelope and of the nuclear pore complexes during mitosis in the *Drosophila* embryo. *Eur. J. Cell Biol.* **34**, 179–189.

Stearns, T., and Botsein, D. (1988). Unlinked non-complementation: Isolation of new conditional-lethal mutations in each of the tubulin genes of *Saccharomyces cerevisiae. Genetics* **119**, 249–260.

Stern, B., Ried, G., Clegg, N. J., Grigliatti, T. A., and Lehner, C. F. (1993). Genetic analysis of the *Drosophila cdc2* homolog. *Development* **117**, 219–232.

Stewart, R. J., Pesavento, P. A., Woerpel, D. N., and Goldstein, L. S. B. (1991). Identification and partial characterisation of six new members of the kinesin superfamily in *Drosophila. Proc. Natl. Acad. Sci. U.S.A.* **88**, 8470–8474.

Sullivan, W., Minden, J. S., and Alberts, B. M. (1990). *daughterless-abo-like*, a *Drosophila* maternal-effect mutation that exhibits abnormal centrosome separation during the late blastoderm divisions. *Development* **110**, 311–323.

Sunkel, C. E., and Glover, D. M. (1988). *polo*, a mitotic mutant of *Drosophila* displaying abnormal spindle poles. *J. Cell Sci.* **89**, 25–38.

Surana, U. H., Amon, A., Dowzer, C., McGrew, J., Byers, B., and Nasmyth, K. (1993). Destruction of the *CDC28/CLB* mitotic kinase is not required for the metaphase to anaphase transition. *EMBO J.* **12**, 1969–1978.

Szabad, J., and Bryant, P. (1982). The mode of action of discless mutations in *Drosophila melanogaster.* Dev. Biol. **93**, 240–256.

Szabad, J., Jursnich, V. A., and Bryant, P. J. (1991). Requirement for cell-proliferation control genes in *Drosophila* oogenesis. *Genetics* **127**, 525–533.

Theurkauf, W. E., and Hawley, R. S. (1992). Meiotic spindle assembly in *Drosophila* females: behavior of nonexchange chromosomes and the effects of mutations in the *nod* kinesin-like protein. *J. Cell Biol.* **116**, 1167–1180.

Tobin, S. L., Zulauf, E., Sanchez, F., Craig, E. A., and McCarthy, B. J. (1980). Multiple actin-related sequences in the *Drosophila melanogaster* genome. *Cell* **19**, 121–131.

Tokuyasu, K. T. (1974a). Dynamics of spermiogenesis in *Drosophila melanogaster*: IV. Nuclear transformation. *J. Ultrastruct. Res.* **48**, 284–303.

Tokuyasu, K. T. (1974b). Dynamics of spermiogenesis in *Drosophila melanogaster*: III. Relation between axoneme and mitochondrial derivates. *Exp. Cell Res.* **84**, 239–250.

Tokuyasu, K. T. (1974c). Spoke heads in sperm tail of *Drosophila melanogaster. J. Cell Biol.* **63**, 334–337.

Tokuyasu, K. T. (1975). Dynamics of spermiogenesis in *Drosophila melanogaster* VI. Significance of "onion" nebenkern formation. *J. Ultrastruct. Res.* **53**, 93–112.

Troelstra, C., Hesen, W., Bootsma, D., and Hoeijmakers, J. H. J. (1993). Strucure and expression of the excision repair gene *ERCC6,* involved in the human disorder Cockaynes syndrome group-B. *Nucleic Acids Res.* 21, 419–426.

Tsuda, L., Indue, Y. H., Yoo, M., Mizuno, M., Hata, M, Lim, Y., Adachi-Yamada, T., Ryo, H., Masamune, Y., and Nishida, Y. (1993). A protein kinase similar to MAP kinase activator acts downstream of the *raf* kinase in *Drosophila. Cell* 72, 407–414.

Tudor, M., Lobocka, M., Goodell, M., Pettit, J., and O'Hare, K. (1992). The *pogo* transposable element family of *Drosophila melanogaster. Mol. Gen. Genet.* 232, 126–134.

Uemura, T., Ohkura, H., Adachi, Y., Morino, K., Shiozaki, K., and Yanagida, M. (1987). DNA topoisomerase II is required for condensation and separation of mitotic chromosomes in *S. pombe. Cell* 50, 917–925.

Uemura, T., Shiomi, K., Togashi, S., and Takeichi, M. (1993). Mutation of *twins* encoding a regulator of protein phosphatase 2A leads to pattern duplication in the *Drosophila* imaginal disc. *Genes Dev.* 7, 429–440.

Vale, R. D., Reese, T. S., and Sheetz, M. P. (1985). Identification of a novel force-generating protein, kinesin, involved in microtubule-based motility. *Cell* 42, 39–50.

Wald, H. (1936). Cytologic studies on the abnormal development of the eggs of the *claret* mutant type of *Drosophila simulans. Genetics* 21, 264–281.

Walker, R. A., Salmon, E. D., and Endow, S. A. (1990). The *Drosophila claret* segregation protein is a minus-end directed motor molecule. *Nature (London)* 347, 780–782.

Warn, R. M., Magrath, R., and Webb, S. (1984). Distribution of F-actin during cleavage of the *Drosophila* syncytial blastoderm. *J. Cell Biol.* 98, 156–162.

Weil, C. F., Oakley, C. E., and Oakley, B. R. (1986). Isolation of *mip* (*microtubule interacting protein*). mutations of *Aspergillus nidulans. Mol. Cell. Biol.* 6,2963–2968.

Wharton, R. P., and Struhl, G. (1991). RNA regulatory elements mediate control of *Drosophila* body pattern by the posterior morphogen *nanos. Cell* 67, 955–967.

White-Cooper, H., Alphey, L., and Glover, D. M. (1993). The *cdc25* homologue *twine* is required for only some aspects of the entry into meiosis in *Drosophila. J. Cell Sci.* 106, 1035–1044.

Whitfield, W. G. F., Millar, S. E., Saumweber, H., Frasch, M., and Glover, D. M. (1988). Cloning of a gene encoding an antigen associated with the centrosome in *Drosophila. J. Cell. Sci.* 89, 467–480.

Whitfield, W. G. F., González, C., Sánchez-Herrero, E., and Glover, D. M. (1989). Transcripts from one of two *Drosophila* cyclin genes become localized in pole cells during embryogenesis. *Nature (London)* 338, 337–340.

Whitfield, W. G. F., González, C., Maldonado-Codina, G., and Glover, D. M. (1990). The A- and B-type cyclins of *Drosophila* are accumulated and destroyed in temporally distinct events that define separable phases of the G2-M transition. *EMBO J.* 9, 2563–2572.

Wieschaus, E., and Sweeton, D. (1988). Requirements for X-linked zygotic gene activity during cellularisation of early *Drosophila* embryos. *Development* 104, 483–493.

Williams, B. C., Karr, T. L., Montgomery, J. M., and Goldberg, M. L. (1992). The *Drosophila zw10* gene product, required for accurate mitotic chromosome segregation, is redistributed at anaphase onset. *J. Cell Biol.* 118, 759–773.

Wilson, P. G., Heck, M., and Fuller, M. T. (1992). Monoastral spindles are generated by mutation in *urchin,* a *bimC* homologue in *Drosophila. Mol. Biol. Cell* 3, 343a.

Wood, J. S., and Hartwell, L. H. (1982). A dependent pathway of gene functions leading to chromosome segregation in *Saccharomyces cerevisiae. J. Cell. Biol.* 94, 718–726.

Woods, D. F., and Bryant, P. J. (1991). The *discs-large* tumor suppressor gene of *Drosophila* encodes a guanylate kinase homolog localised at septate junctions. *Cell* 66, 451–464.

Wynes, D. R., and Henikoff, S. (1992). Somatic instability of a *Drosophila* chromosome. *Genetics* **131**, 683–691.

Xiong, Y., Connolly, T., Futcher, B., and Beach, D. (1991). Human D-type cyclin. *Cell* **65**, 691–699.

Yamaguchi, M., Nishida, Y., Moriuchi, T., Hirose, F., Hui, C.-C., Suzuki, Y., and Matsukage, A. (1990). *Drosophila* proliferating cell nuclear antigen (cyclin). gene: structure, expression during development and specific binding of homeodomain proteins to its 5'-flanking region. *Mol. Cell. Biol.* **10**, 872–879.

Yamamoto, A. H., Komma, D. J., Shaffer, C. D., Pirrotta, V., and Endow, S. A. (1989). The *claret* locus in *Drosophila* encodes products required for eye color and for meiotic chromosome segregation. *EMBO J.* **8**, 3543–3552.

Yang, J. T., Saxton, M., and Goldstein, L. S. B. (1988). Isolation and characterisation of the gene encoding the heavy chain of *Drosophila* kinesin. *Proc. Natl. Acad. Sci. U.S.A.* **85**, 1864–1868.

Young, P. E., Richman, A. M., Ketchum, A. S., and Kiehart, D. P. (1993). Morphogenesis in *Drosophila* requires nonmuscle myosin heavy-chain function. *Genes Dev.* **7**, 29–41.

Zalokar, M., and Erk, I. (1976). Division and migration of nuclei during early development of *Drosophila* eggs. *J. Microsc. Biol. Cell.* **25**, 97–106.

Zhang, P., and Hawley, R. S. (1990). The genetic analysis of distributive segregation in *Drosophila melanogaster*. II. Further genetic analysis of the *nod* locus. *Genetics* **125**, 115–127.

Zhang, P., Knowles, B. A., Goldstein, L. S. B., and Hawley, R. S. (1990). A kinesin-like protein required for distributive chromosome segregation in *Drosophila*. *Cell* **62**, 1053–1062.

Zheng, Y., Jung, M.K., and Oakley, B. R. (1991). Gamma tubulin is present in *Drosophila melanogaster* and *Homo sapiens* and is associated with the centrosome. *Cell* **65**, 817–823.

Zindy, F., Lamas, E., Chenivesse, X., Sobczak, J., Wang, J., Fesquet, D., Henglein, B., and Bréchot, C. (1992). Cyclin A is required in S-phase in normal epithelial cells. *Biochem. Biophys. Res. Commun.* **182**, 1144–1154.

Zulauf, E., Sanchez, F., Tobin, L., Rdest, U., and McCarthy, B. J. (1981). Developmental expression of a *Drosophila melanogaster* actin gene encoding actin I. *Nature (London)* **292**, 556–558.

GENETIC AND MOLECULAR ANALYSIS OF
Drosophila BEHAVIOR

C. P. Kyriacou* and Jeffrey C. Hall†

*Department of Genetics, University of Leicester, Leicester, LE1 7RH,
United Kingdom; and †Department of Biology, Brandeis University,
Waltham, Massachusetts 02254

I. Introduction

The genetic analysis of *Drosophila* behavior is an area which has attracted some attention over the past three decades, but in the past 6 or 7 years, there has been an explosion in the number of workers in this

ADVANCES IN GENETICS, Vol. 31

field. Most of these scientists would probably say that they had little interest in behavior *per se,* and that they were primarily concentrating analyzing the nervous system using genetic and molecular methods. It is understandable perhaps that most *bona fide* geneticists and molecular biologists working in the area would prefer to work with much simpler phenotypes (e.g., gels), rather than the highly complex behavioral patterns that the fly can produce. Nevertheless, their efforts are beginning to reveal some of the underlying mechanisms that determine how the nervous system functions, with obvious implications for those who are more interested in behavior. In this chapter we attempt to highlight some of the more recent studies in the field using examples in which progress has been made in understanding both complex and relatively simpler behavioral phenotypes. The work to be reviewed will include studies on the visual system, olfaction and contact chemoreception, learning, courtship behavior, and biological rhythms. The overlap among these systems will become evident as we progress through the literature.

II. Visual System Genes

A. SEVENLESS

A number of laboratories have been using sophisticated genetic and molecular techniques to analyze the development of the visual system. In doing so they have often relied on genes which were initially isolated by screening for behavioral mutants. A classic example of this approach is the work with *sevenless (sev),* in which the mutant *sev* fly does not show the normal phototactic response to ultraviolet light (Harris *et al.,* 1976). The behavioral response to ultra-violet light is mediated by receptor cell R7 in the ommatidium. Subsequent work has shown that the *sev* gene encodes a transmembrane protein (Banerjee *et al.,* 1987; Tomlinson *et al.,* 1987) with a cytoplasmic tyrosine kinase domain (Hafen *et al.,* 1987; Simon *et al.,* 1989; Basler and Hafen, 1988). A ligand produced by the neighboring receptor cell R8 (Reinke and Zipursky, 1988) binds to the extracellular domain of the SEV protein (Krämer *et al.,* 1991), and a phosphorylation signal is generated inside the cell. This signal results eventually in the R7 progenitor cell becoming the R7 receptor cell.

In the *sev* mutant, however, the R7 progenitor becomes a nonneural cone cell, suggesting that *sev* is a cell-specific homeotic gene. A number of other genes play a role in the induction of the R7 cell fate (Krämer *et al.,* 1991; Rogge *et al.,* 1991; Carthew and Rubin, 1990), and the *sevenless* story has now become a classic within the field of the molecular and cellular control of development. It is sometimes hard to believe that the *sev* gene was in fact originally isolated as a behavioral mutant. This highlights the fact that the more one learns about a "behavioral gene," the less of a behavioral gene it will become, especially if it ends up having an apparently specific developmental function, as in the case of *sev*.

However, the picture is not quite as clear-cut as it seems, because *sev* is pleiotropic. The *sev* mutants prefer green light to UV, unlike the wild type (Heisenberg and Buchner, 1977). A screen for second site suppressors for this *sev*-induced behavioral defect produced a dominant mutation, *Photophobe* (*Ppb*), which had an allele-specific interaction with a *sev* mutant (Ballinger and Benzer, 1988). In combination with the *sev^LY3* allele, *Ppb* reverses the mutant's color preference so that the double mutant favors UV light. This occurs in the absence of receptor cell R7 in *sev^LY3*;*Pp^b*/+ flies. The implication is that *sev* also plays a role in this behavioral pathway, but the pathway is not mediated by receptor cell R7. This epistatic interaction of the two loci may be a reflection of the fact that *sev* is expressed also in the adult brain, that is, is not solely restricted to the developing eye discs (Banerjee *et al.*, 1987; Hafen *et al.*, 1987). This nervous system expression of *sev* may be important for determining the behavioral choice of UV versus green light.

B. *OPTOMOTOR-BLIND*

Other genes that play a role in complex visual behaviors have been identified both by behavioral screens and by mutations that affect brain morphology. One of these genes is *optomotor-blind* (*omb*), which was originally isolated as a behavioral mutant (Heisenberg and Gotz, 1975) and was also, incidentally, the first mutant to be isolated on the basis of a brute-force screen for mutants in the structure of the brain (Heisenberg and Böhl, 1979). The original *optomotor-blind* mutation, *omb^H31*, is associated with a large inversion of the X chromosome, and the mutant appears almost oblivious to rotating vertical stripes in its environment (Heisenberg *et al.*, 1978). The defect is especially severe when the fly is walking, but is less marked during flight. The mutant is missing giant neuronal fibers of the lobula plate in the optic lobe of the visual system. These fibers have been shown to be important for motion detection in larger dipterans (Strausfeld and Bacon, 1983). However other visually guided behaviors are apparently normal in *omb^H31* mutants, including the landing response (Heisenberg *et al.*, 1978).

The gene has been cloned (Pflugfelder *et al.*, 1990) and appears to be one of the few *Drosophila* genes that is spread over 100 kb of genomic DNA. Not surprisingly, the molecular genetics of *omb* appear very complex with more than 20 mRNAs transcribed from this region (Pflugfelder *et al.*, 1990). The distal part of the *omb* gene appears to overlap with other complementation units, *bifid* (*bi*), *l(1)bifid* [*l(1)bi*], *Quadroon* (*Qd*), and *lacquered*^{gls} (*lac*). Adult wing and abdominal morphology is affected by the three nonvital genes and as yet no nervous system dysfunction has been associated with these viable mutants. A lethal *omb* mutation, *l(1)omb*, fails to complement *bi* and *Qd*. The *l(1)omb* gene encodes a protein of 974 amino acids, which has a repeated occurrence of a serine-proline motif (Pflugfelder *et al.*, 1992a) that is found in DNA-binding proteins (Suzuki, 1989). Consequently the OMB protein appears to be a nuclear regulatory protein; this is supported by the discovery that it shares a common domain with the mouse gene *Brachyury*, which can bind DNA (Pflugfelder *et al.*, 1992b).

The 3' regulatory region of *l(1)omb* has been studied (Brunner *et al.*, 1992): mutations which remove progressively more downstream material cause, not surprisingly, progressively more severe neuroanatomical and behavioral defects. Based on these 3' deletions, the regulatory region has been divided into three domains. Removal of the area furthest downstream results in the lobula plate phenotype seen in all *omb* mutants (except *Qd* and *bi*) when placed in trans with *l(1)omb*. Removal of the areas adjacent 5' to this region causes additional anatomical defects, which are sometimes observed between the medulla and lobula neuropil. Further loss of the area 5' to this region enhances the penetration of this latter additional defect. Deletion of the two most 3' regions also causes progressively enhanced optomotor behavior defects (Brunner *et al.*, 1992).

Two *l(1)omb* mutations have also been mapped to one of the many transcripts (called T3) from this locus (Poeck *et al.*, 1993a). The T3 mRNA expression pattern during optic lobe development and its altered expression pattern in a mutant *In(1)omb*^{H31}, suggests that this transcript encodes *omb* function (Poeck *et al.*, 1993b). The two mutations both cause premature translational stops in the T3 transcript (Poeck *et al.*, 1993a).

C. SMALL-OPTIC-LOBES

Another mutation that was isolated by Heisenberg and Böhl's (1979) pioneering work was *small-optic-lobes* (*sol*). The mutant shows severe cell degeneration and an approximately 50% reduction in the pupal optic lobes (Fischbach and Heisenberg, 1981; Fischbach and Technau,

1984). Many cell types are thus severely reduced in number and some classes are totally eliminated; yet the optomotor yaw response is manifestly normal. However, the landing responses, orientation behavior, and figure–ground discrimination are defective (Fischbach and Heisenberg, 1981). Comparison of wild-type and *sol* brains has thus led to a preliminary identification of cells that are and are not necessary for different components of normal visual behaviors (Fischbach and Heisenberg, 1981).

Recently the *sol* locus has been cloned (Delaney *et al.*, 1991) and mapped to a 14-kb region of genomic DNA. Two major transcripts are produced, a 5.8- and 5.2-kb species, and transformation has revealed that a 10.3-kb genomic DNA fragment is sufficient to restore normal brain morphology to a *sol*^{KS58} mutant (Delaney *et al.*, 1991). Northern blotting analysis reveals that both *sol* transcripts are detected at all developmental stages, but the abundance of the mRNA is reduced during the larval stages. The longer 5.8-kb mRNA encodes a conceptual 1597-amino-acid protein, which bears in its amino terminal domain six motifs resembling the zinc fingers found in steroid hormone receptors. The carboxyl terminal of the protein has a 300-amino-acid domain, which appears similar to vertebrate calcium-activated neutral proteases (calpains).

Perhaps then, the SOL protein has protease activity, as well as DNA binding functions. Further transformation and *in vitro* mutagenesis experiments may reveal the functional relevance of each of the SOL protein domains. The smaller *sol* transcript encodes the first 393 amino acids plus two further residues from the carboxyl terminus. A frame shift is responsible for the severe truncation of the protein. In the words of Delaney *et al.* (1991), the function, if any, of the small *sol* product is "enigmatic."

Thus the brain mutations *l(1)omb* and *sol* are being dissected molecularly and will inevitably contribute to our understanding of how such genes "encode" complex visual behavioral phenotypes. We will return to genes that determine components of the visual system in a different context.

III. Olfaction

A. METHODS

Drosophila appear to have the ability to detect and discriminate among a large number of different volatile chemicals. They can do this at both the larval and adult stages. A variety of experimental procedures have been used over the years to isolate olfactory mutants. An early attempt made use of a simple Y-maze olfactometer (Rodrigues and Siddiqi, 1978; also reported in Ayyub *et al.*, 1990) in which one arm of the maze contained the experimental odor and the other arm contained the control air. Flies were placed at the start tube and allowed to distribute themselves in either of the two arms in response to the control air and odorants sucked through the apparatus.

Helfand and Carlson (1989) later described a different type of T-maze, similar to the olfactory learning apparatus of Tully and Quinn (1985), which again measured the distribution of flies in two arms of the maze in response to a control and an experimental odor. McKenna *et al.* (1989) described a simple olfactory jump assay where odorant is passed through a tube containing a fly. If the fly jumps within 3 sec of the stimulus, this is scored as a positive escape response and permits an individual rather than a population of flies to be measured.

Larval olfaction has also been described in detail by Monte *et al.* (1989). Third-instar larvae were placed in the center of a petri dish containing agarose and two filter discs placed opposite each other; one contained odorant and the other the control (the solution in which the odorant was dissolved). The distribution of larvae in the dish after 5 min was examined (Fig. 1). Since the behavioral response of the larvae was rapid, this strongly suggests that the migration is mediated by airborne stimulants.

B. ANATOMY OF OLFACTION

Surgical experiments on adult flies suggest that the olfactory response in a T-maze (Helfand and Carlson, 1989) depends on input from sense organs on the antenna (Helfand and Carlson, 1989). These organs are primarily located in the third antennal segment (Barrows, 1907; Bolwig, 1946; Venard and Pichon, 1981). Ayer and Carlson (1992) have also suggested that the maxillary palp has olfactory function. Venkatesh and Singh (1984) divided the sensory hairs in the third antennal segment into three types, two of which have pores through which airborne chemicals can pass. Neurons from each sensory hair project to the two antennal lobes, from which further projections connect to other parts of the brain, including the mushroom bodies (Stocker *et al.*, 1990). As we will see in a later section, the mushroom bodies are heavily implicated as the neuroanatomical focus for smell-driven learning (Davis and Dauwalder, 1991).

Given that perhaps one thousand neurons convey the receptor signals to the brain, it is not surprising that the fly has such a rich rep-

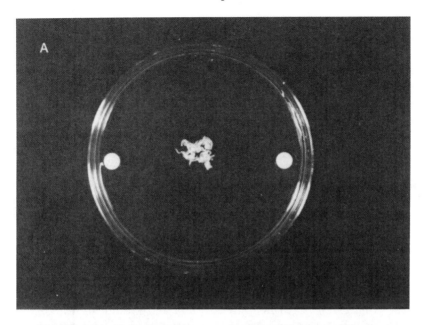

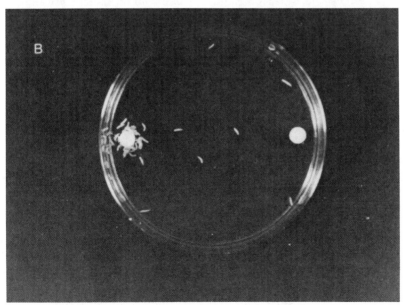

FIG. 1. The larval olfactory assay. The disk on the left contains the odorant, the disk on the right, the control. (A) The clumps of larvae are placed between the two disks. (B) Five minutes later the larvae have migrated toward the odorant. (Reprinted with permission from Monte *et al.*, 1989.)

ertoire of olfactory-driven behavior directed to the large range of chemicals it will experience in its environment. Therefore it is probable that a large number of genes will determine the development of the olfactory receptors, the neurons, the signal transduction cascades, and the final motor output which will build this integrated system.

C. OLFACTORY MUTANTS

Rodrigues and Siddiqi (1978), and Ayyub et al. (1990) described the isolation of mutations in 6 X-linked genes of which mutations in four, olfA, olfB, olfE, and olfF, caused behavioral defects in the presence of aldehydes. Another gene, olfC, could be mutated to give abnormal responses to acetate esters. A sixth mutant olfD^{x9}, produced a general anosmia, and is allelic to smell-blind (sbl) (Lilly and Carlson, 1989), a mutant which failed to learn to avoid electric shock when the shock was associated with odor (Aceves-Piña and Quinn, 1979). This olf mutant is now designated sbl^{olfD} (Lilly and Carlson, 1989). Mutants in all six of these genes affected both adult and larval olfactory behavior. This is interesting, given that the adult antenna and the larval antennal organ, which is the anterior structure believed to be the preimaginal olfactory organ, have quite different developmental origins. This suggests that these genes might encode processes common to the workings of both types of receptors, for example, the signal transduction pathways in both systems, rather than to the building of the structures.

Larvae carrying different mutant alleles of the sbl locus show defective olfactory and contact chemosensory behavior (Lilly and Carlson, 1989). Their phototactic responses, which use the same kind of motor behavior as in the olfactory and contact chemosensory assays, are not impaired. This demonstrates that the mutants (sbl^1 and sbl^{olfD}) are not generally "sick." The two mutations are also heat-sensitive lethals. Four other lethal sbl mutations have now been described, which suggests that the sbl gene either has pleiotropic effects outside the olfactory system, or that viability must depend on a reasonably intact olfactory system early in development. We favor the former scenario, especially since Carlson (1991) has mentioned preliminary results that sbl may encode a component which interacts with sodium channels. If this proves to be the case, then the defects observed in both sbl larvae and adults are more readily understood.

Of the olf genes, only olfE has been cloned so far (Hasan, 1990). Transformation of olfE mutant flies with a 14-kb genomic fragment partially rescued the defective Y-maze and olfactory jump responses of the mutants. Two transcripts were detected—a major one of 5.4 kb and a minor one of 1.7 kb, the latter disappearing during the larval stages. Given that *olfE* mutants also affect larval olfaction, Hasan (1990) has suggested that the 1.7-kb mRNA may not be directly involved in determining the larval olfactory phenotype. The two transcripts are probably alternatively spliced mRNAs from the *olfE* gene. They are seen in embryos and in the heads and bodies of both wild-type and mutant adults. The widespread distribution of *olfE* mRNA implies that this gene plays a role outside the olfactory system. DNA sequence analysis may clarify the possible role of *olfE* in olfaction and contact chemoreception.

D. OLFACTORY-TRAP ABNORMAL AND ABNORMAL CHEMOSENSORY JUMP MUTANTS

Using an ingeniously simple device to trap flies that were attracted to an odorant, Woodard *et al.* (1989) isolated a series of X-linked mutants that showed abnormal behavioral responses. These were termed *olfactory-trap-abnormal* (*ota*). One mutant, *ota1,* was found to be an allele of the *retinal-degeneration-B* (*rdgB*) gene originally isolated by Harris and Stark (1977).

Conversely, other *rdgB* mutations were observed to show olfactory trap defects, as well as abnormalities in their electroantennogram responses (EAGs), which represent a global picture of the receptor potentials of antennal neurons (Woodard *et al.*, 1992). The *rdgB* gene has been cloned and sequenced (Vihtelic *et al.*, 1991) and encodes a putative protein with a transmembrane domain and a calcium binding site. This protein appears to have six potential membrane spanning domains, which suggests that it may be an ion channel or a Ca^{2+} transporter. Antibodies to an RDG-B fusion protein label not only the retina but also regions in the brain (Vihtelic *et al.*, 1991). Consequently, these results are consistent with the view that the role of the *rdgB* gene is not limited solely to the visual system. Clearly if *rdgB* encodes an ion channel protein, it could be involved in several, perhaps many, behavioral systems.

The *abnormal chemosensory jump* (*acj*) mutants have been isolated on the basis of a defective olfactory jump response (McKenna *et al.*, 1989). One mutant, *acj6,* has a defective EAG, suggesting a generally reduced antennal sensitivity to odorants (Ayer and Carlson, 1991). The mutation also affects larval olfaction but does not impair visual behavior. Consequently *acj6* may be a gene that encodes specific properties of the olfactory system, unlike *rdgB*. This is supported by the observation that *acj6* reduces the maxillary palp EAG responses to all odorants except benzaldehyde (Ayer and Carlson, 1992). Thus the benzaldehyde-mediated response pathway appears intact in these mutants, which suggests that the wild-type *acj6* allele encodes some olfactory specificity.

A concerted effort has been made to identify and isolate genes that function in the olfactory system. Some of the hard work of behavioral screening and genetic analysis has been done, and the molecular analysis can be expected to follow. Other methods can be used, apart from mutagenesis, to identify olfaction genes. Riesgo-Escovar *et al.* (1992) have reported an analysis of the olfactory system using enhancer traps. A number of lines show reporter gene expression in different regions of the antennal–maxillary complex, and some are sex and age specific. Imprecise excision of the enhancer trap insertions may lead to interesting and interpretable olfactory phenotypes, including some that may affect sexual behavior; any interesting loci should be amenable to molecular analysis. Another strategy that has been used successfully is that taken by Buck and Axel (1991), who isolated G-protein-coupled receptors from the rat olfactory system using a polymerase chain reaction (PCR)-based approach. A similar analysis using *Drosophila* head cDNA libraries could also prove fruitful.

IV. Learning

A. GENETICS

Drosophila have the ability to associate electric shock with odor (Quinn *et al.*, 1974; Tully and Quinn, 1985), and several X-linked mutant strains have been induced (using chemical mutagenesis) which perform poorly in these conditioning tests (Dudai *et al.*, 1976; Aceves-Pina and Quinn, 1979; Quinn *et al.*, 1979). Genetic mapping experiments confirmed that the *dunce* (*dnc*), *amnesiac* (*amn*), and *rutabaga* (*rut*) strains resulted from single-gene mutations at three separate loci (Aceves-Piña and Quinn, 1979; Booker and Quinn, 1981; Byers *et al.*, 1981; Livingstone, 1985; Quinn *et al.*, 1979; Tully and Gergen, 1986). More recently, two transposon-tagged mutations, *latheo* (Boynton and Tully, 1992) and *linotte* (Dura *et al.*, 1993), have been identified, both situated on chromosome 2.

dnc, amn, and *rut* mutants all show defects in other behavioral tests which purport to measure some aspect of learning (Siegel and Hall, 1979; Booker and Quinn, 1981; Duerr and Quinn, 1982; Folkers, 1982; Tempel *et al.*, 1983; Kyriacou and Hall, 1984; Gailey *et al.*, 1982, 1984; Wittekind and Spatz, 1988; Corfas and Dudai, 1989). These assays involve a number of different sensory modalities, not just the olfactory system. Thus it seems likely that these genes must be acting centrally, rather than peripherally, to affect the learning process. Furthermore, *dnc* and *amn* adults, unlike the wild type, do not remember training that was imposed during their larval period (Tully *et al.*, 1994). Careful behavioral studies of *dnc* also suggest that the mutation disrupts acquisition of learning or the formation of short-term memory (Tully and Gold, 1993). However *dnc* mutants have at least one other defect unrelated to learning. All the known mutant *dnc* alleles give female sterility either in homozygous or in heteroallelic combinations (Kiger and Golanty, 1977; Kiger *et al.*, 1981).

B. MOLECULAR BIOLOGY AND BIOCHEMISTRY OF *DUNCE* AND *RUTABAGA*

Biochemical and genetic analysis reveals that *dnc* is the structural gene for a cAMP-specific phosphodiesterase, PDEII (Byers *et al.*, 1981). This has been confirmed molecularly in that the predicted amino acid sequence of the wild-type *dnc* gene resembles that of bovine calcium/calmodulin-dependent phosphodiesterase and of yeast cAMP-dependent phosphodiesterase (Chen *et al.*, 1986).

The *dnc* gene is huge by *Drosophila* standards; it extends over 140 kb of genomic DNA (Davis and Davidson, 1986; Chen *et al.*, 1986; Davis and Dauwalder, 1991; Qui *et al.*, 1991). A large number of *dnc* transcripts can be detected, resulting from multiple transcription initiation sites, alternative splicing of exons, and differential processing of downstream sequences (Davis and Dauwalder, 1991; Qiu *et al.*, 1991). Remarkably, several other genes are encoded within the introns of *dnc*. One of these is the *Sgs-4* gene, which produces a protein that the larva uses to attach itself to the substrate during pupariation. This gene is transcribed in the same direction as *dnc*. However the *Pig-1* gene is also found within this intron (*Pig-1* is the preintermolt gene, which, like *Sgs-4*, produces a larval salivary-gland protein) and is transcribed in the opposite direction (Chen *et al.*, 1986). Other *Sgs-* and *Pig*-like genes also appear to be transcribed within the introns of *dnc* (Davis and Dauwalder, 1991), and the *sperm-amotile* (*sam*) complementation group also falls within the *dnc* region (Chen *et al.*, 1986). This type of nested gene organization provokes some interesting evolutionary speculations as to how it arose.

Biochemical analysis of the *rut* gene (and its mutants) indicates that it encodes a calcium-stimulated adenylyl cyclase (Livingstone *et al.*, 1984; Dudai *et al.*, 1985; Livingstone, 1985). A *Drosophila* homolog of a bovine Ca^{2+} on calmodulin-sensitive anadenylyl cyclase gene has also been mapped by *in situ* hybridization to the position occupied by *rut* on the polytene chromosomes (Krupinski *et al.*, 1989). Molecular cloning of the *rut* locus confirmed the biochemical-genetic data (Levin *et al.*, 1992). Furthermore, the single nucleotide change in *rut¹* mutants has been identified and shown to cause a complete loss of cyclase activity *in vitro* (Levin *et al.*, 1992).

Thus both *dnc* and *rut* encode components of the cAMP second messenger signaling system. This is particularly interesting given the prominent position assigned to the cAMP pathway by Kandel and colleagues in their studies of learning in *Aplysia* (see, e.g., Abrams and Kandel, 1988). In addition, experiments with mammalian memory on the D1 dopamine receptor have demonstrated how the receptor requires the stimulation of adenylyl cyclase and the regulation of the cAMP system (Sawaguchi and Goldman-Rakic, 1991).

Finally, transformation experiments in *Drosophila* have shown that inhibitors of both cAMP-dependent protein kinase (Drain *et al.*, 1991) and calcium/calmodulin-dependent protein kinase (Griffith *et al.*, 1993) produce learning defects in different types of conditioning tests. Similar findings have also been reported for mice mutant for α-calcium/calmodulin-dependent protein kinase II, where both long-term hippocampal potentiation (Silva *et al.*, 1992a) and spatial learning (Silva *et al.*, 1992b) are disrupted.

C. EXPRESSION PATTERNS OF *DNC* AND *RUT*

The *dnc* gene and gene products are expressed in the adult brain, and seem to be concentrated in the neuropil rather than the nerve cell bodies (Nighorn *et al.*, 1991). These *dnc*-staining synaptic regions are especially pronounced in the mushroom bodies, long thought to be the "higher centers" of the brain in social insects. Surgical ablation of these regions in wood ants (Vowles, 1964) and bees (Erber *et al.*, 1980) leads to learning defects in these hymenopterans. Furthermore, *Drosophila* mutants identified initially on the basis of defective mushroom bodies (Fischbach and Heisenberg, 1984) also show poor olfactory learning ability (Heisenberg *et al.*, 1985). Thus, information has accumulated over the years indicating that the mushroom bodies may be regions where learning and "memory traces" might be stored. Since several sensory modalities also have pathways which converge on the mushroom bodies (Heisenberg, 1980), it is not difficult to imagine that these structures could be sites for the integration of sensory input.

In addition, Technau (1984) has demonstrated that the number of new Kenyon fibers in the mushroom bodies rises during the first week of adult life. These dynamic changes in brain structure may have something to do with the fly's sensory experiences. This is because flies raised in isolated, "deprived" sensory environments show highly significant reductions in the number of fibers found in the stalks of the mushroom bodies (Technau, 1984). Moreover, *dnc* and *rut* mutants do not show the experience-dependent modulation in fiber number when raised in enriched and deprived environments (Balling *et al.*, 1987). This interweaving of what appears to be "critical periods" early in an adult fly's life, when the fly may be especially receptive to adaptive experiences, and the effects of the *dnc* and *rut* mutants further underscore the intimate association between learning and the mushroom bodies. How alterations in cAMP cascade might produce changes in synaptic function and how neuronal excitability—which has recently also been found to be affected in both *dnc* and *rut* mutants (Zuong and Wu, 1991)—affect the learning process has yet to be elucidated.

The particular set of neurons that may be involved in specific types of learning is unknown but may prove tractable, given recent developments in molecular methods for delivering cytotoxic agents to subsets of cells (Kunes and Steller, 1991; Moffat *et al.*, 1992; Bellen *et al.*, 1992). Alternatively or in addition, by using molecular extensions to the well-tried and trusted technique of making genetic mosaics, it may be possible to remove gene function, for example *dnc* or *rut,* from various "enhancer-trapped" neurons (O'Kane and Gehring, 1987) and observe the behavioral correlates of such interventions (O'Kane and Moffat, 1992). Therefore, after a period of some stagnation, the pace of progress in determining the molecular basis of *Drosophila* learning has quickened.

V. Courtship Behavior

A. BEHAVIORAL SEQUENCES

Sexual behavior in *Drosophila* has been studied from a genetic perspective since 1915, when Sturtevant performed some pioneering experiments with cuticular mutants (Sturtevant, 1915). The first detailed ethological study using a mutant was that of Bastock (1956), who focused on *yellow* in *D. melanogaster*. Most of the ethological experimental work has revolved around this particular species, which has an easily recognized sequence of behaviors that ethologists term "fixed action patterns." These behavioral elements are stereotyped and occur in a regular sequential pattern (see Burnet and Connolly, 1974, for review).

The male orients to the female, taps her abdomen, follows her while extending the wing nearest her head, and vibrates it to produce a "love song." When the male is close to the female, he will lick her genitalia and attempt to copulate by grasping her abdomen. He then curls his abdomen to achieve genital contact. The sequence of behavior—orientation, vibration, licking, and attempted copulation—continues until copulation is achieved or the courtship interaction is broken. The male therefore displays the more prominent elements, whereas virgin females generally decamp, or fend off and kick the male, and fertilized females also extrude their ovipositor (Connolly and Cook, 1973).

Given that courtship behavior is complex, utilizing most or all of the sensory modalities, it is not surprising that mutations which disrupt vision, olfaction, and hearing produce behavioral defects. Thus, blind males cannot follow the female, deaf females cannot respond to the song, and males and females whose sense of smell is abnormal, tend to take longer to initiate mating (see reviews in Tompkins, 1984; Hall, 1986). Rather than review these rather obvious cause-and-effect cases, we turn to the study of mutations that cause more surprising or interesting effects on courtship.

B. Auditory Behavior

1. cacophony and dissonance

The song of male *D. melanogaster* is produced by wing vibration and consists of two components: a pulse song, with interpulse intervals (IPIs) varying on average between 30 and 40 ms; and a hum component with a frequency of approximately 160 Hz (von Schilcher, 1976a; Wheeler *et al.*, 1988, 1989). Two mutations, *cacophony* (*cac*) and *dissonance* (*diss*), both located on the X chromosome, lead to abnormal pulse songs (von Schilcher, 1976b, 1977; Kulkarni and Hall, 1987; Kulkarni *et al.*, 1988). In *cac* mutants, the pulses are polycyclic and have a large amplitude, while in *diss,* the pulses at the beginning of a burst may be monocyclic, but usually become polycyclic (and higher in amplitude) toward the end of a song burst (see Fig. 2).

Interestingly, both the *cac* and *diss* mutations map to loci where there already exist mutant alleles that affect visual behavior. In the case of *cac,* complementation tests reveal that *cac* and a mutation called *night-blind-A* (*nbA*) are mutant alleles of a vital locus, *l(1)L13* (Kulkarni and Hall, 1987). The *nbA* mutants perform poorly on optomotor and phototactic tests (Heisenberg and Götz, 1975), yet their singing behavior appears normal (Kulkarni and Hall, 1987). However, the *cac* song defect is uncovered when *cac* is placed in trans with *l(1)L13.* The *cac* mutant itself does not appear to have visual defects.

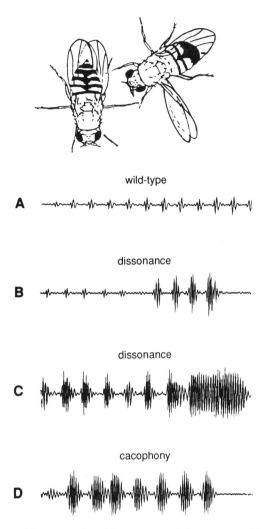

FIG. 2. The courtship songs of wild-type, *cac*, and *diss* mutants. Note how the song pulses of the mutants are usually polycyclic for *cac*, but can be monocyclic for *diss*, especially at the beginning of a song burst. (Reprinted with permission from Kyriacou, 1990b; redrawn from an original by Kulkarni *et al.*, 1988.)

Therefore the *l(1)L13* locus appears to be a complex one, if indeed the *l(1)L13* mutations define a single gene. Preliminary mosaic analysis suggests that the thorax is the site of action for *cac* regarding the abnormal courtship song (Hall *et al.*, 1990).

The *diss* mutant itself has an associated visual defect (Kulkarni *et* al., 1988); diss is formally an allele of the no-on-transient-A (nonA) locus, being designated nonA^{diss} (Rendahl et al., 1992). The nonA mutants have defective electroretinograms (ERGs) (Pak et al., 1970), although their phototaxis is not severely impaired (Heisenberg, 1972). Heisenberg and Buchner (1977) suggested on the basis of optomotor tests that the visual pathways associated with photoreceptors R1–6 are affected by the mutation. However, nonA mutants do not show defective courtship song—only the nonA^{diss} allele produces the song abnormality.

Jones and Rubin (1990) cloned and sequenced the nonA gene, and transformation studies confirmed that nonA genomic DNA can rescue the song defect of nonA^{diss} mutants (Rendahl et al., 1992). The locus encodes three transcripts that are found at all developmental stages and distributed ubiquitously in the embryo and adult. Antibody directed against nonA protein revealed a pattern of expression consistent with that of the mRNAs and a nuclear and cytoplasmic localization within most cells of most tissues.

The nonA product has sequence homology to RNA binding proteins (Bandziulis et al., 1989), for example, as encoded by the Sex-lethal (Sxl) gene (Bell et al., 1988), which is critical to sex determination in Drosophila. Speculation as to the possible regulatory role of nonA can be found in Rendahl et al. (1992). In the absence of any clear biochemical data on the function of nonA, it remains something of a surprise that two mutations, cac and nonA^{diss}, which were isolated with respect to abnormalities in song, are both associated with visual functions. The biochemical role of nonA may eventually help to resolve this neurogenetic connection between vision and acoustics.

2. The Clock Gene period

The period (per) gene plays a relatively subtle role in the patterning of the courtship song. Kyriacou and Hall (1980) observed that the average 30–40 ms IPIs are not constant during a courtship, but cycle, with a periodicity of about 55 sec in wild-type D. melanogaster males. This song cycle is shortened to 40 sec in per^s mutants and lengthened to 80 sec in the per^L mutant. The per^{01} mutant males do not show a consistent cyclical pattern in their IPI production. Thus within D. melanogaster, the per gene plays a role in determining the song cycle.

Males of the sympatric sibling species D. simulans also produce a song cycle, but the periodicity is 35– to 40 sec and is superimposed upon generally longer IPIs (Kyriacou and Hall, 1980, 1986; Wheeler et al., 1991). Reciprocal crosses between the two species revealed that the difference in these species-specific song cycles is sex linked (Kyriacou and Hall, 1986), whereas the species difference in IPI appears to be autosomally determined. *D. melanogaster* and *D. simulans* females respond preferentially to simulated songs carrying the conspecific male cycle period and the conspecific male range of IPIs (Kyriacou and Hall, 1982, 1986). However, the correct species-specific cycle superimposed on the incorrect species range of IPIs, or vice versa, did not lead to enhanced mating of females. Female hybrid crosses between *D. melanogaster* and *D. simulans* show the peculiar trait of preferring cycles and IPIs intermediate between the two species (Kyriacou and Hall, 1986). The "peculiarity" is that, because of the sex linkage of the song cycle, hybrid males can only sing one or the other of the two species-specific cycles, depending on which parent supplied their X chromosome. Superficially, however, the enhanced responses of females to intermediate songs suggests a kind of behavioral "coupling" between the hybrid receiver (female) and the hybrid sender (male); this could have a single gene basis.

Such genetic coupling was originally proposed by Alexander (1962), who suggested that if one gene determined both the male output signal and the female's corresponding song input mechanisms, this could favor the rapid evolution of a communication system. However, genetic coupling has never been convincingly demonstrated for the case of a single gene (Butlin and Ritchie, 1990; Boake, 1991).

The hypothesis was critically tested recently by Greenacre *et al.* (1993) with *D. melanogaster per* mutant females. Congenic *per* mutant females were found to prefer 55-sec cycles over 40-sec cycles, indicating that genetic coupling was not in evidence with respect to *per*. The coupling hypothesis would demand, for example, that *per^s* females prefer the 40-sec cycle produced by the *per^s* male, and this clearly was not the case. However, females from a laboratory stock that had been maintained by simple mass transfer for at least 15 years, and in which males and females were both *per^s*, showed quite different responses to song cycles. These *per^s* females gave enhanced mating when stimulated with 40-sec cycles, although their preference for 40-sec cycles over 55-sec cycles was not significant. Nevertheless, the behavioral difference between these *per^s* females and the congenic *per^s* females, where the former show no preference for the 55-sec cycle and the latter significantly preferred 55-sec cycles, *was* highly significant (Greenacre *et al.*, 1993). Thus it appears that in this stock, a coevolution occurred in the receiver (the *per^s* female) to the sender (the *per^s* male), presumably mediated by the selection of genes at other loci. The mutant females had thus been under selection to respond to the mutant 40-sec cycles produced by the males, so the male behavior had "driven" fe-

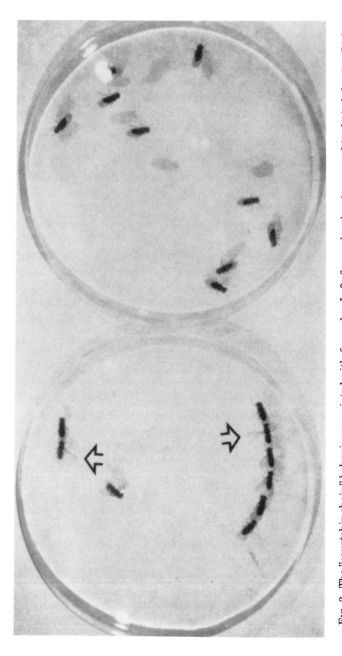

FIG. 3. The "courtship chain" behavior associated with *fru* males. Left, five males showing courtship chain behavior. Right, nine males that are wild-type for this phenotype. (Photograph from Gailey and Hall, 1989.)

male choice, rather than the more usual opposite case in which female preferences are inferred to drive bizarre male behavior and morphology (Kirkpatrick and Ryan, 1991).

It may be possible to map these other genes which have changed the *per*^{s} females' song preference. Since the *per*^{s} males in this *per*^{s} strain had maintained their 40-sec cycle (Greenacre *et al.*, 1993), it shows that the genes selected in the females for responding to 40-sec cycles do not have corresponding effects on the male song cycle; this again goes against the genetic coupling hypothesis.

3. fruitless

Another mutant with an unusual song is *fruitless* (*fru*), which is associated with a small inversion in chromosome 3 (Gailey and Hall, 1989). Historically, *fru* was induced by Gill (1963) using X-rays and is homozygous viable but male-sterile. Hall (1978) listed three behavioral defects in *fru* males:

1. They do not curl their abdomens at females to achieve copulation.
2. They court other *fru* and wild-type males. Furthermore, *fru* males court in long "courtship chains" (see Fig. 3).
3. They also stimulate normal males to court them.

Solvent extracts from *fru* males will stimulate intermale courtship, which suggests that a volatile substance in *fru* males provides "sex-appeal" for other males (Tompkins *et al.*, 1980). More recently, Wheeler *et al.* (1989) have demonstrated that the courtship song of *fru* males is abnormal, in that the median IPI values are higher than those of the wild type. Other more subtle defects in the song pulses of *fru* males have also been detected (Bernstein *et al.*, 1992).

The three nonacoustic defects in *fru* have somewhat surprisingly been mapped to different breakpoints on chromosome 3 (Gailey and Hall, 1989). The behavioral sterility maps to the 91A-B polytene band boundary, as does the courtship chain phenotype. However, the ability to induce courtship from wild-type males maps to the other nearby breakpoint in the *fru* inversion (Gailey and Hall, 1989). Another defect in some *fru* mutants is the near absence of a male-specific abdominal muscle (Gailey *et al.*, 1991). This muscle spans the fifth abdominal segment of males (Lawrence and Johnston, 1986) and appears to be regulated by the *fru* locus, because deletions of cytogenetic interval 91B remove the muscle in males. However, this removal of the muscle does not correlate with the inability of *fru* mutants to copulate because some *fru* genetic variants are able to curl their abdomens and mate even when the muscle is absent (Gailey *et al.*, 1991). Perhaps other male-specific muscle abnormalities, as yet undetected in *fru* males, also lead to the abnormal songs produced by these males. A transposable element has been inserted in the 91B region which produces *fru*-like behavior (Gailey and Hall, 1989). Consequently, the cloning of this locus should be facilitated, and this might provide some clues as to how *fru* variants lead to such a range of phenotypic effects.

D. OLFACTION AND COURTSHIP

1. Genetic and Molecular Analysis of Female Receptivity

The *fru* mutation has introduced us to the existence of volatile compounds that can stimulate courtship between males. *Drosophila* have both male- and female-specific pheromones that both excite and inhibit courtship. Needless to say, olfactory mutants such as *olfD* and *sbl* in males and females produce abnormal smell-driven behavior. For example, the female decamping response appears to be arrested just before copulation, in that the wild-type females (but not mutant ones) tend to reduce their locomotor activity over the course of courtship (Tompkins *et al.*, 1982). However, *sbl* and *olfD* females do not show such slowing down, suggesting that the response is mediated by olfaction (Tompkins *et al.*, 1982; Gailey *et al.*, 1986). Also, *sbl* and *olfD* mutant males show higher levels of courtship to fertilized females than wild-type males (Tompkins *et al.*, 1983; Gailey *et al.*, 1986). This suggests that the males are not sensing an inhibitory pheromone produced in the fertilized females.

The inhibition of male courtship by fertilized females is at least partly regulated by a sex peptide (SP) that is transferred from the male to the female during copulation. The peptide is a secreted component of the male accessory gland; when injected into the abdominal cavity of virgin *D. melanogaster* females, SP causes a reduction of sexual receptivity and induction of ovulation (Chen *et al.*, 1988). The mature peptide consists of 36 amino acids and is synthesized as a precursor with a signal sequence of 19 amino acids at its N-terminal end (Chen *et al.*, 1988).

Ectopic expression of this peptide in flies transformed with an *hsp70* promotor fused to a cDNA encoding the SP produces the changes in receptivity and corresponding rejection of males seen in mated females (Aigaki *et al.*, 1991). However, the ectopically expressed peptide only produces the effect for about 24 hr, which is considerably less than the effects of natural mating (Manning, 1967). By adding an enhancer element from the *yolk-protein* (*yp1*) gene into the *hsp70-SP* construct, Aigaki *et al.* (1991) were able to generate a constitutive level of rejection behavior by females. The *ypl* enhancer directs the SP to the female fat body, apparently stabilizing it. What is not quite clear from these studies is whether the change in female rejection behavior that is regulated by SP also causes the production of inhibitory pheromones which reduce the male's approaches to the female (Vander Meer *et al.*, 1986; Scott, 1986).

D. *melanogaster* males produce a male-inhibitory pheromone, 7-tricosene, which is transferred to females through direct cuticular contact during mating (Scott, 1986). However, fertilized females later produce 7-tricosene themselves (Scott, 1986). Whether the SP is directly relevant to the females' production of 7-tricosene is unknown.

Finally, we should consider the role of *cis*-vaccenyl acetate (cVA), a male-specific lipid transferred to the female during mating (Butterworth, 1969), which also has antiaphrodisiac properties. cVA was considered for some time to be the major male inhibiting factor in fertilized females (Zawistowski and Richmond, 1986). However, while cVA may play a role, it appears to be more important as an aggregation pheromone because when it is passed from the male to the female during mating, it is quickly lost from the female and deposited onto the food substrate (Bartelt *et al.*, 1985).

2. Experience-Dependent Effects on Mating Behavior

Siegel and Hall (1979) observed that males paired with fertilized females, and then placed with virgin females, subsequently showed reduced responsiveness to virgin females. This modified behavior toward virgin females persisted for 2–3 hr. Interestingly, males carrying the memory mutation *amn* recovered their normal response after only 30 min (Siegel and Hall, 1979). The result suggests that the aftereffect in wild-type males could be due to a general debilitation of behavior (i.e., olfactory receptors "clogged up" by the inhibitory pheromone of fertilized females), and that the *amn* males are simply less responsive to such stimuli. More likely, however, some process related to learning is involved in the wild-type males' modified behavior, and the *amn* mutant is defective in this process. Moreover, wild-type males that had been previously paired with fertilized females were subsequently responsive to immature males (young males elicit courtship from older males; Gailey *et al.*, 1982). This shows clearly that a general behavioral depression is not produced by the pairing of wild-type males with fertilized females (Gailey *et al.*, 1984).

Mature males which have been previously paired with immature males subsequently express little or no courtship to immobilized young males (Gailey *et al.*, 1982). However, males previously paired with another mature male, or kept in isolation, will respond to young males with vigorous courtship (Gailey et al., 1982). The amn mutation also leads to a decrease in the ability of mature males to become "conditioned" by immature males (Gailey et al., 1982).

These examples of the modification of courtship behavior or "courtship conditioning" (emphasizing the learning aspect) show how mating behavior can be altered by previous experience. Similar effects are seen with artificial courtship songs, which can "prime" females to be more responsive to subsequent male courtship (von Schilcher, 1976a; Kyriacou and Hall, 1984).

The courtship conditioning of males with mated females is mediated by volatile chemicals. An extract from fertilized females can reduce the courtship a male performs for a virgin female (Tompkins and Hall, 1981). Control extracts from virgin females and mature males did not subsequently produce courtship modifications (Tompkins et al., 1983). Some of the behavioral subtleties that are evident in these male–female and male–male interactions, and their possible evolutionary implications are reviewed in Siegel et al. (1984).

3. Neurogenetic Control of Courtship Behavior

Hotta and Benzer (1976) performed a gynandromorph experiment to identify the anatomical loci responsible for male and female courtship behavior. Using the fate-mapping technique based on body-surface landmarks, they suggested that genetically male tissue was required in the brain for orientation and wing extension. Courtship song required a male focus in the thoracic ganglion (von Schilcher and Hall, 1979), and licking and attempted copulation had more diffuse foci (Hall, 1977, 1979; von Schilcher and Hall, 1979). Female "sex appeal" depends on a mosaic having abdominal regions that are female and presumably able (by virtue of their XX genotype) to release the aphrodisiac pheromone, 7,11-heptacosadiene (Jallon, 1984). Elements of female courtship behavior, however, require a diplo-X female brain (Tompkins and Hall, 1983); these tissues would presumably integrate incoming olfactory visual and acoustic stimuli from the male.

Implicit in these types of experiments are two not completely mutually exclusive possibilities: Either a female anatomically carries male foci, presumably neural tissues, which are not activated during courtship because these foci would produce male-specific behavior; or there may exist sexual dimorphisms in the brain. A similar suggestion can be made for males, which by analogy, would not activate the "female centers" they may possess. Such sexual dimorphisms in brain structure are documented for larger insects (Strausfeld, 1980; Schneiderman *et al.*, 1982). Technau (1984) has observed in *D. melanogaster* that in the mushroom body there are about 10% more neurons in the peduncular lobe of females. Since Hall (1979) has implicated the region of the dorsal brain of the male near the mushroom bodies as being of critical importance for initiating the early stages of male courtship, perhaps this sexual dimorphisms has functional meaning. The evidence available at present is negative, however, in that *mushroom-body-deranged* (*mbd*) mutants (Heisenberg and Böhl, 1979), which have severe reductions in the number of fibers in the mushroom bodies, do not show any detectable defects in male behavior (Heisenberg, 1980). Even the more subtle mated-female aftereffects (Siegel and Hall, 1979) are intact in *mbd* males (Heisenberg, 1980). It may be that the sex-specific mushroom body region is not affected in the *mbd* mutant used in these behavioral tests. In any event, no clear sex-specific function has yet been associated with this brain region.

One way of analyzing brain regions, such as those pinpointed by gynandromorph studies as being important for sex-specific behavior, is to produce mosaics in which that part of the brain has been lesioned. Greenspan *et al.* (1980) attempted this by using an *Acetyl-cholinesterase* (*Ace*) mutation in gynanders to neurochemically ablate parts of the nervous system. Currently, the enhancer-trap methodology (O'Kane and Gehring, 1987), allied to the development of techniques for killing particular cells described in Section IV,C (Kunes and Steller, 1991), and the FLP recombinase system (Golic and Lindquist, 1989) for making mosaic lesions within an enhancer-trap line (O'Kane and Moffat, 1992), may permit quite subtle ablations of regions of the nervous system. These types of experiments (Moffat *et al.*, 1992) have the advantage of being able to produce many almost identical mosaics from a single cross and are a significant improvement over the usual method of obtaining a small percentage of uniquely mosaic individuals. These elegant molecular methods could substantially enhance existing mosaic technology and boost our knowledge of structure–function and brain–behavior relationships in *Drosophila*.

VI. Biological Rhythms

A. THE *PERIOD* GENE: BEHAVIOR AND GENETICS

The identification of the *period* (*per*) gene by Konopka and Benzer (1971) has proved to be a major advance in the study of biological cycles. The *per* work has been reviewed many times (see, e.g., Rosbash and Hall, 1989; Kyriacou, 1990a; Hall, 1990), so we cover the older literature only briefly before moving on to current developments. The three original *per* mutants isolated by Konopka and Benzer (1971) were *per^s*, *per^L*, and *per^{01}*. These three mutations shortened to 19 hr, lengthened to 29 hr, or obliterated, respectively, the circadian cycles in pupal-adult eclosion, a population assay of the biological clock. Similar changes in the circadian cycle were observed in individual flies carrying each of these mutations when locomotor activity cycles were examined. These mutant genes also affected the ultradian (shorter-than-24-hr) cycle which is observed in the male's courtship song and was described in Section V,B,2 (Kyriacou and Hall, 1980, 1989). Consequently, as in the case of the *fru* gene, we have yet another example of gene pleiotropy, in which at least two behavioral phenotypes, courtship song and circadian cycles, appear to be influenced by the same gene. The *per* gene also plays a role in the timing of development, in that *per^s* flies develop faster from egg to adult, and *per^L* flies develop more slowly than the wild type (Kyriacou *et al.*, 1990). The arrhythmic *per^{01}* mutants give rather erratic developmental timing results. Thus the *per* gene affects behavioral and nonbehavioral phenotypes. It is noticeable that in all these phenotypes there is a temporal component which seems to be altered in the mutants in a predictable direction, irrespective of the temporal domain (ultradian, circadian, infradian). Thus the *per* gene appears to encode a fundamental property of biological cycles.

B. MOLECULAR BIOLOGY OF *PER*: THE THREONINE–GLYCINE REPEATS

The *per* gene has been cloned by two groups of workers, one based at Rockefeller University and one at Brandeis University (Bargiello and Young, 1984; Reddy *et al.*, 1984). The *per* DNA sequence predicts a large protein of approximately 1200 amino acids (Jackson *et al.*, 1986; Citri *et al.*, 1987). The three *per* mutants have been molecularly mapped: *per^{01}* encodes a translational stop after approximately 460 amino acids, whereas *per^s* and *per^L* each produce an amino acid substitution (Baylies *et al.*, 1987; Yu *et al.*, 1987a). These results did not lead to any dramatic insight into the nature of the PER protein itself.

The middle of the conceptual PER protein has a string of about 20 Thr-Gly (Thr-Gly) pairs (Jackson *et al.*, 1986; Citri *et al.*, 1987). The removal of these Thr-Gly repeats by *in vitro* mutagenesis (Yu *et al.*, 1987b) resulted in a significant shortening of the song cycle from 55 to 40 sec in transformants carrying this deleted *per* construct. Encoded just downstream of the Thr-Gly residues is a region which is species specific in the closely related species of the *melanogaster* subgroup (Thackeray and Kyriacou, 1990; Wheeler *et al.*, 1991; Peixoto *et al.*,

1992). Strikingly, one species-specific difference in the song cycle—55 sec in wild-type *D. melanogaster* versus 40 sec in *D. simulans*—appears to be encoded in this region (Wheeler *et al.*, 1991). In fact, between one and four species-specific amino acid differences appear to be all that is required to switch the song cycle of *D. simulans* to that of *D. melanogaster* (Wheeler *et al.*, 1991). One of these amino acid differences falls in a region that at least theoretically could be a phosphorylation site (Kemp and Pearson, 1990). Perhaps, then, the species-specific song cycle difference between *D. melanogaster* and *D. simulans* is due to a single difference in the phosphorylation profile of the PER protein in the two species.

Since courtship song cycles have been shown (at least in the laboratory) to play a role in choice of mate (Kyriacou and Hall, 1982, 1986; Greenacre *et al.*, 1993), it is conceivable that *per* may even be a gene that is intimately associated with the speciation process itself (Coyne, 1992). Such a role for *per* may not be restricted to its control of courtship song cycles. Species-specific locomotor activity cycles can be transferred from one species to another simply by moving the *per* gene from the donor into the host species. This was demonstrated by Petersen *et al.* (1988), who transformed the *D. pseudoobscura per* gene into a *D. melanogaster per⁺* genetic background. The resulting transformants took on the circadian locomotor pattern of the *D. pseudoobscura* donor species. Consequently, somewhere in the regions of the *per* gene that differ in sequence between species (Colot *et al.*, 1988) will be areas that affect the species specificity of the circadian locomotor phenotype. Since the activity patterns of flies will also influence the extent to which they interact when in sympatry, *per* may be an important speciation gene for this reason, in addition to its determination of song cycles.

The Thr-Gly region is also highly polymorphic in length within *D. melanogaster*. Different strains taken from the laboratory or the wild show variation in the number of pairs of Thr-Glys encoded (Yu *et al.*, 1987b; Costa *et al.*, 1991; Peixoto *et al.*, 1992). Some strains of *D. melanogaster* have only 14 Thr-Gly repeats, whereas others have as many as 23 (Costa *et al.*, 1991, 1992; Peixoto *et al.*, 1992). Recently Costa *et al.* (1992) have shown that this length is clinally distributed. Populations in the south of Europe have shorter Thr-Gly repeats than populations in the north of Europe.

A cline traditionally suggests that the character being examined is under some kind of natural selection, presumably thermal (Anderson and Oakeshott, 1984; David *et al.*, 1989). Perhaps the Thr-Gly region plays a role in maintaining the integrity of the PER protein under different thermal environments. It is therefore interesting that removing the Thr-Gly repeat as discussed above (Yu *et al.*, 1987b) not only changes the courtship song cycle, but also the apparent thermostability of the PER protein, as inferred from the temperature sensitivity of the circadian locomotor phenotype of the *in vitro*-deleted transformants (Ewer *et al.*, 1990). Thus, temperature stability, an important property of circadian cycles (Pittendrigh, 1954), is relatively poor in such Thr-Gly deleted mutants. Therefore two different lines of evidence, one based on the latitudinal cline in the Thr-Gly length polymorphism, and the other based on deletion of the repeat, raise the possibility that the Thr-Gly region confers on its carrier an adaptive response to temperature fluctuation in the environment. Alternatively, the cline in Thr-Gly length could also be related to the different photoperiodism found at different latitudes.

C. PER PROTEIN AND TRANSCRIPT LOCALIZATION

The Thr-Gly repeats are similar to Serine-Glycine repeats found in mammalian proteoglycans (Jackson *et al.*, 1986). Proteoglycans can be found both intra- and extracellularly (for review see Ruoslahti, 1988). Preliminary biochemical results which indicate that PER is a proteoglycan (Reddy *et al.*, 1986; Bargiello *et al.*, 1987) have not been supported by recent work, which uses a new generation of anti-PER antibodies to show that the PER primary product is not extensively modified by carbohydrate. Instead, it appears that PER may be phosphorylated (Edery *et al.*, 1994).

The expression of the PER protein has been studied using both anti-PER antibodies (Siwicki *et al.*, 1988; Saez and Young, 1988; Zerr *et al.*, 1990) and *per*-reporter gene constructs (Liu *et al.*, 1988, 1991). These workers observed that *per* is distributed in a large number of tissues, including the nervous system, gut, Malpighian tubules, reproductive system, ring glands, and salivary glands (see Hall and Kyriacou, 1990, for review). Little *per* expression is observed at the larval stages, even though Kyriacou *et al* (1990) did detect phenotypic differences in larval development times among *per* mutants. In general the patterns of PER protein expression correlate quite well with the patterns of *per* mRNA localization (James *et al.*, 1986; Bargiello *et al.*, 1987; Liu *et al.*, 1988; Saez and Young, 1988). The PER protein in the adult nervous system is found in the photoreceptors of the eyes, in the optic lobes, central brain complex, and thoracic ganglia. Certain glial cells of the CNS express PER, and a group of cells, termed the "lateral neurons," were proposed by Siwicki *et al.* (1988) and Zerr *et al.* (1990) to provide circadian pacemaker function.

Mosaic analyses (Konopka *et al.*, 1983; Hall, 1984; Ewer *et al.*, 1992) have revealed that the circadian pacemaker is localized in the head, consistent with Handler and Konopka's (1979) brain transplantation experiments. Ewer *et al.* (1992) have also carried out double labeling experiments in which adult fly sections were stained both with anti-PER antibody and a neuron-specific antibody (Robinow and White, 1988). The results confirmed both the neuronal and glial expression of *per*. Furthermore, Ewer *et al.* (1992) performed a mosaic analysis with *per*$^+$//*per*^{01} individuals which suggested that perhaps glial expression of *per*$^+$ in the subesophageal ganglion may be sufficient to cause weak behavioral rhythmicity. Stronger rhythmicity was obtained when *per*$^+$ was expressed in larger patches, including the lateral neurons. These authors suggest tentatively that those neurons may constitute the pacemaker, while the *per*-expressing glial cells contribute to the fly's circadian rhythmicity.

The expression of *per* is controlled in part by cis-acting sequences. Removal of them results in rather interesting changes in rhythm strength and period (Liu *et al.*, 1991). Transformed flies carrying a *per* gene construct missing much of the 5' material but with all of the *per* coding sequences gave weak rhythms (as measured by the proportion of individuals showing significant rhythmicity), but most flies had periods close to 24 hr (Hamblen *et al.*, 1986). Liu *et al.* (1991) showed that these transformants had generally low levels of *per* expression, except in the region of the lateral neurons. Another strain of transformants, deleted of a large *per* intron, exhibited relatively high "penetrance" (in that a greater proportion of individuals were rhythmic), but the rhythms had appreciably longer than normal periods. These results were interpreted to mean that relatively high penetrance of rhythmicity requires *per* expression in most of its normal brain locations, while normal 24-hr periods can result from fairly robust *per* expression levels in the lateral neurons alone.

Therefore, the critical site required for *per* expression to produce behavioral rhythmicity has been narrowed down to small regions of the brain. A related question concerns the temporal requirement for *per* to produce normal circadian cycles. Is *per* required early in development to "build" structures that will later be necessary for mediating behavioral rhythmicity? The answer appears to be "no." Ewer *et al.* (1988, 1990) used a *heat shock promotor-per* (*hsp-per*) fusion, and transformed this gene into a *per*^{01} genetic background. By giving transformant adults heat pulses, they were able to generate rhythmicity in locomotor behavior. Reducing the temperature resulted in a return to arrhythmicity. Further experiments with these *hsp-per* constructs revealed that the circadian pacemaker could not be entrained in the absence of *per* expression, suggesting that the *per* gene product is a component of the pacemaker (Ewer *et al.*, 1988, 1990).

The *hsp-per* experiments could also suggest that because *per* expression is required only in the adult for behavioral rhythmicity, then the transcription of *per* may merely provide a missing "coupling" component, which then mediates rhythmicity. Coupling has a long history in chronobiology. Briefly, if cells have rhythmic properties that cycle with short periods and these cells are coupled together, a longer circadian period may emerge (see, e.g., Dowse *et al.*, 1987). Bargiello *et al.* (1987) provided experimental evidence that the PER protein may provide a coupling function. Salivary gland cells from *per*[s] larvae were more tightly coupled than those taken from *per*[+], and *per*[01] cells showed very weak coupling. However, this study was retracted recently (Saez *et al.*, 1992) because attempts to replicate it failed (Siwicki *et al.*, 1992; Flint *et al.*, 1993).

D. Circadian Cycling of *per* mRNA and Protein

High-resolution microscopy has revealed that PER can be found predominantly in the nucleus and at lower levels in the cytoplasm, except in the female ovaries, where it appears to be exclusively cytoplasmic (Liu *et al.*, 1992). Siwicki *et al.* (1988) and Zerr *et al.* (1990) have showed that the PER protein cycles in apparent abundance in the fly brain, with a free-running circadian period close to 24 hr. The *per* mRNA has also been found to cycle in the fly's head, again with a circadian period (Hardin *et al.*, 1990). There is at least a 6- to 8-hr difference between the time that the peak *per* mRNA levels are observed and the time that the PER protein peaks occur. These mRNA and protein peak levels occur early and late respectively in the subjective night. In addition, *per*[s] produced about a 20-hr mRNA rhythm (in constant conditions), suggesting that the PER protein feeds back on itself to change its own mRNA oscillation (Hardin *et al.*, 1990). DNA sequences embedded within a 1.3-kb genomic fragment located upstream of the transcription start site appear to be sufficient for mRNA cycling (Hardin *et al.*, 1992a) This strongly suggests that *per* mRNA cycles are transcriptionally regulated.

In vitro mutagenesis of the *per*[s] site reveals that if the serine at position 589, which is altered to an asparagine in *per*[s] (Baylies *et al.*, 1987; Yu *et al.*, 1987a), is changed to any other amino acid apart from the structurally similar threonine, shorter cycles are observed in transformants (Rutila *et al.*, 1992). In fact, mutating any of the amino acids in the vicinity of the *per*[s] site usually produces shorter periods

(Baylies *et al.*, 1992), as does inserting small unrelated peptides near the *per^s* region and into the amino terminal of the PER protein (Rutila *et al.*, 1992). This effect is especially dramatic with the amino-terminal mutations, and since the amino terminal can be an important focus for protein degradation (Varshavsky, 1992), it could be that altering the protein stability might have produced this effect.

If the PER-short protein is more stable than PER-normal, and if, furthermore, the PER negatively regulates its own mRNA, as it appears to do (Hardin *et al.*, 1990; Zweibel *et al.*, 1991), the following scenario can be envisaged: The more stable (mutant) PER-short protein achieves a "critical concentration" earlier than the wild-type protein and prematurely truncates the cycling of the mRNA. Therefore the cycling of the PER protein can be viewed as a negative feedback loop, and the effects of the mutations on protein stability can be incorporated into the model. The "*per^s* domain" may thus be a region that interacts with other proteins, which in the wild type could allow the protein to be degraded. Absence of this site in *per^s*-like mutants might, therefore, result in a more stable protein.

E. SEQUENCE SIMILARITIES

Embedded within the conceptional PER protein are two repeats with approximately 50 amino acids. These sequences are also found in the *Drosophila single-minded* (*sim*) gene (Crews *et al.*, 1988) and the mammalian *aryl hydrocarbon receptor nuclear translocator* (*arnt*) gene, as well as the aryl hydrocarbon receptor (*Ahr*) gene (Hoffman *et al.*, 1991; Burbach *et al.*, 1992), giving rise to the term the "PAS region" (*per-arnt-sim*). Both the *sim* and *arnt* genes encode nuclear proteins. *sim* plays a pivotal role in the development of cells that lie along the midline of the embryonic CNS, including neurons, glia, and other cells (Nambu *et al.*, 1990). *sim* appears to be required for the transcription of other genes that are necessary for the development of these midline cells (Nambu *et al.*, 1991). The *arnt* gene encodes a protein that is needed to translocate the protein encoded by the *aryl hydrocarbon receptor* (*Ahr*) gene, (which binds toxins such as dioxin), from the cytosol to the nucleus after binding ligands (Hoffman *et al.*, 1991). The receptor and *arnt* products thus mediate the carcinogenic effects of such agents, and their association could be mediated by their PAS domains.

Regions of the *arnt*, *Ahr*, and *sim* genes encode peptides that show high conformity to the consensus sequences for the basic helix–loop–helix motif (bHLH) of certain DNA binding proteins, which bind as homodimers or heterodimers (Benezra *et al.*, 1990; Weintraub *et al.*,

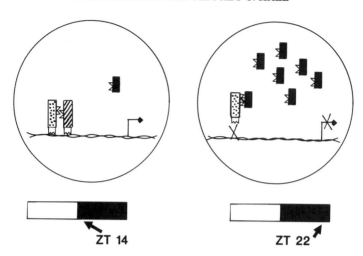

Fɪɢ. 4. Simple model of how PER might negatively regulate its own transcription. Early in the night phase (Zeitgeber time, ZT14), *per* is transcribed by a putative dimeric transcription factor complex (striped and stippled boxes), both subunits of which interact via PAS regions and have bHLH regions to bind to DNA. Late in the night phase (ZT22), high concentrations of PER protein antagonize the effects of the transcription factors by binding to one or both via their mutual PAS regions.

1991). Thus, the *sim, arnt,* and *Ahr* genes appear to encode transcription factors or components of transcription complexes. This in turn may suggest that the PER protein also acts as a transcription factor, switching on or off downstream genes which are needed for the expression of rhythmic behavior. Certainly the (partly) nuclear localization of PER (Liu *et al.,* 1992) is consistent with this view.

If PER binds to another factor, producing a heterodimer, then this complex may act as the transcription factor that could regulate other genes. This would bypass the problem of PER being thought of as a transcription factor, though it has no obvious DNA binding domains.

Recently, Huang *et al.* (1993) demonstrated that the PAS region allows PER to bind to itself, and mutations at or near the PAS domain markedly reduced this dimerization. One of these mutations is *per^L*. Furthermore, the PAS region also could mediate PER-SIM and PER-ARNT interactions. Thus, we predict that perhaps a bHLH-PAS protein in *Drosophila* acts as a partner to PER, and such an interaction could prevent the ability of the PER–partner complex to bind *per*-regulatory DNA (Fig. 4). This would be consistent with PER seemingly influencing transcription of its own gene by negative feedback. Mutation in such hypothetical PER-partner genes would be expected to seriously disrupt the *per* mRNA cycling and biological clock phenotypes as well. Additional clock mutants have indeed been discovered, and it is to these that the discussion now turns.

F. OTHER CLOCK MUTANTS

1. Clock

The *Clock* (*Clk*) mutation also maps to near the *per* region and produces shorter than normal cycles of approximately 22.5 hr (Dushay *et al.*, 1990). Genetic mapping suggested that *Clk* is a *per* mutation, and recent transformation experiments involving a *per* fragment cloned from the *Clk* mutant confirmed this guess (Dushay *et al.*, 1992). Consequently, *Clk* is now renamed *per^{Clk}*. It is not known whether *per^{Clk}* is a coding or regulatory mutant of *per*. If *per^{Clk}* is a regulatory mutant of *per*, then this may explain why courtship song cycles are nearly normal in *per^{Clk}* mutant males (Dushay *et al.*, 1990). *per^{Clk}* is the only mutation known that affects the circadian phenotype but apparently not the song cycle. However, since the difference in circadian period between the wild type and *per^{Clk}* is only on the order of 5%, a proportionately similar 3–4-sec difference in song period would be difficult to detect.

2. Andante and ebony

The *Andante* (*And*) mutant has longer than normal cycles of approximately 25.5 hr and was originally mapped very close to the *miniature-dusky* (*m-dy*) locus (Konopka *et al.*, 1991). Newby *et al.* (1991) demonstrated that *And* is an allele of the *m-dy* gene, and these authors suggested that the *And* mutation should be called *dy^{And}*. The *dy^{And}* mutation is an allele of a complex locus whose mutant alleles influence not only circadian period but also wing size and male and female fertility (Newby *et al.*, 1991).

The body-color mutation, *ebony* (*e*), also has defects in circadian behavior. The circadian locomotor activity of *ebony* mutants is essentially arrhythmic, especially at lower temperatures, but the eclosion rhythms are normal (Newby and Jackson, 1991). *ebony* is one of the classic morphological mutations that received much attention from the early behavioral geneticists (see, e.g., Rendel, 1951; Jacobs, 1960). A more recent study reported that *ebony* mutants were hyperactive in an open-field situation (Kyriacou, 1981). From the results of Newby and Jackson (1991), it could be that this apparent hyperactivity was caused by the defective circadian locomotor activity of this mutant. Thus *ebony* is an interesting gene because mutations affect the locomotor activity cycle but not the circadian eclosion cycle. A considerable amount of biochemical evidence has accumulated over the years on the possible defects in *ebony* mutants (Black, 1988). The gene has also been cloned (Caizzi *et al.*, 1987) and the gene product tentatively identified as β-alamine-dopamine synthetase (BAS). The mutant does not incorporate β-alamine into its cuticle (Jacobs and Brubaker, 1963) and has elevated dopamine levels in 1-day-old adults (Hodgetts and Konopka, 1973). The mutation also causes a number of other interesting behavioral alterations, which may be related to its unusual population biology (Kyriacou *et al.*, 1978; Kyriacou, 1985). However, at present none of the biochemical defects in *ebony* mutants appear to give any firm clues as to how these lesions may be affecting circadian rhythms.

3. *disconnected*

A particularly interesting mutation that affects circadian rhythms is *disconnected* (*disco*). This mutant usually has its eyes disconnected from the optic lobes of the brain, producing optic lobes which exhibit severe anatomical defects (Steller *et al.*, 1987). The mutants behave practically arrhythmically both in their eclosion profiles and in their free-running cycles of locomotor activity (Dushay *et al.*, 1989; Dowse *et al.*, 1989). However in light-dark cycles, an abnormal pattern of rhythmic locomotor behavior can be observed, which suggests that despite their disconnected phenotype, these mutants can respond to light (Dushay *et al.*, 1989). Light input may be necessary to "start" the locomotor activity clock (Dowse and Ringo, 1989; see however Seghal *et al.*, 1992). Even so, blind *norp-A* (*no-receptor-potential*) flies and the eyeless, ocelliless, *sol;so* (*small optic lobes; sine oculis*) double mutants exhibit quasi-normal circadian rhythmicity (Helfrich, 1986; Dushay *et al.*, 1989). Thus it appears unlikely that the disconnection of the eyes from the brain in *disco* mutants can explain the arrhythmic locomotor activity phenotype.

Recent evidence has shown that the *disco* mutant circadian oscillator appears to be working normally, in that the *per* mRNA cycling is present in *disco* mutants (Hardin *et al.*, 1992b). This suggests that the defect in *disco* is in the output signals from the oscillator to the effector organs that determine rhythmic behavior (Hardin *et al.*, 1992b). Steller *et al.* (1987) have reported that *disco* produces defects in "peripheral neurons" in the thorax and abdomen, and so it may be that the *disco* lesion that causes arrhythmicity will be found in the anatomical region. However, it is already known that the lateral neurons, which express Per protein, are largely undetectable in *disco* mutant; the *per*-expressing glia are present in this mutant (Zerr *et al.*, 1990). Thus, glial expression of *per* is not sufficient for rhythmicity in the absence of the lateral neurons, at least in these mutants (see also Ewer *et al.*, 1992). However, this conclusion must be tempered by the fact that an output pathway may also be defective in *disco* mutants.

4. *aj42*

Sehgal *et al.* (1991) have isolated a number of new mutant lines using a transposon tagging scheme, one of which, *aj42*, shows arrhythmic eclosion and locomotor profiles, and a striking "nocturnal" reversal of locomotor activity in a light-dark cycle. These two phenotypes are caused by two different genetic lesions, one causing the day–night activity reversal, and the other the arrhythmicity (Young, 1992).

VII. Conclusions, Perspectives, and Prospects

We have taken some of the more recent examples of the genetic and molecular analysis of complex behavior in *Drosophila* and seen how these methods can be applied to ask questions at different biological levels. The hard-core "clone and sequence" molecular approach has been extremely successful in moving forward our understanding of, for example, how the R7 cell develops in the visual system, with obvious implications for visual behavior (see Section II). However, a nagging question arises with such examples: Is *sevenless* really a "behavioral" gene? The easy answer is that a gene of this sort is one that students of behavior wish to study; and, after all, the *sev* mutation was isolated in a behavioral screen (Harris *et al.*, 1976). Our conclusion about *sev* is that, given its interesting interaction with *Ppb* (Ballinger and Benzer, 1988), and the fact that *sev* may be expressed in the nervous system, then it could be considered to be just as much of a behavioral gene as *per*. Without this added dimension to *sev*, we would have been tempted to classify it as a mere "developmental" gene, because its function initially appeared to be limited to specifying the fate of cell R7 during the late larval period.

The *per* gene would of course meet most criteria for being a behavioral gene. Complete removal of the gene does not result in dead insects (Smith and Konopka, 1981), but ones with excellent viability, which are arrhythmic and hyperactive (Hamblen *et al.*, 1986; Hamblen-Coyle *et al.*, 1989). The *cac* and *diss* mutations affect song characteristics, which are likely to be used in species recognition

(Wheeler *et al.,* 1989). The cloning and sequencing of *cac* and *diss* in different species, and the transformation of these genes from *D. simulans* into *D. melanogaster* may reveal, as was the case with *per* (Wheeler *et al.,* 1991), something about the species-specific molecular control of courtship song. Thus, important evolutionary questions, as well as informative "molecular mechanics" can be approached simultaneously by examining such behavioral genes. The attempt to define a behavioral gene is not an empty philosphical exercise. Without such a guideline, any gene which when mutated produces a sick animal (and when more severely mutated, a dead one) could be considered to be "behavioral."

The work with the *per* gene has now been extended into the field of molecular ecology with the discovery of a latitudinal cline in the Thr-Gly length polymorphism (Costa *et al.,* 1992), again reflecting the added dimension that can accrue to behavioral genes. Sokolowski (1980) has produced similar ecologically relevant work, beginning with the behavioral polymorphism she discovered in *D. melanogaster* larvae taken from both laboratory and natural populations. She classified these behavioral phenotypes into "rovers" and "sitters," which describes their locomotor activity. In natural populations, 70% of *D. melanogaster* individuals are rovers (Sokolowski, 1980).

This behavioral polymorphism appears to be due to a single gene difference at the *foraging (for)* locus (de Belle *et al.,* 1989). Once the gene is molecularly cloned, sequenced, and the conceptual protein deduced, this should provide insight into the control of locomotor behavior in larvae. Variation in the gene sequences can be examined both between and within natural populations of different species. There are various mathematical techniques that allow these DNA sequences to be scrutinized and that can provide valuable information on whether natural selection has been operating, for example, on the DNA polymorphism that gives rise to the different larval behavioral strategies (MacDonald and Kreitman, 1991). These techniques could provide an independent assessment, in this case as to whether natural selection was maintaining the rover/sitter gene frequencies. Thus one can do more than analyze how a behavioral gene influences the phenotype via the product it encodes and the tissues in which it is expressed. Behavior is an organism's first response to environmental challenges. Obtaining a full understanding of a behavioral gene at the molecular level can also mean studying that molecule in ways which will determine the selection pressures that have molded it.

The *per* gene may be the current "flagship" example of what can be done with behavioral genes. While it is not clear exactly how *per* determines biological cycles, the species specificity of locomotor activity and song cycles (Petersen *et al.*, 1988; Wheeler *et al.*, 1991) provides a means with which to study possible speciation events. The *for* gene could also be an important player in the speciation process because it is relevant to behaviors involved in habitat choice.

We have come a long way in behavioral genetics from the seemingly entrenched positions of many workers in the field in the late 1960s and early 1970s who appeared to resent studies into the single gene control of behavior (Wilcock, 1969). These workers believed that the study of major genes represented an overly "mechanical" and "nonevolutionary" approach, and they advocated studying "polygenically determined" traits only. We believe that the examples reviewed here reveal how incorrect this older view was. In fact, the roving and sitting phenotypes appear at first glance to meet the criteria of a quantitative polygenic trait (Sokolowski, 1980), by analogy to the different calling strategies of "caller" males versus "satellite" males in crickets (Cade, 1981). However, such "complex" behavioral differences turn out to be under the control of one or two genes in both cases (de Belle *et al.*, 1989; Cade, 1981). Similarly, the *fru* strain, the males of which show bisexual behavior, would suggest, superficially, that such a complex phenotype should be encoded by a polygenic system. In fact two genetic lesions were shown to be responsible for the abnormal behavior seen in the *fru* variants, yet each lesion was readily mappable to a *bona fide* locus.

In conclusion, the genetic and molecular dissection of behavior with single genes provides us with a way, not only to understand how genes can determine behavioral phenotypes in a mechanistic sense, but also how these genes and the behavior which they influence have evolved.

REFERENCES

Abrams, T. W., and Kandel, E. R. (1988). Is contiguity detection in classical conditioning a system or a cellular property? Learning in *Aplysia* suggests a possible molecular site. *Trends Neurosci.* 11, 128–135.
Aceves-Piña, E., and Quinn, W. (1979). Learning in normal and mutant *Drosophila* larvae. *Science* 206, 93–96.
Aigaki, T., Fleischmann, I., Chen, P. S., and Kubli, E. (1991). Ectopic expression of sex peptide alters reproductive behavior of female D. melanogaster. *Neuron* 7, 557–563.
Alexander, B. D. (1962). Evolutionary change in cricket acoustic communication. *Evolution* 16, 443–467.
Anderson, P. R., and Oakeshott, J. G. (1984). Parallel geographical patterns of allozyme variation in two sibling *Drosophila* species. *Nature (London)* 308, 729–731.
Ayer, R. K., Jr., and Carlson, J. (1991). *acj6:* a gene affecting olfactory physiology and behavior in *Drosophila*. *Proc. Natl. Acad. Sci. U.S.A.* 88, 5467–5471.

Ayer, R. K., Jr., and Carlson, J. (1992). Olfactory physiology in the *Drosophila* antenna and maxillary palp: *acj6* distinguishes two classes of odorant pathways. *J. Neurobiol.* 23, 965–982.

Ayyub, C., Paranjape, J., Rodrigues, V., and Siddiqi, O. (1990). Genetics of olfactory behavior in *Drosophila melanogaster*. *J. Neurogenet.* 6, 243–262.

Balling, A., Technau, G. M., and Heisenberg, M. (1987). Are the structural changes in adult *Drosophila* mushroom bodies memory traces? Studies on biochemical learning mutants. *J. Neurogenet.* 4, 64–73.

Ballinger, D. G., and Benzer, S. (1988). *Photophobe (Ppb)*, a *Drosophila* mutant with a reversed sign of phototaxis: the mutation shows an allele specific interaction with *sevenless*. *Proc. Natl. Acad. Sci. U.S.A.* 85, 3960–3964.

Bandziulis, R. J., Swanson, M. S., and Dreyfuss, G. (1989). RNA-binding proteins as developmental regulators. *Genes Dev.* 3, 431–437.

Banerjee, U., Renfranz, P. J., and Pollock, J. A. (1987). Molecular characterization and expression of *sevenless*, a gene involved in neuronal pattern formation in the Drosophila eye. *Cell* 49, 281–291.

Bargiello, T. A., and Young, M. W. (1984). Molecular genetics of a biological clock in *Drosophila*. *Proc. Natl. Acad. Sci. U.S.A.* 81, 2142–2146.

Bargiello, T. A., Saez, T. A., Baylies, M. K., Gasic, G., Young, M. W., and Spray, D. C. (1987). The *Drosophila* gene per affects intercellular junctional communication. *Nature (London)* 328, 686–691.

Barrows, E. (1907). The reactions of the fly, *Drosophila ampelophila loew* to odorous substances. *J. Exp. Zool.* 4, 515–537.

Bartelt, R. J., Schaner, A. M., and Jackson, L. L. (1985). Cis-vaccenyl acetate as an aggregation pheromone in *D. melanogaster*. *J. Chem. Ecol.* 11, 1747–1756.

Basler, K., and Hafen, E. (1988). *sevenless* and *Drosophila* eye development: a tyrosine kinase controls cell fate. *Trends Genet.* 4, 74–79.

Bastock, M. (1956). A gene mutation that changes a behaviour pattern. *Evolution* 10, 421–439.

Baylies, M. K., Bargiello, T. A., Jackson, F. R., and Young, M. W. (1987). Changes in abundance or structure of the per gene product can alter periodicity of the *Drosophila* clock. *Nature (London)* 326, 390–392.

Baylies, M. K., Vosshall, L. B., Sehgal, A., and Young, M. W. (1992). New short period mutations of the Drosophila clock gene *period*. *Neuron* 9, 575–581.

Bell, L. R., Maine, E. M., Schedl, P., and Cline, T. W. (1988). *Sex-lethal*, a Drosophila sex-determination gene, exhibits sex-specific RNA splicing and sequence similarity to RNA binding proteins. *Cell* 5, 1037–1046.

Bellen, H. J., d'Evelyn, D., Harvey, M., and Elledge, S. J. (1992). Isolation of temperature sensitive diptheria toxins in yeast and their effects on *Drosophila* cells. *Development* 114, 787–796.

Benezra, R., Davis, R. L., Lockshon, D., Turner, D. L., and Weintraub, H. (1990). The protein Id: A negative regulator of helix–loop–helix DNA binding proteins. *Cell* 61, 49–59.

Bernstein, A. S., Neumann, E. K., and Hall, J. C. (1992). Temporal analysis of tone pulses within the courtship songs of two sibling *Drosophila* species, their interspecific hybrid, and behavioral mutants of *D. melanogaster*. *J. Insect Behav.* 5, 15–36.

Black, B. C. (1988). Studies of the genetics and biochemistry of catecholamine metabolism using *Drosophila* behavioral mutants. *In* "Progress in Catecholamine Research. Part A: Aspects and Peripheral Mechanisms" (A. Dahlstrom, R. H. Belmaker, and M. Sandler, eds.), pp. 297–302. Alan R. Liss, New York.

Boake, C. R. B. (1991). Coevolution of senders and receivers of sexual signals: Genetic coupling and genetic correlations. *Trends Ecol. Evol.* **6**, 225–227.

Bolwig, N. (1946). Sense and sense organs of the anterior end of the housefly larvae. *Vidensk. Medd. Dan. Naturhist. Foren.* **109**, 81–217.

Booker, R., and Quinn, W. G. (1981). Conditioning of leg position in normal and mutant *Drosophila. Proc. Natl. Acad. Sci. U.S.A.* **76**, 3940–3944.

Boynton, S., and Tully, T. (1992). *latheo*, a new gene involved in associative learning and memory in *Drosophila melanogaster*, identified from P element mutagenesis. *Genetics* **131**, 655–672.

Brunner, A., Wolf, R., Pflugfelder, G. O., Poeck, B., and Heisenberg, M. (1992). Mutations in the proximal region of the *optomotor-blind* locus of *Drosophila melanogaster* reveal a gradient of neuroanatomical and behavioral phenotypes. *J. Neurogenet.* **8**, 43–55.

Buck, L., and Axel, R. (1991). A novel multigene family may encode odorant receptors: a molecular basis for odor recognition. *Cell* **65**, 175–187.

Burbach, K. M., Poland, A., and Bradfield, C. A. (1992). Cloning of the Ah-receptor cDNA reveals a distinctive ligand-activated transcription factor. *Proc. Natl. Acad. Sci. U.S.A.* **89**, 8185–8189.

Burnet, B., and Connolly, K. (1974). Activity and sexual behavior in *Drosophila melanogaster. In* "Genetics of Behavior" (J. H. Evan McLean, ed.), pp. 201–258. Elsevier, New York.

Butlin, R. K., and Ritchie, M. G. (1989). Genetic coupling in mate recognition systems: What is the evidence? *Biol. J. Linn. Soc.* **37**, 237–246.

Butterworth, F. A. (1969). Lipids of *Drosophila:* A newly detected lipid in the male. *Science* **163**, 1356.

Byers, D., Davis, R. L., and Kiger, J. A. (1981). Defect in cyclic AMP phosphodiesterase due to the *dunce* mutation of learning in *Drosophila melanogaster. Nature (London)* **289**, 79–81.

Cade, W. (1981). Alternative male strategies: Genetic differences in crickets. *Science* **212**, 563–564.

Caizzi, R., Ritossa, F., Ryseck, R. P., Richter, S., and Hovemann, B. (1987). Characterization of the *ebony* locus in *Drosophila melanogaster. Mol. Gen. Genet.* **206**, 66–70.

Carlson, J. (1991). Olfaction in *Drosophila:* genetics and molecular analysis. *Trends Neurosci.* **14**, 520–524.

Carthew, R. W., and Rubin, G. M. (1990). *seven in absentia*, the gene required for specification of R7 cell fate in the Drosophila eye. *Cell* **63**, 561–577.

Chen, C.-N., Denome, S., and Davis, R. L. (1986). Molecular analysis of cDNA clones and the corresponding genomic coding sequences of the *Drosophila dunce*[+] gene, the structural gene for cAMP phosphodiesterase. *Proc. Natl. Acad. Sci. U.S.A.* **83**, 9313–9317.

Chen, P. S., Stumm-Zollinger, E., Aigaki, T., Balmer, J., Bienz, M., and Bohlen, P. (1988). A male accessory gland peptide that regulates reproductive behavior of female D. melanogaster. *Cell* **54**, 291–298.

Citri, Y., Colot, H. V., Jacquier, A. C., Yu, Q., Hall, J. C., Baltimore, D., and Rosbash, M. (1987). A family of unusually spliced and biologically active transcripts is encoded by a *Drosophila clock* gene. *Nature (London)* **326**, 42–47.

Colot, H. V., Hall, J. C., and Rosbash, M. (1988). Interspecific comparisons of the *period* gene of *Drosophila* reveals large blocks of non-conserved coding DNA. *EMBO J.* **7**, 3929–3937.

Connolly, K., and Cook, R. (1973). Rejection behaviours by female *Drosophila melano-*

gaster: their ontogeny, causality and effects upon the behaviour of the courting male. *Behaviour* **52,** 155–171.

Corfas, G., and Dudai, Y. (1989). Habituation and dishabituation of a cleaning reflex in normal and mutant *Drosophila. J. Neurosci.* **9,** 56–62.

Costa, R., Peixoto, A. A., Thackeray, J. T., Dalgleish, R., and Kyriacou, C. P. (1991). Length polymorphism in the Threonine-Glycine-Encoding repeat region of the *period* gene in *Drosophila. J. Mol. Evol.* **32,** 238–246.

Costa, R., Peixoto, A. A., Barbujani, G., and Kyriacou, C. P. (1992). A latitudinal cline in a *Drosophila* clock gene. *Proc. R. Soc. London, Ser. B* **250,** 43–49.

Coyne, J. A. (1992). Genetics and speciation. *Nature (London)* **355,** 511–515.

Crews, S. T., Thomas, J. B., and Goodman, C. S. (1988). The Drosophila *single-minded* gene encodes a nuclear protein with sequence similarity to the *per* gene product. *Cell* **52,** 143–151.

David, J. R., Alonso-Moraga, A., Borai, F., Capy, P., Merçot, H., McEvey, S. F., Munoz-Serrano, A., and Tsakas, S. (1989). Latitudinal variation of *Adh* gene frequencies in *Drosophila melanogaster:* a Mediterranean instability. *Heredity* **62,** 11–16.

Davis, R. L., and Dauwalder, B. (1991). The *Drosophila dunce* locus: learning and memory genes in the fly. *Trends Genet.* **7,** 224–229.

Davis, R. L., and Davidson, N. (1986). The memory gene *dunce*⁺ encodes a remarkable set of RNAs with internal heterogeneity. *Mol. Cell. Biol.* 1464–1470.

de Belle, J. S., Hilliker, A. J., and Sokolowski, M. B. (1989). Genetic localisation of *foraging (for):* A major gene for larval behavior in *Drosophila melanogaster. Genetics* **123,** 157–164.

Delaney, S. J., Hayward, D. C., Barleben, F., Fischbach, K.-F., and Miklos, G. L. G. (1991). Molecular cloning and analysis of *small optic lobes,* a structural brain gene of *Drosophila melanogaster. Proc. Natl. Acad. Sci. U.S.A.* **88,** 7214–7218.

Dowse, H. B., and Ringo, J. M. (1989). Rearing *Drosophila* in constant darkness produces phenocopies of *period* circadian clock mutants. *Physiol. Zool.* **62,** 785–803.

Dowse, H. B., Hall, J. C., and Ringo, J. M. (1987). Circadian and ultradian rhythms in *period* mutants of *Drosophila melanogaster. Behav. Genet.* **17,** 19–35.

Dowse, H. B., Dushay, M. S., Hall, J. C., and Ringo, J. M. (1989). High-resolution analysis of locomotor activity rhythms in *disconnected,* a visual-system mutant of *Drosophila melanogaster. Behav. Genet.* **19,** 529–542.

Drain, P., Folkers, E., and Quinn, W. G. (1991). cAMP-dependent protein kinase and the disruption of learning in transgenic flies. *Neuron* **6,** 71–82.

Dudai, Y., Jan, Y.-N., Byers, D., Quinn, W., and Benzer, S. (1976). *dunce,* a mutant of *Drosophila* deficient in learning. *Proc. Natl. Acad. Sci. U.S.A.* **73,** 1684–1688.

Dudai, Y., Sher, B., Segal, D., and Yovell, Y. (1985). Defective responsiveness of adenylate cyclase to forskolin in the *Drosophila* memory mutant *rutabaga. J. Neurogenet.* **2,** 356–380.

Duerr, J. S., and Quinn, W. G. (1982). Three *Drosophila* mutations that block associative learning also affect habituation and sensitization. *Proc. Natl. Acad. Sci. U.S.A.* **79,** 3646–3650.

Dura, J.-M., Preat, T., and Tully, T. (1993). Identification of *linotte,* a new gene affecting learning and memory in *Drosophila melanogaster. J. Neurogenet.* **9,** 1–15.

Dushay, M. S., Rosbash, M., and Hall, J. C. (1989). The *disconnected* visual system mutations in *Drosophila melanogaster* drastically disrupt circadian rhythms. *J. Biol. Rhythms* **4,** 1–27.

Dushay, M. S., Konopka, R. J., Orr, E., Greenacre, M. L., Kyriacou, C. P., Rosbash, M., and Hall, J. C. (1990). Phenotypic and genetic analysis of *Clock,* a new circadian rhythm mutant in *Drosophila melanogaster. Genetics* **127,** 557–578.

Dushay, M. S., Rosbash, M., and Hall, J. C. (1992). Mapping the *Clock* rhythm mutation to the *period* locus of *Drosophila melanogaster* by germline transformation. *J. Neurogenet.* **8,** 173–179.

Erber, J., Masuhr, R., and Menzel, R. (1980). Localisation of short-term memory in the bee (*Apis mellitera*). *Physiol. Entomol.* **5,** 343–358.

Ewer, J., Rosbash, M., and Hall, J. C. (1988). An inducible promotor fused to the *period* gene in *Drosophila* conditionally rescues adult *per*-mutant arrhythmicity. *Nature (London)* **333,** 82–84.

Ewer, J., Rosbash, M., and Hall, J. C. (1990). Requirement for *period* gene expression in the adult and not during development for locomotor activity rhythms of imaginal *Drosophila melanogaster*. *J. Neurogenet.* **7,** 31–73.

Ewer, J., Frisch, B., Hamblen-Coyle, M., Rosbash, M., and Hall, J. C. (1992). Expression of the *period* clock gene in different cell types in the brain of *Drosophila* adults and mosaic analysis of these cells' influence on circadian behavioral rhythms. *J. Neurosci.* **12,** 3321–3349.

Fischbach, K. F., and Heisenberg, M. (1981). Structural brain mutant of *Drosophila melanogaster* with reduced cell number in the medulla cortex and with normal optomotor yaw response. *Proc. Natl. Acad. Sci. U.S.A.* **78,** 1105–1109.

Fischbach, K. F., and Heisenberg, M. (1984). Neurogenetics and behaviour in insects. *J. Exp. Biol.* **112,** 65–93.

Fischbach, K. F., and Technau, G. (1984). Cell degeneration in the developing optic lobes of the *sine oculis* and *small-optic-lobes* mutants of *Drosophila melanogaster*. *Dev. Biol.* **104,** 219–239.

Flint, K. K., Rosbash, M., and Hall, J. C. (1993). Transfer of dye among salivary gland cells is not affected by genetic variations of the *period* clock gene in *Drosophila melanogaster*. *J. Membr. Biol.* **136,** 333–342.

Folkers, E. 1982. Visual learning and memory of *Drosophila melanogaster* wild-type C-S and the mutants *dunce, amnesiac, turnip* and *rutabaga*. *J. Insect. Physiol.* **28,** 535–539.

Gailey, D. A., and Hall, J. C. (1989). Behavior and cytogenetics of *fruitless* in *D. melanogaster:* Different courtship defects caused by separate, closely linked lesions. *Genetics* **121,** 773–785.

Gailey, D. A., Jackson, F. R., and Siegel, R. W. (1982). Male courtship in *Drosophila:* the conditioned response to immature males and its genetic control. *Genetics* **102,** 771–782.

Gailey, D. A., Jackson, F. R., and Siegel, R. W. (1984). Conditioning mutations in *D. melanogaster* affect an experience-dependent behavioral modification in courting males. *Genetics* **106,** 613–623.

Gailey, D. A., Lacaillade, R. C., and Hall, J. C. (1986). Chemosensory elements of courtship in normal and mutant, olfaction-deficient *Drosophila melanogaster*. *Behav. Genet.* **16,** 375–405.

Gailey, D. A., Taylor, B. J., and Hall, J. C. (1991). Elements of the *fruitless* locus regulate development of the muscle of Lawrence, a male-specific structure in the abdomen of *Drosophila melanogaster* adults. *Development* **113,** 879–890.

Gill, K. S. (1963). A mutation causing abnormal courtships and mating behavior in male *Drosophila melanogaster*. *Am. Zool.* **3,** 507.

Golic, K., and Lindquist, S. (1989). The FLP recombinase of yeast catalyses site-specific recombination in the *Drosophila* genome. *Cell* **59,** 499–509.

Greenacre, M. L., Ritchie, M. G., Byrne, B. C., and Kyriacou, C. P. (1993). Female song preference and the *period* gene in *Drosophila*. *Behav. Genet.* **23,** 85–90.

Greenspan, R. J., Finn, J. A., Jr., and Hall, J. C. (1980). Acetylocholinesterase mutants in *Drosophila* and their effects on the structure and function of the central nervous system. *J. Comp. Neurol.* **189**, 741–774.

Griffith, L. C., Verselis, L. M., Aitken, K. M., Kyriacou, C. P., Danho, W., and Greenspan, R. J. (1993). Inhibition of calcium/calmodulin-dependent protein kinase in Drosophila disrupts behavioral plasticity. *Neuron* **10**, 501–509.

Hafen, E., Basler, K., Edstroem, J. E., and Rubin, G. M. (1987). *Sevenless* a cell specific homeotic gene of *Drosophila* encodes a putative transmembrane receptor with a tyrosine kinase domain. *Science* **236**, 55–63.

Hall, J. C. (1977). Portions of the central nervous system controlling reproductive behavior in *Drosophila melanogaster. Behav. Genet.* **7**, 291–312.

Hall, J. C. (1978). Courtship among males due to a male-sterile mutation in *Drosophila melanogaster. Behav. Genet.* **8**, 125–141.

Hall, J. C. (1979). Control of male reproductive behavior by the central nervous system of *Drosophila:* dissection of a courtship pathway by genetic mosaics. *Genetics* **92**, 437–457.

Hall, J. C. (1984). Complex brain and behavioral functions disrupted by mutations in *Drosophila. Dev. Genet.* **4**, 355–378.

Hall, J. C. (1986). Learning and rhythms in courting mutant *Drosophila. Trends Neurosci.* **9**, 414–418.

Hall, J. C. (1990). Genetics of circadian rhythms. *Annu. Rev. Genet.* **24**, 659–697.

Hall, J. C., and Greenspan, R. J. (1979). Genetic analysis of *Drosophila* neurobiology. *Annu. Rev. Genet.* **13**, 127–195.

Hall, J. C., and Kyriacou, C. P. (1990). Genetics of biological rhythms in *Drosophila. Adv. Insect Physiol.* **22**, 221–298.

Hall, J. C., Kulkarni, S. J., Kyriacou, C. P., Yu, Q., and Rosbash, M. (1990). Genetic and molecular analysis of neural development and behavior in *Drosophila*. In "Developmental Behavior Genetics" (M. G. Hahn, J. K. Hewitt, N. D. Henderson, and R. Benno, eds.), pp. 100–112. Oxford Univ. Press, New York.

Hamblen, M., Zehring, W. A., Kyriacou, C. P., Reddy, P., Yu, Q., Wheeler, D. A., Zwiebel, L. J., Konopka, R. J., Rosbash, M., and Hall, J. C. (1986). Germ-line transformation involving DNA from the *period* locus in *Drosophila melanogaster:* Overlapping genomic fragments that restore circadian and ultradian rhythmicity to *per⁰* and *per⁻* mutants. *J. Neurogenet.* **3**, 249–291.

Hamblen-Coyle, M., Konopka, R. J., Zwiebel, L. J., Colot, H. V., Dowse, H. B., Rosbash, M., and Hall, J. C. (1989). A new mutation at the *period* locus of *Drosophila melanogaster* with some novel effects on circadian rhythms. *J. Neurogenet.* **5**, 229–256.

Handler, A. M., and Konopka, R. J. (1979). Transplantation of a circadian pacemaker in *Drosophila. Nature (London)* **279**, 236–238.

Hardin, P. E., Hall, J. C., and Rosbash, M. (1990). Feedback of the *Drosophila period* gene product on circadian cycling of its messenger RNA levels. *Nature (London)* **343**, 536–540.

Hardin, P. E., Hall, J. C., and Rosbash, M. (1992a). Circadian oscillation in *period* gene mRNA levels are transcriptionally regulated. *Proc. Natl. Acad. Sci. U.S.A.* **89**, 11711–11715.

Hardin, P. E., Hall, J. C., and Rosbash, M. (1992b). Behavioral and molecular analyses suggest that circadian output is disrupted by *disconnected* mutants in *D. melanogaster. EMBO J.* **11**, 1–6.

Harris, W. A., and Stark, W. S. (1977). Hereditary retinal degeneration in *Drosophila melanogaster:* a mutant defect associated with the phototransduction process. *J. Gen. Physiol.* **69**, 261–291.

Harris, W. A., Stark, W. S., and Walker, J. A. (1976). Genetic dissection of the photoreceptor system in the compound eye of *Drosophila melanogaster. J. Physiol. (London)* **255,** 415–439.

Hasan, G. (1990). Molecular cloning of an olfactory gene from *Drosophila melanogaster. Proc. Natl. Acad. Sci. U.S.A.* **87,** 9037–9041.

Heisenberg, M. (1972). Comparative behavioral studies on two visual mutants of *Drosophila. J. Comp. Physiol.* **80,** 119–136.

Heisenberg, M. (1980). Mutants of brain structure and function. What is the significance of the mushroom bodies for behavior? *In* "Development and Neurobiology of *Drosophila*" (O. Siddiqi, P. Babu, R. M. Hall, and J. C. Hall, eds.), pp. 373–390. Plenum, New York.

Heisenberg, M., and Böhl, K. (1979). Isolation of anatomical brain mutants of *Drosophila* by histological means. *Z. Naturforsch. C* **34C,** 143–147.

Heisenberg, M., Borst, A., Wagner, S., and Byers, D. (1985). *Drosophila* mushroom body mutants are deficient in olfactory learning. *J. Neurogenet.* **2,** 1–30.

Heisenberg, M., and Buchner, E. (1977). The role of retinula cell types in visual behavior of *Drosophila melanogaster. J. Comp. Physiol.* **117,** 127–162.

Heisenberg, M., and Götz, K. G. (1975). The use of mutations for the partial degradation of vision in *Drosophila melanogaster. J. Comp. Physiol.* **98,** 217–241.

Heisenberg, M., Wonneberger, R., and Wolf, R. (1978). optomotor blind[H31]—a *Drosophila* mutant of the lobula plate giant neurons. *J. Comp. Physiol.* **124,** 287–296.

Helfand, S. L., and Carlson, J. R. (1989). Isolation and characterization of an olfactory mutant in *Drosophila* with a chemically specific defect. *Proc. Natl. Acad. Sci. U.S.A.* **86,** 2908–2912.

Helfrich, C. (1986). The role of the optic lobes in the regulation of the locomotor activity rhythms in *Drosophila melanogaster:* behavioral analysis of neural mutants. *J. Neurogenet.* **3,** 321–343.

Hodgetts, R. B., and Konopka, R. J. (1973). Tyrosine and catecholamine metabolism in wild-type *Drosophila melanogaster* and a mutant *ebony. J. Insect Physiol.* **19,** 1211–1220.

Hoffman, E. C., Reyes, H., Chu, F.-F., Sander, F., Conley, L. H., Brooks, B. A., and Hankinson, O. (1991). Cloning of a factor required for activity of the Ah (Dioxin) receptor. *Science* **252,** 954–958.

Hotta, Y., and Benzer, S. (1976). Courtship in *Drosophila* mosaics: sex specific loci for sequential action patterns. *Proc. Nat. Acad. Sci. U.S.A.* **73,** 4154–4158.

Huang, Z. J., Edery, I., and Rosbash, M. (1993). PAS is a dimerization domain common to *Drosophila* Period and several transcription factors. *Nature (London)* **364,** 259–262.

Jackson, F. R., Bargiello, T. A., Yun, S. H., and Young, M. W. (1986). Product of *per* of *Drosophila* shares homology with proteoglycans. *Nature (London)* **320,** 185–188.

Jacobs, M. E. (1960). Influence of light on mating of *D. melanogaster. Ecology* **41,** 182–188.

Jacobs, M. E., and Brubaker, K. K. (1963). β-alanine utilization of *ebony* and non-*ebony Drosophila melanogaster. Science* **139,** 1282.

Jallon, J.-M. (1984). A few chemical words exchanged by *Drosophila* during courtship and mating. *Behav. Genet.* **14,** 441–478.

James, A. A., Ewer, J., Reddy, P., Hall, J. C., and Rosbash, M. (1986). Embryonic expression of the *period* clock gene in the central nervous system of *Drosophila melanogaster. EMBO J.* **5,** 2313–2320.

Jones, R. K., and Rubin, G. M. (1990). Molecular analysis of *no-on-transient A,* a gene required for normal vision in Drosophila. *Neuron* **4,** 711–723.

Kemp, B. E., and Pearson, R. B. (1990). Protein kinase recognition sequence motifs. *Trends Biochem. Sci.* **15**, 342–346.

Kiger, J. A., Jr., and Golanty, E. (1977). A cytogenetic analysis of cyclic nucleotide phosphodiesterase activities in *Drosphila. Genetics* **85**, 609–622.

Kiger, J. A., Jr., Davis, R. L., Salz, H., Fletcher, T., and Bowling, M. (1981). Genetic analysis of cyclic nucleotide phophodiesterases in *Drosophila melanogaster. Adv. Cyclic Nucleotide Res.* **14**, 273–288.

Kirkpatrick, M., and Ryan, M. J. (1991). The evolution of mating preferences and the paradox of the lek. *Nature (London)* **350**, 33–38.

Konopka, R. J., and Benzer, S. (1971). Clock mutants of *Drosophila melanogaster. Proc. Natl. Acad. Sci. U.S.A.* **68**, 2112–2116.

Konopka, R. J., Wells, S., and Lee, T. (1983). Mosaic analysis of a *Drosophila* clock mutant. *Mol. Gen. Genet.* **190**, 284–288.

Konopka, R. J., Smith, R. F., and Orr, D. (1991). Characterization of *Andante,* a new *Drosophila* clock mutant and its interaction with other clock mutants. *J. Neurogenet.* **7**, 103–114.

Krämer, H., Cagan, R. L., and Zipursky, L. (1991). Interaction of *bride* of *sevenless* membrane-bound ligand and the *sevenless* tyrosine-kinase receptor. *Nature (London)* **352**, 207–212.

Krupinski, J., Coussen, F., Bakalyer, H. A., Tang, W. J., Feinstein, P. G., Orth, K., Slaughter, C., Reed, R. R., and Gilman, A. G. (1989). Adenylyl cyclase amino acid sequence: possible channel- or transporter-like structure. *Science* **244**, 1558–1564.

Kulkarni, S. J., and Hall, J. C. (1987). Behavioral and cytogenetic analysis of the *cacophony* courtship song mutant and interacting genetic variants in *Drosophila melanogaster. Genetics* **115**, 461–475.

Kulkarni, S. J., Steinlauf, A. F., and Hall, J. C. (1988). The *dissonance* mutant of courtship song in *Drosophila melanogaster:* isolation, behavior and cytogenetics. *Genetics* **118**, 267–285.

Kunes, S., and Steller, H. (1991). Ablation of *Drosophila* photoreceptor cells by conditional expression of a toxin gene. *Genes Dev.* **5**, 970–983.

Kyriacou, C. P. (1981). The relationship between locomotor activity and sexual behaviour in *ebony* strains of *Drosophila melanogaster. Anim. Behav.* **29**, 462–471.

Kyriacou, C. P. (1985). Long term *ebony* polymorphisms a comparison of the contributions of behavioral and non-behavioral fitness characters. *Behav. Genet.* **15**, 165–180.

Kyriacou, C. P. (1990a). The molecular ethology of the *period* gene in *Drosophila. Behav. Genet.* **20**, 191–211.

Kyriacou, C. P. (1990b). Genetic and molecular analysis of eukaryotic behaviour. *Semin. Neurosci.* **2**, 217–229.

Kyriacou, C. P., and Hall, J. C. (1980). Circadian rhythm mutations in *Drosophila* affect short-term fluctuations in the male's courtship song. *Proc. Natl. Acad. Sci. U.S.A.* **77**, 6929–6933.

Kyriacou, C. P., and Hall, J. C. (1982). The function of courtship song rhythms in *Drosophila. Anim. Behav.* **30**, 794–801.

Kyriacou, C. P., and Hall, J. C. (1984). Learning and memory mutations impair acoustic priming of mating behaviour in *Drosophila. Nature (London)* **308**, 62–65.

Kyriacou, C. P., and Hall, J. C. (1986). Inter-specific genetic control of courtship song production and reception in *Drosophila. Science* **232**, 494–497.

Kyriacou, C. P., and Hall, J. C. (1989). Spectral analysis of *Drosophila* courtship song rhythms. *Anim. Behav.* **37**, 850–859.

Kyriacou, C. P., Burnet, B., and Connolly, K. (1978). The behavioural basis of over dominance in competitive mating success at the *ebony* locus of *Drosophila melanogaster*. *Anim. Behav.* **26**, 1195–1206.

Kyriacou, C. P., Oldroyd, M., Wood, J., Sharp, M., and Hill, M. (1990). Clock mutants alter developmental timing in *Drosophila*. *Heredity* **64**, 395–401.

Lawrence, P. A., and Johnston, P. (1986). The muscle pattern of a segment of Drosophila may be determined by neurons and not by contributing myoblasts. *Cell* **45**, 505–513.

Levin, L. R., Han, P.-L., Hwang, P. M., Feinstein, P. G., Davis, R. L., and Reed, R. R. (1992). The Drosophila learning and memory gene *rutabaga* encodes a Ca^{2+}/calmodulin-responsive adenylyl cyclase. *Cell* **68**, 479–489.

Lilly, M., and Carlson, J. (1989). *smellblind:* a gene required for *Drosophila* olfaction. *Genetics* **124**, 293–302.

Liu, X., Lorenz, L., Yu, Q., Hall, J. C., and Rosbash, M. (1988). Spatial and temporal expression of the *period* gene in *Drosophila melanogaster*. *Genes Dev.* **2**, 228–238.

Liu, X., Yu, Q., Huang, Z., Zwiebel, L. J., Hall, J. C., and Rosbash, M. (1991). The strength and periodicity of D. melanogaster circadian rhythms are differentially affected by alterations in *period* gene expression. *Neuron* **6**, 753–766.

Liu, X., Zweibel, L. J., Hinton, D., Benzer, S., Hall, J. C., and Rosbash, M. (1992). The *period* gene encodes a predominately nuclear protein in adult *Drosophila*. *J. Neurosci.* **12**, 2735–2744.

Livingstone, M. S. (1985). Genetic dissection of *Drosophila* adenylate cyclase. *Proc. Natl. Acad. Sci. U.S.A.* **82**, 5992–5996.

Livingstone, M. S., Sziber, P. P., and Quinn, W. G. (1984). Loss of calcium/calmodulin responsiveness in adenylase cyclase of *rutabaga*, a Drosophila learning mutant. *Cell* **37**, 205–215.

MacDonald, J. H., and Kreitman, M. E. (1991). Adaptive protein evolution at the *Adh* locus in *Drosophila*. *Nature (London)* **351**, 652–654.

Manning, A. (1967). The control of sexual receptivity in female *Drosophila*. *Anim. Behav.* **15**, 239–250.

McKenna, M., Monte, P., Helfand, S. L., Woodward, C., and Carlson, J. (1989). A simple chemosensory response in *Drosophila* and the isolation of *acj* mutants in which it is affected. *Proc. Natl. Acad. Sci. U.S.A.* **86**, 8118–8122.

Moffat, K. G., Gould, J. H., Smith, H. K., and O'Kane, C. J. (1992). Inducible cell ablation in *Drosophila* in cold-sensitive *ricin-A* chain. *Development* **114**, 681–687.

Monsma, S. A., and Wolfner, M. F. (1990). Synthesis of two *Drosophila* male accessory gland proteins and their fate after transfer to the female during mating. *Dev. Biol.* **142**, 465–475.

Monte, P., Woodard, C., Ayer, R., Lilly, M., Sun, H., and Carlson, J. (1989). Characterization of the larval olfactory response in *Drosophila* and its genetic basis. *Behav. Genet.* **19**, 267–283.

Nambu, J. R., Franks, R. G., Hu, S., and Crews, S. T. (1990). The *single-minded* gene of Drosophila is required for the expression of genes important for the development of CNS midline cells. *Cell* **63**, 63–75.

Nambu, J. R., Lewis, J. O., Whorton, K. A., Jr., and Crews, S. T. (1991). The Drosophila *single-minded* gene encodes a helix–loop–helix protein that acts as a master regulator of CNS midline development. *Cell* **69**, 1157–1168.

Newby, L. M., and Jackson, F. R. (1991). *Drosophila ebony* mutants have altered circadian activity rhythms but normal eclosion rhythms. *J. Neurogenet.* **7**, 85–101.

Newby, L. M., White, L., DiBartolomeis, S. M., Walker, B. J., Dowse, H. B., Ringo, J. M. Khuda, N., and Jackson, F. R. (1991). Mutational analysis of the *Drosophila min-*

iature-dusky (*m-dy*) locus: effects on cell size and circadian rhythms. *Genetics* **128**, 571–582.

Nighorn, A., Healy, M. J., and Davis, R. L. (1991). The cyclic AMP phosphodiesterase encoded by the Drosophila dunce gene is concentrated in the mushroom body neuropil. *Neuron* **6**, 455–467.

O'Kane, C. J., and Gehring, W. J. (1987). Detection *in situ* of genomic regulatory elements in *Drosophila. Proc. Nat. Acad. Sci. U.S.A.* **84**, 9123–9127.

O'Kane, C. J., and Moffat, K. G. (1992). Selective cell ablation and cell surgery. *Curr. Opinion Genet. Dev.* **2**, 602–607.

Pak, W. L., Grossfield, J., and Arnold, S. (1970). Mutants of the visual pathway of *Drosophila melanogaster. Nature (London)* **227**, 518–520.

Peixoto, A. A., Costa, R., Wheeler, D. A., Hall, J. C., and Kyriacou, C. P. (1992). Evolution of the Threonine-Glycine repeat region of the *period* gene in the *melanogaster* species subgroup. *J. Mol. Evol.* **35**, 411–419.

Petersen, G., Hall, J. C., and Rosbash, M. (1988). The *period* gene of Drosophila carries species-specific behavioral instructions. *EMBO J.* **7**, 3939–3947.

Pflugfelder, G. O., Schwarz, H., Roth, H., Poeck, B., Sigl, A., Kerscher, S., Jonschker, B., Pak, W. L., and Heisenberg, M. (1990). Genetic and molecular characterization of the *optomotor-blind* gene locus in *Drosophila melanogaster. Genetics* **126**, 91–104.

Pflugfelder, G. O., Roth, H., Poeck, B., Kerscher, S., Schwarz, H., Jonschiker, B., and Heisenberg, M. (1992a). The *l(1)optomotor-blind* gene of *Drosophila melanogaster* is a major organizer of optic lobe development: Isolation and characterization of the gene. *Proc. Natl. Acad. Sci. U.S.A.* **89**, 1199–1203.

Pflugfelder, G. O., Roth, H., and Poeck, B. (1992b). A homology domain shared between *Drosophila optomotor-blind* and mouse *Brachyury* is involved in DNA-binding. *Biochem. Biophys. Res. Commun.* **186**, 918–925.

Pittendrigh, C. S. (1954). On temperature independence in the clock system controlling emergence time in *Drosophila. Proc. Natl. Acad. Sci. U.S.A.* **40**, 1018–1029.

Poeck, B., Balle, J., and Pflugfelder, G. O. (1993a). Transcript identification in the *optomotor-blind* locus of *Drosophila melanogaster* by intragenic mapping and PCR-aided sequence analysis of lethal point mutations. *Mol. Gen. Genet.* **238**, 325–332.

Poeck, B., Hofbauer, A., and Pflugfelder, G. O. (1993b). Expression of the *optomotor-blind* gene transcript in neuronal and glial cells of the developing nervous system. *Development* **117**, 1017–1029.

Qiu, Y., Chen, C.-N., Malone, T., Richker, L., Beckendorf, S. K., and Davis, R. L. (1991). Characterisation of the memory gene *dunce* of *Drosophila melanogaster. J. Mol. Biol.* **222**, 553–565.

Quinn, W. G., Harris, W. A., and Benzer, S. (1974). Conditioning behavior in *Drosophila melanogaster. Proc. Natl. Acad. Sci. U.S.A.* **71**, 708–712.

Quinn, W. G., Sziber, P. P., and Booker, R. (1979). The *Drosophila* memory mutant *amnesiac. Nature (London)* **277**, 212–214.

Reddy, P., Jacquier, A. C., Abovich, N., Peterson, G., and Rosbash, M. (1986). The *period* clock locus of D. melanogaster codes for a proteoglycan. *Cell* **46**, 53–61.

Reddy, P., Zehring, W. A., Wheeler, D. A., Pirrotta, V., Hadfield, C., Hall, J. C., and Rosbash, M. (1984). Molecular analysis of the *period* locus in Drosophila melanogaster and identification of a transcript involved in biological rhythms. *Cell* **38**, 701–710.

Reinke, R., and Zipursky, S. L. (1988). Cell–cell interactions in the Drosophila retina: the *bride-of-sevenless* gene is required in photoreceptor cell R8 for R7 cell development. *Cell* **55**, 321–330.

Rendahl, K. G., Jones, K. R., Kulkarni, S. J., Bagully, S. H., and Hall, J. C. (1992). The *dissonance* mutation at the *no-on-transient-A* locus of *D. melanogaster:* genetic control of courtship song and visual behaviors by a protein with putative RNA-binding motifs. *J. Neurosci.* **12,** 390–407.

Rendel, J. (1951). Mating of *ebony, vestigial* and wild-type *D. melanogaster* in light and dark. *Evolution* **5,** 226–230.

Riesgo-Escovar, J., Woodard, C., Gaines, P., and Carlson, J. (1992). Development and organisation of the *Drosophila* olfactory system: an analysis using enhancer traps. *J. Neurobiol.* **23,** 947–964.

Robinow, S., and White, K. (1988). The locus *elav* of *Drosophila melanogaster* is expressed in neurons at all developmental stages. *Dev. Biol.* **126,** 294–303.

Rodrigues, V., and Siddiqi, O. (1978). Genetic analysis of a chemosensory pathway. *Proc. Indian Acad. Sci., Sect. B* **87B,** 147–160.

Rogge, R. D., Karlovich, C. A., and Banerjee, U. (1991). Genetic dissection of a neurodevelopmental pathway: *sevenless* functions downstream of the *sevenless* and EGF receptor tyrosine kinases. *Cell* **64,** 38–48.

Rosbash, M., and Hall, J. C. (1989). The molecular biology of circadian rhythms. *Neuron* **3,** 387–398.

Ruoslahti, E. (1988). Structure and biology of proteoglycans. *Annu. Rev. Cell Biol.* **4,** 229–255.

Rutila, J. E., Edery, I., Hall, J. C., and Rosbash, M. (1992). The analysis of new short-period circadian rhythm mutants suggests features of *D. melanogaster period* gene function. *J. Neurogenet.* **8,** 101–113.

Saez, L., and Young, M. W. (1988). *In situ* localisation of the per clock protein during development of *Drosophila melanogaster. Mol. Cell. Biol.* **8,** 5378–5385.

Saez, L., Young, M. W., Baylies, M. K., Gasic, G., Bargiello, T. A., and Spray, D. C. (1992). *Per*—no link to gap junctions. *Nature (London)* **360,** 342.

Sawaguchi, T., and Goldman-Rakic, P. S. (1991). D1 dopamine receptors in prefrontal cortex: involvement in working memory. *Science* **251,** 947–950.

Schneiderman, A. M., Matsumato, S. G., and Hildebrand, J. G. (1982). Transsexually grafted antennae influence development of sexually dimorphic neurons in moth brain. *Nature (London)* **298,** 844–846.

Scott, D. (1986). Sexual mimicry regulates the attractiveness of mated *D. melanogaster* females. *Proc. Natl. Acad. Sci. U.S.A.* **83,** 8429–8433.

Scott, D., and Jackson, L. (1988). Interstrain comparisons of male predominant aphrodisiacs in *D. melanogaster. J. Insect Physiol.* **34,** 863–871.

Scott, D., and Richmond, R. C. (1987). Evidence against an anti-aphrodisiac role for cis-vaccenyl acetate in *D. melanogaster. J. Insect Physiol.* **33,** 363–369.

Sehgal, A., Man, B., Price, J. L., Visshall, L. B., and Young, M. W. (1991). New clock mutations in *Drosophila. Ann. N.Y. Acad. Sci.* **618,** 1–10.

Sehgal, A., Price, J., and Young, M. W. (1992). Ontogeny of a biological clock in *Drosophila melanogaster. Proc. Natl. Acad. Sci. U.S.A.* **89,** 1423–1427.

Siegel, R. W., and Hall, J. C. (1979). Conditioned responses in courtship behavior of normal and mutant *Drosophila. Proc. Natl. Acad. Sci. U.S.A.* **76,** 3430–3434.

Siegel, R. W., Hall, J. C., Gailey, D. A., and Kyriacou, C. P. (1984). Genetic elements of courtship in *Drosophila:* mosaics and learning mutants. *Behav. Genet.* **14,** 383–410.

Silva, A. J., Stevens, C. F., Tonegawa, S., and Wang, Y. (1992a). Deficient hippocampal long-term potentiation in α-calcium-calmodulin kinase II mutant mice. *Science* **257,** 201–206.

Silva, A. J., Paylor, R., Wehner, J. M., and Tonegawa, S. (1992b). Impaired spatial learning in α-calcium-calmodulin kinase II mutant mice. *Science* **257,** 206–211.

Simon, M. A., Bowell, D. D., and Rubin, G. M. (1989). Structure and activity of the sevenless protein: a protein tyrosine kinase receptor required for photoreceptor development in *Drosophila. Proc. Natl. Acad. Sci. U.S.A.* **86**, 8333–8337.

Siwicki, K. K., Eastman, C., Petersen, G., Rosbash, M., and Hall, J. C. (1988). Antibodies to the *period* gene product of Drosophila reveal diverse tissue distribution and rhythmic changes in the visual system. *Neuron* **1**, 141–150.

Siwicki, K. K., Flint, K. K., Hall, J. C., Rosbash, M., and Spray, D. C. (1992). The *Drosophila period* gene and dye coupling in larval salivary glands: a re-evaluation. *Biol. Bull.* **183**, 340–341.

Smith, R., and Konopka, R. (1981). Circadian clock phenotypes of chromosome abberations with a breakpoint at the *per* locus. *Mol. Gen. Genet.* **183**, 243–251.

Sokolowski, M. B. (1980). Foraging strategies of *Drosophila melanogaster:* a chromosomal analysis. *Behav. Genet.* **10**, 291–302.

Steller, H., Fischbach, K.-F., and Rubin, G. M. (1987). *disconnected:* a locus required for neuronal pathway formation in the visual system of Drosophila. *Cell* **50**, 1139–1153.

Stocker, R., Lienhard, M., Borst, A., and Fischbach, K. F. (1990). Neural architecture of antennal lobe in *Drosophila melanogaster. Cell Tissue Res.* **262**, 9–34.

Strausfeld, N. J. (1980). Male and female visual neurons in dipterous insects. *Nature (London)* **283**, 381–383.

Strausfeld, N. J., and Bacon, J. P. (1983). Multimodal convergence in central nervous systems of insects. *Fortschr. Zool.* **28**, 47–76.

Sturtevant, A. H. (1915). Experiments on sex recognition and the problem of sexual selection in *Drosophila. Anim. Behav.* **5**, 351–366.

Suzuki, M. (1989). SPXX, a frequent sequence motif in gene regulatory proteins. *J. Mol. Biol.* **207**, 61–84.

Technau, G. M. (1984). Fiber number in the mushroom bodies of adult *Drosophila melanogaster* depends on age, sex and experience. *J. Neurogenet.* **1**, 113–126.

Tempel, B. L., Bonini, N., Dawson, D. R., and Quinn, W. G. (1983). Reward learning in normal and mutant *Drosophila. Proc. Natl. Acad. Sci. U.S.A.* **80**, 1482–1486.

Thackeray, J. R., and Kyriacou, C. P. (1990). Molecular evolution in the *Drosophila yakuba period* locus. *J. Mol. Evol.* **31**, 389–401.

Tomlinson, A., Boewtell, D. D., Hafen, E., and Rubin, G. M. (1987). Localization of the *sevenless* protein a putative receptor for positional information in the eye imaginal disk of Drosophila. *Cell* **51**, 143–150.

Tompkins, L. (1984). Genetic analysis of sex appeal in *Drosophila. Behav. Genet.* **14**, 411–440.

Tompkins, L., and Hall, J. C. (1981). The different effects on courtship of volatile compounds from mated and virgin *Drosophila* females. *J. Insect Physiol.* **27**, 17–21.

Tompkins, L., and Hall, J. C. (1983). Identification of brain sites controlling female receptivity in mosaics of *Drosophila melanogaster. Genetics* **103**, 179–195.

Tompkins, L., Hall, J. C., and Hall, L. M. (1980). Courtship simulating volatile compounds from normal and mutant *Drosophila. J. Insect Physiol.* **26**, 689–697.

Tompkins, L., Gross, A. C., Hall, J. C., Gailey, D. A., and Siegal, R. W. (1982). The role of female movement in the sexual behavior of *Drosophila melanogaster. Behav. Genet.* **12**, 295–307.

Tompkins, L., Siegal, R. N., Gailey, D. A., and Hall, J. C. (1983). Conditioned courtship in *Drosophila* and its mediation by chemical cues. *Behav. Genet.* **13**, 565–578.

Tully, T., and Gergen, J. P. (1986). Deletion mapping of the *Drosophila* memory mutant *amnesiac. J. Neurogenet.* **3**, 33–47.

Tully, T., and Gold, D. (1993). Differential effects of *dunce* mutations on associative learning and memory in *Drosophila. J. Neurogenet.* **9**, 55–71.

Tully, T., and Quinn, W. G. (1985). Classical conditioning and retention in normal and mutant *Drosophila melanogaster*. *J. Comp. Physiol.* **157**, 263–277.

Tully, T., Cambiazo, V., and Kruse, L. (1994). Memory through metamorphosis in normal and mutant *Drosophila. J. Neurosci.* **14**, 68–74.

Vander Meer, R. K., Obin, M. S., Zawistowski, S., Sheehan, K. B., and Richmond, R. C. (1986). A reevaluation of the role of cis-vaccenylacetate, cis-vaccenol and esterase 6 in the regulation of mated female sexual attractiveness in *Drosophila melanogaster. J. Insect Physiol.* **32**, 681–686.

Varshavsky, A. (1992). The N-end rule. *Cell* **69**, 725–735.

Venard, R., and Pichon, Y. (1981). Étude electro-antennogrophique de al response peripherique de l'antenne de *Drosophila melanogaster* a des stimulations odorantes. *C. R. Sci., Acad. Hebd. Seances Ser. D* **293**, 839–842.

Venkatesh, S., and Singh, R. N. (1984). Sensilla on the third antennal segment of *Drosophila melanogaster* Meigen (*Diptera-Drosophilidae*). *Int. J. Insect Morphol. Embryol.* **13**, 51–63.

Vihtelic, T. S., Hyde, D. R., and O'Tousa, J. E. (1991). Isolation and characterization of the *Drosophila retinal degeneration B (rdgB)* gene. *Genetics* **127**, 761–768.

von Schilcher, F. (1976a). The function of sine song and pulse song in the courtship of *Drosophila melanogaster. Anim. Behav.* **24**, 622–625.

von Schilcher, F. (1976b). The behavior of *cacophony,* a courtship song mutant in *Drosophila melanogaster. Behav. Biol.* **17**, 187–196.

von Schilcher, F. (1977). A mutation which changes courtship song in *Drosophila melanogaster. Behav. Genet.* **7**, 251–259.

von Schilcher, F., and Hall, J. C. (1979). Neural topography of courtship song in sex mosaics of *Drosophila melanogaster. J. Comp. Physiol. A* **129**, 85–95.

Vowles, D. M. (1964). Olfactory learning and brain lesions in the wood ant (*F.rufa*). *J. Comp. Physiol. Psychol.* **158**, 104–111.

Weintraub, H., Davis, R., Tapscott, S., Thayer, M., Krause, M., Benezra, R., Blackwell, T. K., Turner, D., Rupp, R., Hollenberg, S., Zhuang, Y., and Lasser, A. (1991). The *myoD* gene family: nodal point during specification of the muscle cell lineage. *Science* **251**, 761–766.

Wheeler, D. A., Fields, W. L., and Hall, J. C. (1988). Spectral analysis of *Drosophila* courtship: *Drosophila melanogaster, Drosophila simulans* and their interspecific hybrid. *Behav. Genet.* **18**, 675–703.

Wheeler, D. A., Kulkarni, S. J., Gailey, D. A., and Hall, J. C. (1989). Spectral analysis of courtship songs in behavioral mutants of *Drosophila melanogaster. Behav. Genet.* **19**, 503–528.

Wheeler, D. A., Kyriacou, C. P., Greenacre, M. L., Yu, Q., Rutila, J. E., Rosbash, M., and Hall, J. C. (1991). Molecular transfer of a species-specific courtship behavior from *Drosophila simulans* to *Drosophila melanogaster. Science* **251**, 1082–1085.

Wilcock, J. (1969). Gene action and behaviour. An evaluation of major gene pleiotropism. *Psychol. Bull.* **72**, 1–29.

Wittekind, W. C., and Spatz, H. C. (1988). Habituation of the landing response of *Drosophila. In* "Modulation of Synaptic Transmission and Plasticity in Nervous Systems" (G. Hertting and H. C. Spatz, eds.), pp. 351–368. Springer-Verlag, Berlin.

Woodard, C., Huang, T., San, H., Helfand, L., and Carlson, J. (1989). Genetic analysis of olfactory behavior in *Drosophila:* a new screen yields the *ota* mutants. *Genetics* **123**, 315–326.

Woodard, C., Alcorta, E., and Carlson, J. (1992). The *rdgB* gene of *Drosophila:* a link between vision and olfaction. *J. Neurogenet.* **8**, 17–32.

Young, M. W. (1992). *Drosophila's* biological clocks. *Discuss. Neurosci.* **8**, 93–98.

Yu, Q., Jacquier, A. C., Citri, Y., and Colot, H. M. (1987a). Molecular mapping of point mutations in the *period* gene that stop or speed up biological clocks in *Drosophila melanogaster. Proc. Natl. Acad. Sci. U.S.A.* **84**, 784–788.

Yu, Q., Colot, H. V., Kyriacou, C. P., Hall, J. C., and Rosbash, M. (1987b). Behaviour modification by *in vitro* mutagenesis of a variable region within the *period* gene of *Drosophila. Nature (London)* **326**, 765–769.

Zawistowski, S., and Richmond, R. C. (1986). Inhibition of courtship and mating of *Drosophila melanogaster* by the male produced lipid, cis-vaccenyl acetate. *J. Insect Physiol.* **32**, 189–192.

Zerr, D. N., Rosbash, M., Hall, J. C., and Siwicki, K. K. (1990). Circadian rhythms of period protein immunoreactivity in the CNS and the visual system of *Drosophila. J. Neurosci.* **10**, 2749–2762.

Zhong, Y., and Wu, C. F. (1991). Altered synaptic plasticity in *Drosophila* memory mutants with a defective cyclic AMP cascade. *Science* **251**, 198–201.

Zweibel, L. J., Hardin, P. E., Liu, X., Hall, J. C., and Rosbash, M. (1991). A post-transcriptional mechanism contributes to circadian cycling of a per-β-galactosidase fusion protein. *Proc. Natl. Acad. Sci. U.S.A.* **88**, 3882–3886.

GENETICS AND MOLECULAR GENETICS OF SULFUR ASSIMILATION IN THE FUNGI

George A. Marzluf

Department of Biochemistry, The Ohio State University, Columbus, Ohio 43210

I. Introduction

In the filamentous fungi, complex regulatory circuits control entire areas of metabolism; for example, the expression of catabolic enzymes involved in nitrogen, carbon, phosphorus, or sulfur metabolism, in each case, is controlled by a set of regulatory genes and specific metabolic repressors or inducers. Both positive-acting and negative-acting regulatory genes mediate the global controls which integrate the biochemical pathways within each of these major areas of metabolism. Various sulfur compounds, especially the sulfur-containing amino

187

acids, methionine and cysteine, are required for the growth and activation of all cells, whether prokaryotic or eukarytoic, plant or animal.
The sulfur-containing amino acid methionine initiates the synthesis
of nearly all proteins in all organisms; cysteine plays a critical role
in the structure, stability, and catalytic function of many proteins,
whereas S-adenosylmethionine plays a pivotal role in methyl group
transfer and in polyamine biosynthesis. Sulfur is an essential component of many other biologically important compounds or macromolecules. Thus, it is not surprising that considerable enzymatic machinery
and regulatory circuitry is devoted to ensuring that cells secure and
assimilate a steady supply of sulfur.

In the case of eukaryotic organisms, the most complete understanding of the assimilation of sulfur and its regulation is with filamentous
fungi, particularly Neurospora crassa, which was utilized in the pioneering studies of Robert Metzenberg and his colleagues (Metzenberg
and Ahlgren, 1971; Metzenberg et al., 1964; Metzenberg and Parson,
1966). The sulfur source most frequently encountered, inorganic sulfate, is utilized by a well-defined assimilatory pathway in Neurospora,
Aspergillus, and other fungi as well as by plants (Fig. 1). After transport into the cells, inorganic sulfate is first activated via ATP in two
enzymatic steps to generate 3'-phosphoadenosine-5'-phosphosulfate
(PAPS), reduced to sulfite, and then to sulfide, which is condensed with
0-acetylserine to generate cysteine. Many of the enzymes of the pathway were studied extensively two decades ago by Martin Flavin and
his colleagues at the National Institutes of Health (Nagai and Flavin,
1967; Flavin and Slaughter, 1964, 1967). Sulfite reductase is a particularly interesting and very complex enzyme; it contains an iron-sulfur
cluster and the very unusual cofactor, siroheme, which is also found in
nitrite reductase (Exley et al., 1993).

Four conserved cysteine residues are believed to be involved in binding the iron-sulfur cubane and siroheme prosthetic groups in nitrite
reductases (Exley et al., 1993). The N. crassa and A. nidulans structural genes for nitrite reductase have been cloned and sequenced
(Johnstone et al., 1990; Exley et al., 1993), but the gene(s) encoding
sulfite reductase has not yet been isolated for any filamentous fungus.
Cysteine is incorporated into proteins and is also a precursor for the
synthesis of methionine. Moreover, unlike the situation in bacteria, in
Neurospora and Aspergillus this pathway is fully reversible owing to
the presence of enzymes required for reverse trans-sulfuration, and
methionine or homocysteine represent excellent sulfur sources for
Neurospora (Flavin and and Slaughter, 1964, 1967). Mutants of Neurospora are available which define the enzymatic steps of the sulfur
assimilatory pathway (Fig. 1). A similar set of mutants has been iden-

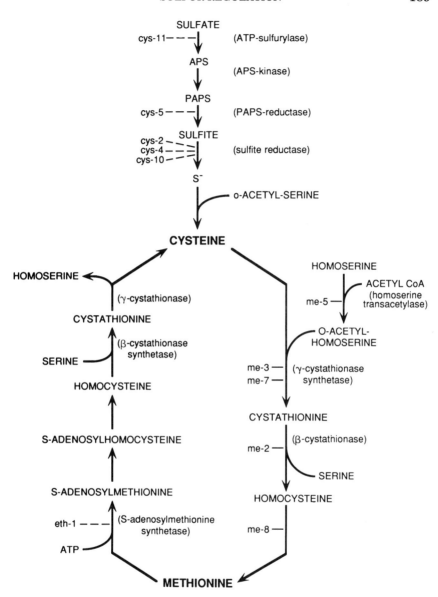

FIG. 1. The sulfur assimilatory pathway. The pathway leading from inorganic sulfate that gives rise to cysteine and methionine is shown, and mutants which affect specific steps in *Neurospora crassa* are identified. Several key enzymes of the pathway are shown in parenthesis.

tified in *Aspergillus nidulans* (Arst, 1968; Paszewski and Grabski, 1974; Natorff *et al.*, 1993).

Sulfur repression by methionine appears to highly regulate the expression of some of the enzymes of this assimilatory pathway but does not affect the synthesis of others. Thus, the amount of Γ-cystathionase increased 30-fold upon sulfur limitation whereas the levels of homoserine transacetylase, β-cystathionase, cystathionine-β-synthase, and cystathionine-Γ-synthase remained unchanged (Flavin and Slaughter, 1967). Similarly, derepression led to approximately a 5-fold increase in sulfite reductase (Metzenberg and Parson, 1966).

II. The Sulfur Regulatory Circuit

A. DESIGN AND PHYSIOLOGICAL BASIS OF THE CIRCUIT

The sulfur regulatory circuit of *Neurospora crassa* operates to ensure that favored sources of sulfur are utilized. Control occurs primarily at the level of entry of various potential sulfur sources into the assimilatory pathway. *Neurospora* preferentially utilizes certain favored sulfur sources, such as methionine. However, this fungus can use various secondary sulfur sources, for example, inorganic sulfate, choline-0-sulfate, aromatic sulfates, glucose-6-sulfate, cysteic acid, or proteins, but their utilization requires *de novo* synthesis of various sulfur catabolic enzymes. A set of unlinked but coregulated structural genes specifies these sulfur catabolic enzymes. Synthesis of this entire family of enzymes, which includes two distinct sulfate permease species, a methionine-specific permease, choline sulfate permease, choline sulfatase, an aromatic sulfate permease, aryl sulfatase, and an extracellular alkaline protease, only occurs during conditions of sulfur limitation (Hanson and Marzluf, 1975; Marzluf, 1970; Marzluf and Metzenberg, 1968; Pall, 1971).

B. GENETIC CONTROLS

The expression of these sulfur catabolic enzymes is controlled by three regulatory genes, *scon-1, scon-2,* and *cys-3* (Marzluf and Metzenberg, 1968; Burton and Metzenberg, 1972; Paietta, 1990). *Cys-3* mutants are completely deficient in the entire set of catabolic enzymes and fail to grow on any of the various secondary sulfur sources. The *cys-3* + gene encodes a trans-acting regulatory protein which acts in a positive way to turn on the expression of the sulfur catabolic enzymes

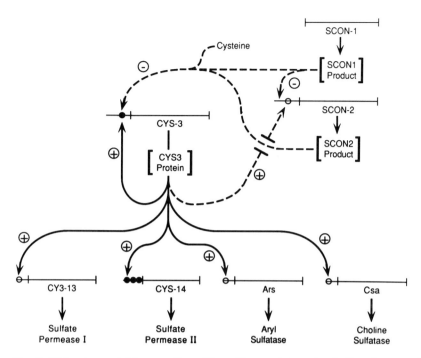

FIG. 2. Molecular model for operation of the sulfur regulatory circuit. Regulatory and representative structural genes are shown. The exact function of the *scon-1* and *scon-2* genes is unknown but it is suspected that they code for products which negatively control the expression of the *cys-3* regulatory gene. *Cys-3* is expressed only when the cells become limited for sulfur. CYS3 is a trans-acting DNA-binding protein which increases its own expression via autogenous control and also turns on *cys-14, ars,* and other structural genes of the circuit. Three CYS3 binding sites in the 5' promoter region of *cys-14* and a single duplex site upstream of *cys-3* have been identified and are shown as filled circles; open circles represent predicted CYS3 sites in other genes. Dashed lines identify possible control signals which have not yet been established. +, positive control; −, negative control.

(Fig. 2). In contrast, both *scon-1* and *scon-2* appear to act in a negative fashion, and mutants of either of them show constitutive synthesis of the sulfur catabolic enzymes.

C. STRUCTURAL GENES OF THE CIRCUIT

Several genes of the sulfur circuit of *N. crassa* have recently been cloned, including two structural genes, *cys-14,* which encodes sulfate permease II, and *ars,* which encodes aryl sulfatase, as well as the positive-acting *cys-3* and the negative-acting *scon-2* regulatory genes

(Paietta *et al.*, 1987; Ketter and Marzluf, 1988; Fu *et al.*, 1989; Fu and Marzluf, 1990; Paietta, 1989, 1990). The cellular content of *ars* mRNA and of *cys-14* mRNA is highly regulated by sulfur repression and by the *cys-3* and *scon-1* control genes. Both of these mRNAs are present in the wild type only upon sulfur limitation. These two mRNAs are both expressed in a constitutive fashion in *scon-1* mutants, and are missing entirely in *cys-3* mutants. Nuclear run-on experiments demonstrated that *ars* expression is controlled at the level of initiation of transcription (Paietta, 1989). Thus, it appears that *cys-14* and *ars,* and presumably each of the other structural genes of the sulfur circuit, are subject to transcriptional control.

The genes which encode ATP-sulfurylase and PAPS-reductase of both *Aspergillus nidulans* and *Penicillium chrysogenum* have recently been cloned (G. Turner, unpublished observations). These two structural genes are adjacent in both of these fungal species; the entire nucleotide sequence of the *A. nidulans* PAPS-reductase gene has been determined, and that of the *A. nidulans* ATP sulfurylase gene is nearly completed (G. Turner, unpublished observations). It will be of considerable interest to determine whether these two related, closely linked genes are controlled by a common promoter region or are simply adjacent to one another but independently regulated.

III. Aryl Sulfatase and Utilization of Aromatic Sulfates

Aromatic sulfate esters, for example, tyrosine-0-sulfate, serve as secondary sulfur sources for *Neurospora* (Fig. 3). Tyrosine-0-sulfate is transported into *Neurospora* cells as an intact molecule, followed by its intracellular hydrolysis by aryl sulfatase, to yield internal pools of inorganic sulfate and tyrosine (D. Plock and G. A. Marzluf, unpublished observations). Mutant strains incapable of the transport of any inorganic sulfate nevertheless transport exogenous tyrosine-0-sulfate at a wild-type rate and use this aromatic sulfate ester as an excellent sulfur source. Tyrosine-0-sulfate accumulates within the cells of the *ars* mutant, which, however, cannot use it as a source of sulfur or tyrosine. Synthesis of the uptake system for tyrosine-0-sulfate is strongly repressed by methionine and is completely missing in *cys-3* mutant strains (D. Plock and G. A. Marzluf, unpublished observations). These features imply that a transport system for aromatic sulfates is a member of the sulfur regulatory circuit of *Neurospora*.

The enzyme aryl sulfatase has been the focus of much attention in

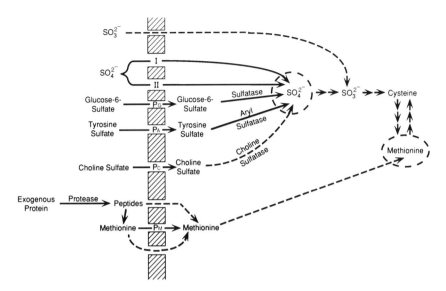

FIG. 3. Uptake and utilization of environmental sulfur sources. The assimilatory pathway is abbreviated; the transport and metabolism of various sulfur sources are shown. I, II, P_G, P_A, P_C, and P_M represent specific sulfur-controlled permeases which transport the respective metabolites as shown. Methionine is transported by both a specific sulfur-controlled permease, P_M, and also by a general amino transport system(s). An extracellular enzyme, the protease, and intracellular sulfur catabolic enzymes, such as aryl sulfatase and choline sulfatase, are also subject to a high degree of sulfur control.

studies of sulfur control because it is stable and extremely easy to assay; moreover, its activity increases approximately 1000-fold upon sulfur derepression (Metzenberg and Parson, 1966). Mutants of the aryl sulfatase structural gene, *ars,* have been isolated which completely lack this enzyme (Metzenberg and Ahlgren, 1970). Other *N. crassa* strains, cleverly obtained by introgression of an *ars* gene from *N. tetrasperma,* possess aryl sulfatase with an altered electrophoretic mobility (Metzenberg and Ahlgren, 1971). Metzenberg and Ahlgren (1971) used these electrophoretic variants of aryl sulfatase to provide an unequivocal demonstration that the *cys-3*$^+$ gene acts in a positive fashion. In *cys-3*$^+$/*-3*$^-$ heterokaryons, the wild-type *cys-3* gene not only turned on the *ars* gene present in the same nucleus but also activated an *ars* gene that encoded an electrophoretic variant of the enzyme in nuclei containing the *cys-3* mutant gene.

Electrophoretic variants of aryl sulfatase were similarly employed to examine the characteristics of the *scon-1*$^+$ gene. The *scon-1* mutation leads to the constitutive expression of aryl sulfatase, choline sulfatase, sulfatase permease, and other sulfur-controlled enzymes, that is, the synthesis of these enzymes in scon-1 mutants is insensitive to high levels of methionine, which strongly represses their synthesis in the wild type (Burton and Metzenberg, 1972). Remarkably, the effect of scon-1 upon ars gene expression was found to be nucleus limited in heterokaryons, suggesting that the scon-1 gene product was maintained within, or, at least, in very close proximity to the nucleus in which it was encoded (Burton and Metzenberg, 1972). The nature of the scon-1 gene product and its mode of action remain unknown. Molecular cloning of scon-1 would be an invaluable step to understanding this fascinating gene.

Paietta (1989) cloned the ars+ gene by chromosome walking. This gene is transcribed to give a 2.3-kb mRNA whose content parallels that of aryl sulfatase enzyme activity upon derepression of wild-type cells. The cys-3 mutant lacks any detectable ars mRNA and enzyme activity, whereas the scon-1 mutant displays constitutive expression of both the messenger RNA and enzyme (Paietta, 1989). Neither the nucleotide sequence of the ars gene nor the amino acid sequence of aryl sulfatase have yet been reported.

IV. Dual Sulfate Permeases

In N. crassa, two unlinked genes, cys-13 and cys-14, appear to specify distinct sulfate permease species which can be readily distinguished by significantly different K_m values for the sulfate uptake (Marzluf, 1970). Either single mutant strain grows readily on inorganic sulfate and retains one sulfate uptake system. However, the cys-13 cys-14 double mutant lacks all sulfate transport activity and cannot utilize sulfate. Both of these sulfate permease species are subject to sulfur catabolic repression and are controlled by the cys-3+ gene; cys-3 mutants lack all capacity for sulfate transport. Moreover, the dual sulfate permeases are developmentally regulated, permease I predominating in conidiospores, whereas permease II is the major form in mycelia (Marzluf, 1970). Sulfate uptake is energy dependent but does not require the conversion of the sulfate ion to adenosine phosphosulfate (APS); cys-11 mutants, which lack ATP sulfurylase, transport sulfate at a wild-type rate (Marzluf, 1974). The sulfate permeases are regulated by dynamic turnover, with a functional half-life of approximately 2 hr, whereas the enzyme aryl sulfatase is very stable (Marzluf, 1972). Upon sulfur derepression, significant accumulation of cys-14 mRNA requires about 2 hr, indicating that this gene is only slowly turned on

(Jarai and Marzluf, 1991); in contrast, *cys-14* mRNA turns over relatively rapidly, with a half-life of approximately 15 min (Ketter *et al.*, 1991).

The *cys-14*⁺ gene, which encodes sulfate permease II, has been cloned and conceptual translation of its nucleotide sequence indicates that it encodes a protein of approximately 90 kDa (Ketter *et al.*, 1991). The amino acid sequence of CYS14 reveals 12 highly hydrophobic helical regions, which appear to represent membrane-spanning domains. Antibodies specific for CYS14 were employed to demonstrate that this protein is indeed found in *Neurospora* and is localized within the plasma membrane fraction (Jarai and Marzluf, 1991). As expected, *cys-3* mutants lack any detectable CYS14. In wild types, the cellular content of CYS14 is highly regulated by *de novo* synthesis and turnover, which is in good agreement with the kinetics of appearance and disappearance of sulfate transport activity. These features thus strongly suggest that CYS14 functions as a membrane-bound sulfate transporter. The utilization of inorganic sulfate is highly regulated at the level of its entry into the cell by controlling the transcription of the sulfate permease genes and limited mRNA stability, and also in the dynamic turnover of both sulfate permease species, which have a half-life of approximately 2 hr (Marzluf, 1972).

V. Choline-O-Sulfate Utilization

Choline-O-sulfate occurs widely throughout the plant kingdom and is stored in many fungi; its synthesis involves the transfer of sulfate from 3′-phosphoadenosine-5′-phosphosulfate to choline via the enzyme PAPS-choline sulfotransferase. *Neurospora crassa* utilizes choline-O-sulfate as an excellent secondary sulfur source. Exogenous choline-O-sulfate is transported as an intact molecule via a specific permease into the cells, within which it is hydrolyzed by choline sulfatase to yield an internal pool of inorganic sulfate (Fig. 3). As expected, *cys-13 cys-14* double mutant cells, which cannot transport inorganic sulfate or utilize it for growth, transport choline-O-sulfate at a wild-type rate and readily use it as a sole sulfur source for growth. Both choline-O-sulfate permease and choline sulfatase are subject to a high degree of genetic and metabolic control as members of the sulfur regulatory circuit. The expression of choline-O-sulfate permease and of choline sulfatase is strongly repressed by methionine in wild types, but is synthesized in a constitutive fashion in *scon^c*, whereas both activities are completely missing under all growth conditions in *cys-3* mutants. Arst (1971) isolated mutants of *A. nidulans* which cannot utilize exogenous choline-*0*-sulfate for growth but which have no other apparent pleiotropic effects. The characteristics of these *csu* mutants suggest that they are deficient in choline sulfatase, choline sulfate permease, or both.

VI. Other Secondary Sulfur Metabolites

A. ALTERNATIVE SULFUR SOURCES

N. crassa and other filamentous fungi are able to utilize many other compounds as sole sulfur sources, and the use of most of these requires the synthesis of one or more enzymes or permeases. One particularly intriguing case is the use of an external protein as a sulfur source. Under conditions of sulfur limitation where a protein is available, *cys-3* + strains, but not *cys-3* mutants, synthesize and secrete an alkaline protease into the growth medium (Hanson and Marzluf, 1975). This same extracellular protease can be turned on in an independent fashion, even in *cys-3* mutants, when the cells are instead limited for nitrogen or for carbon (Hanson and Marzluf, 1975). Peptides released from the exogenous protein by the extracellular protease are taken into the cells and their cysteine and methionine residues used to satisfy the growth requirements of the *Neurospora* cells. Other sulfur-containing compounds, for example, glucose-6-sulfate, cysteic acid, and taurine can be used as sole sulfur sources for growth of *Neurospora* and other fungi.

B. METABOLIC REPRESSORS

The exact identity of the metabolite which exerts sulfur catabolite repression in *Neurospora* has never been firmly established. Growth of wild-type *Neurospora* cells with adequate amounts of either inorganic sulfate or methionine leads to full repression of aryl sulfatase and many other related sulfur catabolic enzymes. Yet neither of these compounds represents the actual repressor metabolite. Mutants which lack ATP-sulfurylase can accumulate the sulfate ion but cannot assimilate it and do not exhibit sulfur catabolite repression. Jacobson and Metzenberg (1977) demonstrated that aryl sulfatase synthesis was not repressed by methionine in a leaky serine autotroph unless this strain was provided with exogenous serine, when repression was complete. This result implies that methionine itself is not the repressor and strongly suggests that cysteine, or a compound closely related to or derived from cysteine, represents the true repressing metabolite.

An irreparable, temperature-sensitive, ethionine-resistant mutant, *eth-1^R*, of *Neurospora* was found to be insensitive to repression by methionine of aryl sulfatase (Metzenberg *et al.*, 1964). Selenomethionine resistance also arises by mutation at the *eth-1* locus (Chen and Metzenberg, 1974). The *eth-1* gene appears to encode S-adenosylmethionine synthetase, which catalyzes the first step in the conversion of methionine to cysteine (Chen and Metzenberg, 1974). These results also indicate that methionine itself exerts sulfur repression but requires activity of the reverse transsulfuration, that is, conversion to cysteine. The *eth-1^r* mutant had another property of special interest, namely, it is osmotic-remedial. All of the pleiotropic defects associated with the *eth-1^r* mutation could be phenotypically repaired by simply raising the osmolarity of the growth medium (Metzenberg, 1968).

VII. The Sulfur Regulatory Genes

A. POSITIVE AND NEGATIVE CONTROL GENES

At least three distinct regulatory genes, *scon-1*, *scon-2*, and *cys-3*, appear to control the expression of the structural genes which specify sulfur catabolic enzymes (Marzluf and Metzenberg, 1968; Burton and Metzenberg, 1972; Paietta, 1990). Both *scon-1* and *scon-2* appear to act in a negative fashion, since mutation of either leads to constitutive expression of aryl sulfatase, sulfate permease, and other members of the sulfur circuit. They may be part of a sequential control network and act to control the *cys-3^+* regulatory gene, which in turn, directly controls expression of the structural genes. *Cys-3^+* transcripts are constitutively expressed during conditions of either sulfur excess or sulfur limitation in both *scon-1* and *scon-2* mutant strains. Very little molecular detail is known concerning *scon-1*. The *scon-2* gene has been isolated and partially characterized although its nucleotide sequence has not yet been reported nor has that of a potential encoded protein (Paietta, 1990). Interestingly, a 2.6-kb *scon-2^+* transcript was present in wild-type strains only during sulfur-limited growth conditions, but was constitutive in *scon-1* mutant strains. Moreover, *scon-2* transcripts were present in reduced amounts in the *cys-3* mutant, which suggests a complex feedback loop in which *scon-2* controls *cys-3* which, in turn, along with *scon-1*, also regulates *scon-2* (see Fig. 2). It is unknown whether such regulation occurs by promoter activation or inhibition or by some type of direct protein–protein interaction. It will be interesting to determine whether the promoter of the *scon-2* gene contains CYS3 protein binding sites and whether the *scon-1* gene is regulated by *cys-3*, *scon-2*, or in any other fashion.

Natorff *et al.* (1993) have recently reported identification in *Asper-*

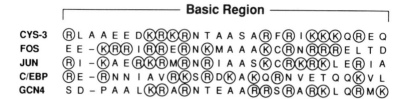

FIG. 4. The bzip region of CYS3 and other trans-acting regulatory proteins. The leu-cine zipper and adjacent amino-terminal basic region of these regulatory proteins appear to constitute a bipartite DNA binding domain. The basic region contacts DNA directly whereas the leucine zipper is responsible for dimerization of these proteins. Methionine instead of leucine is found in one zipper position of CYS3; a mutant CYS3 protein that has methionine in all zipper positions retains significant function (see text).

gillus nidulans of four distinct negative-acting sulfur control genes, *sconA, sconB, sconC,* and *sconD;* the characteristics of mutants of these genes are similar to those of *scon-1* and *scon-2* mutants of *N. crassa.* It will be important to obtain unequivocal evidence which con-vincingly demonstrates that each of these four *scon* genes actually has a regulatory function and to determine the gene product of each of these putative control genes. No positive-acting regulatory gene ho-mologous to *cys-3* has yet been found in *A. nidulans.*

B. THE *CYS-3* REGULATORY GENE

The *cys-3*⁺ gene of *N. crassa* encodes a positive-acting regulatory protein which turns on the expression of *cys-14* and *ars* and pre-sumably each of the other unlinked but coregulated structural genes. The *cys-3* null mutants fail to express any *cys-14* or *ars* mRNA, and lack all sulfate permease and aryl sulfatase activity. Furthermore, temperature-sensitive *cys-3* mutants possess mRNAs and enzyme ac-tivities at the permissive temperature but not at the conditional tem-perature. The protein CYS3 is a member of the bzip class of DNA-binding proteins (Fig. 4) which includes the yeast protein GCN4 and the mammalian proteins C/EBP-1, FOS, and JUN (Fu and Marzluf, 1990; Fu *et al.,* 1989; Kanaan and Marzluf, 1991; Kanaan *et al.,* 1992). The CYS3 DNA-binding domain is bipartite, consisting of a leucine zipper which functions in the dimerization of two CYS3 monomers plus an adjacent upstream basic region which makes direct contact with DNA (Fu and Marzluf, 1990; Kanaan and Marzluf, 1991; Kanaan *et al.,* 1992).

Chemical cross-linking studies have shown that CYS3 exists as a dimer in the presence or absence of DNA (Kanaan *et al.,* 1992). A trun-

┌─────────────────────── Leucine Zipper ───────────────────────┐

A ⌈L⌉ E K S A K E ⌈M⌉ S E K V T Q ⌈L⌉ E G R I Q A ⌈L⌉ E T E N K W ⌈L⌉
T ⌊L⌋ Q A E T D Q ⌊L⌋ E D E K S A ⌊L⌋ Q T E I A N ⌊L⌋ L K E K E K ⌊L⌋
Ⓡ ⌊L⌋ E E K V K T ⌊L⌋ K A Q N S E ⌊L⌋ A S T A N M ⌊L⌋ R E Q V A Q ⌊L⌋
E ⌊L⌋ T S D N D R ⌊L⌋ R K R V E Q ⌊L⌋ S R E L D T ⌊L⌋ R G I F R Q ⌊L⌋
Q ⌊L⌋ E D K V E E ⌊L⌋ L S K N Y H ⌊L⌋ E N E V A R ⌊L⌋ K K L V G E R

FIG. 4. —*Continued*

cated CYS3 protein, consisting only of the bzip region, is also capable
of dimer formation and DNA binding. The CYS3 dimers appear to be
quite stable and not to dissociate and exchange subunits to a detectable extent, even when subjected to freezing and thawing. However,
when coexpressed *in vivo* or *in vitro,* hybrid dimers, consisting of one
full-length monomer and one truncated CYS3 monomer, are formed
and are active in DNA binding (Kanaan and Marzluf, 1991; Kanaan *et
al.,* 1992).

C. DNA BINDING BY THE CYS3 PROTEIN

The CYS3 protein expressed in *E. coli* has been used to examine
DNA binding with gel shift and with DNA footprint experiments (Fu
and Marzluf, 1990). CYS3 binds *in vitro* to three distinct sites upstream of *cys-14,* which encodes sulfate permease II, and to a single
site in the 5′ promoter region of the *cys-3* gene itself. The truncated
CYS3 protein which contains primarily the bzip domain binds to these
same sites with the same specificity and affinity as the full-length protein. Thus, DNA recognition is determined entirely by the bzip domain
and does not require other regions of the CYS3 protein (Kanaan and
Marzluf, 1991).

The three CYS3 binding sites upstream of *cys-14* and the single site
upstream of *cys-3* have very limited sequence homology. Each site contains two or more repeats of a CAT sequence, which may provide a
limited dyad symmetry and represent the central core of a CYS3 binding site. The proximal CYS3 binding site upstream of *cys-14* has two
central CAT elements, and alteration of either of these sequences results in approximately an 80% loss in recognition by the CYS3 protein
(Ketter *et al.,* 1991). Since CYS3 is responsible for turning on an entire
family of coregulated genes, presumably with different kinetics and to
different extents, it is perhaps not surprising that CYS3 recognizes
different sequence elements. The single binding site in the promoter of
the *cys-3* gene appears to be responsible for autogenous control and

actually is apparently a duplex site which is bound by CYS3 with high affinity (Fu and Marzluf, 1990).

D. MUTATION STUDIES OF CYS3

In a *cys-3* null mutant, two basic amino acids within the DNA binding domain of the CYS3 protein were replaced by glutamine (Fu *et al.*, 1989). This mutant protein, expressed in *E. coli*, was incapable of DNA binding (Fu and Marzluf, 1990). A temperature-sensitive *cys-3* mutant encodes a protein which has a glutamine in place of an arginine residue in the bzip motif; this mutant CYS3 protein displays temperature-dependent DNA binding *in vitro*.

Several conserved amino acids in the bzip domain of CYS3 have been changed by site-directed mutagenesis to determine whether they are essential for function *in vivo* or for CYS3 dimerization or DNA binding *in vitro*. Replacement of basic amino acids within the positively charged region upstream of the zipper by glutamine abolished *cys-3* function *in vivo* (Kanaan and Marzluf, 1991). These mutant CYS3 proteins formed dimers but were completely deficient in DNA binding. These results support the concept that the basic region constitutes the DNA contact surface and that substitution of even one critical amino acid in this region entirely eliminates CYS3 DNA-binding activity.

Immediately downstream of the basic region, the CYS3 protein contains a heptad repeat motif, the hallmark of a "leucine zipper," an alpha helical motif with an extended hydrophobic surface which allows dimer formation. The effect of amino acid substitutions within this putative CYS3 zipper structure depends upon the position within the zipper and the nature of the substitutions; some replacements prevent dimer formation and thus totally eliminate DNA binding *in vitro* and *cys-3* function *in vivo*, whereas other amino acid substitutions resulted in functional CYS3 proteins (Kanaan and Marzluf, 1991).

The introduction of valine at any two positions within the zipper led to a nonfunctional CYS3 protein, and a single charged amino acid at any zipper position and even certain single valine substitutions eliminated CYS3 function. In contrast, CYS3 proteins with as many as three methionine residues in zipper positions possessed full *cys-3* activity. Surprisingly, a CYS3 protein which had a pure "methionine zipper" was functional *in vivo* and displayed significant, although reduced, dimer formation and DNA binding *in vitro*. Thus, although leucine, with its extended hydrophobic side chain, appears to be a particularly functional residue in a so-called "leucine zipper," certain other amino acids, for example, methionine or isoleucine, can also function in a zipper motif, thus explaining the natural occurrence of methionine in a zipper position of CYS3 (Fig. 4).

E. CONTROL OF THE cys-3 REGULATORY GENE

The cys-3 regulatory gene is itself highly regulated by sulfur catabolite repression and by at least two other unlinked control genes. The cys-3 gene is only barely expressed when the cells have a sufficient sulfur source; cys-3 expression is greatly increased during conditions of sulfur limitation (Fu et al., 1989). The products of the two "sulfur controller" genes, scon-1 + and scon-2 +, appear together to negatively control cys-3, possibly at the level of transcription, when an adequate cellular supply of sulfur is available. When the cys-3 gene is expressed by way of a heterologous promoter in cells subject to repressing levels of sulfur, the ars gene is turned on, implying that the positive function of the CYS3 protein is not prevented via a protein–protein interaction with a hypothetical SCON protein (Paietta, 1992). This result supports the suggestion that cys-3 is controlled primarily at the level of transcription, and that either an SCON1 and SCON2 protein, or even an SCON1–SCON2 heteromultimer, might be the element that recognizes the repressing sulfur metabolite.

F. AUTOGENOUS REGULATION OF CYS3

It seems virtually certain that the CYS3 protein autogenously regulates its own expression in a positive fashion. It appears that when the cys-3 + gene is turned on by sulfur limitation, the resulting CYS3 protein binds at an element in the promoter of its own structural gene and strongly enhances its own expression (Fu and Marzluf, 1990; Fu et al., 1989). Mobility shift and DNA footprinting experiments have demonstrated that CYS3 binds in vitro with high affinity to this promoter element. Moreover, cys-3 missense mutants which encode a nonfunctional CYS3 protein are greatly deficient in synthesis of both cys-3 mRNA and CYS3 protein. This result can best be interpreted to indicate that the mutant CYS3 protein cannot activate its own expression, which supports the concept of positive autogenous control.

G. CYS3: A NUCLEAR PROTEIN

Western blotting analyses conducted with anti-CYS3 polyclonal antibodies have demonstrated that nuclear extracts derived from cys-3 + cells grown under limiting sulfur conditions possess a protein of the

size predicted for CYS3 (Kanaan and Marzluf, 1993). Moreover, mobility shift experiments conducted with the wild-type nuclear extract
revealed a specific DNA–protein complex at the same position as obtained with the CYS3 protein expressed in *E. coli.* Nuclear extracts
prepared from *cys-3* ⁺ cells which were instead grown under sulfur excess conditions lack this protein. Moreover, nuclear extracts of a *cys-3*
missense mutant lack any detectable CYS3 protein (Kanaan and Marzluf, 1993). Thus, these results reveal that CYS3 is a nuclear DNA-
binding protein and also provide additional evidence for the concept
that *cys-3* expression is subject to positive autogenous regulation. The
mutant CYS3 protein is presumably not expressed in *Neurospora* at a
detectable level because it cannot turn on its own synthesis. However,
the possibility that the mutant CYS3 protein is unstable and degraded
in vivo has not yet been excluded.

H. MOLECULAR DISSECTION OF THE CYS3 PROTEIN

The CYS3 protein consists of 236 amino acid residues and has a
number of potentially important regions in addition to the bzip DNA
binding domain. The CYS3 protein has several distinct regions which
are highly enriched in one or more amino acids, a feature that has been
observed with other regulatory proteins. A threonine-serine-rich segment is found near the amino terminus of CYS3, which is followed by
a proline-rich domain and a short acidic region, then the bzip DNA-
binding domain, and finally, an alanine-rich segment. Any of these
regions might be important for proper folding, stability, or function of
the CYS3 protein in trans-activation or in recognition of sulfur catabolite signals.

Various deleted forms of the *cys-3* gene were constructed to identify
regions of the protein which were important for CYS3 function *in vivo;*
the function of the various mutated genes was analyzed by transforming each of them into a *cys-3* mutant strain, and conducting aryl sulfatase assays on the transformants. A CYS3 protein lacking all residues C-terminal to the bzip domain, including the alanine-rich region,
retains partial function *in vivo.* All deletions examined which removed
segments of CYS3 amino terminal to the bzip region totally eliminated
cys-3 function, suggesting that the structural integrity of this entire
region may be important for proper function of the activation domains
or for stability of the CYS3 protein. Alanine saturation mutagenesis,
used to identify functionally important amino acids in this region of
CYS3, demonstrated that the acidic region is essential but only as a
structural or spacer domain; its acidic nature is not important for

CYS3 function. The proline-rich region, which contains three closely spaced pairs of histidine residues, appears to constitute a CYS3 activation domain. The substitution of alanine for different proline residues in this region completely abolished CYS3 function, and alanine replacements of the histidine residues reduced the activation of aryl sulfatase by approximately 90% (Kanaan and Marzluf, 1993). In sharp contrast, alanine substitutions for 6 threonine or serine residues near the amino terminus of CYS3 resulted in approximately a 10-fold increase in gene activation, perhaps suggesting that phosphorylation can modulate the activity of the CYS3 protein.

VIII. Sulfur Regulation in Yeast

In the yeast *Saccharomyces cerevisiae,* the expression of the entire set of genes which encode enzymes required for sulfate assimilation requires the lifting of sulfur catabolite repression and a functional *met-4⁺* gene; *met-4* mutants lack the entire set of enzymes required for assimilation of inorganic sulfate. MET4 is a positive-acting regulatory protein and its DNA-binding domain, which is composed of a bzip leucine zipper motif, shows striking homology to that of the *Neurospora* CYS3 protein. MET4 binds to recognition elements with the sequence 5′-TGGCAAATG-3′, which resembles the core sequences responsible for CYS3 binding (Thomas *et al.,* 1992).

A surprising finding is that transcription of some, but not all, of the sulfur-related structural genes of *S. cerevisiae* also require a second DNA-binding protein, known as centromere-binding factor 1 (CBF1). The expression of the *met-16* is reduced to an almost undetectable level in a *cbf1* mutant, although *met-3* expression is not affected. CBF1 appears to be a member of a class of multifunctional DNA-binding proteins of yeast which bind to elements found throughout the genome, such as centromeres, telemeres, and replication origins. CBF1 binds to a sequence element present in all yeast centromeres and is required for proper chromosome segregation.

Why then is CFB1 implicated in the expression of certain sulfur-controlled structural genes? Thomas *et al.* (1992) have proposed the interesting possibility that CBF1 plays a cooperative role in transcription activation by increasing the accessibility of promoter regions to MET4, perhaps by preventing nucleosome formation. It is intriguing that *met-3*, which is fully activated by MET4 in the complete absence of CBF1, contains a long poly(AT) tract near its TATA box. This poly(AT) tract may exclude nucleosomes, thus opening the *met-3* promoter for the positive-acting MET4 protein, without the need for a supporting role by CBF1. These features suggest that the *met-3* promoter may possess constitutive hypersensitive sites, whereas CBF1 may be required for the development of such hypersensitive sites in *met-16*.

IX. The Sulfur Circuit: Some Answers, More Questions

Many intriguing questions concerning the sulfur regulatory circuit of fungi remain unanswered. These include the precise identification of the metabolite which mediates sulfur repression and the manner in which this molecular signal is detected and transmitted to control the expression of the *cys-3* gene. It seems plausible that upon binding of the repressor metabolite, the product of one or both of the negative-acting *scon* genes is activated and then represses *cys-3* expression. It is also not known whether the *scon-1* and *scon-2* products act in concert or independently or even sequentially. Molecular cloning and characterization of *scon-1* and *scon-2* genes and determination of the nature and subcellular localization of their respective products will provide critical new information to help address these questions, including a necessary concern whether both *scon-1* and *scon-2* represent genuine regulatory genes.

When the fungal cells experience sulfur limitation, the negative control preventing *cys-3* expression is lifted and *cys-3* transcription is turned on. The resulting CYS3 protein, via positive autogenous control, apparently further increases *cys-3* expression, yielding an increasing pool of CYS3 protein which binds to recognition DNA elements in the promoters of the various unlinked structural genes, resulting in turning their expression from off to on. The exact nature of what constitutes a DNA binding element for CYS3 is still unknown.

Another point of interest is to understand how cells in which the sulfur circuit is fully activated due to the synthesis of a large pool of the CYS3 protein are able to turn off structural gene expression in a timely fashion if a sulfur excess is suddenly realized. Much new information will be required to address such questions. For example, it is important to determine the stability of the *cys-3* mRNA and protein and to determine whether the function of the CYS3 protein might be modulated by phosphorylation or some similar protein modification. Although many significant questions remain to be addressed, the recent breakthroughs in our understanding of the operation of the sulfur regulatory circuit represent truly major advances. The future promises to be equally rewarding.

ACKNOWLEDGMENTS

Research in the author's laboratory is supported by Grant GM-23367 from the National Institutes of Health. I thank Geoff Turner for sharing results prior to their publication.

REFERENCES

Arst, H. N. (1968). Genetic analysis of the first steps of sulphate metabolism in *Aspergillus nidulans*. *Nature* **219**, 268–270.

Arst, H. N. (1971). Mutants of *Aspergillus nidulans* unable to use cholin-0-sulfate. *Genet. Res.* **17**, 273–277.

Burton, E. G., and Metzenberg, R. L. (1972). Novel mutation causing derepression of several enzymes of sulfur metabolism in *Neurospora crassa*. *J. Bacteriol.* **109**, 140–151.

Chen, G. S., and Metzenberg, R. L. (1974). Isolation and properties of selenomethionine-resistant mutants of *Neurospora crassa*. *Genetics* **77**, 627–638.

Exley, G. E., Colandene, J. D., and Garrett, R. H. (1993). Molecular cloning, characterization, and nucleotide sequence of *nit-6*, the structural gene for nitrite reductase in *Neurospora crassa*. *J. Bacteriol.* **175**, 2379–2392.

Flavin, M., and Slaughter, C. (1964). Cystathionine cleavage enzymes of *Neurospora*. *J. Biol. Chem.* **239**, 2212–2219.

Flavin, M., and Slaughter, C. (1967). The derepression and function of enzymes of reverse trans-sulfuration in *Neurospora*. *Biochim. Biophys. Acta* **132**, 406–411.

Fu, Y. H., and Marzluf, G. A. (1990). *cys-3*, the positive-acting sulfur regulatory gene of *Neurospora crassa*, encodes a sequence-specific DNA binding protein. *J. Biol. Chem.* **265**, 11942–11947.

Fu, Y. H., Paietta, J. V., Mannix, D. G., and Marzluf, G. A. (1989). *Cys-3*, the positive-acting sulfur regulatory gene of *Neurospora crassa*, encodes a protein with a putative leucine zipper DNA-binding element. *Mol. Cell. Biol.* **9**, 1120–1127.

Hanson, M. A., and Marzluf, G. A. (1975). Control of the synthesis of a single enzyme by multiple regulatory circuits in *Neurospora crassa*. *Proc. Natl. Acad. Sci. U.S.A.* **72**, 1240–1244.

Jacobson, E. S., and Metzenberg, R. L. (1977). Control of arylsulfatase in a serine auxotroph of *Neurospora*. *J. Bacteriol.* **130**, 1397–1398.

Jarai, G., and Marzluf, G. A. (1991). Sulfate transport in *Neurospora crassa*: Regulation, turnover, and cellular localization of the CYS14 protein. *Biochemistry* **30**, 4768–4773.

Johnstone, I. L., McCabe, P. C., Greaves, P., Gurr, S. I., Cole, G. E., Brow, M. A. D., Unkles, S. E., Clutterbuck, A. J., Kinghorn, J. R., and Innis, M. A. (1990). The isolation and characterization of the crnA–niiA–niaD gene cluster for nitrate assimilation in the filamentous fungus *Aspergillus nidulans*. *Gene* **90**, 181–192.

Kanaan, M., and Marzluf, G. A. (1991). Mutational analysis of the DNA-binding domain of the CYS3 regulatory protein of *Neurospora crassa*. *Mol. Cell. Biol.* **11**, 4356-4362.

Kanaan, M., and Marzluf, G. A. (1993). The positive-acting sulfur regulatory protein CYS3 of *Neurospora crassa*: nuclear localization, autogenous control, and regions required for transcriptional activation. *Mol. Gen. Genet.* **239**, 334–344.

Kanaan, M., Fu, Y. H., and Marzluf, G. A. (1992). The DNA-binding domain of the CYS3 regulatory protein of *Neurospora crassa* is bipartite. *Biochemistry* **31**, 3197–3203.

206   GEORGE A. MARZLUF

Ketter, J. S., and Marzluf, G. A. (1988). Molecular cloning and analysis of the regulation of cys-14, a structural gene of the sulfur regulatory circuit of Neurospora crassa. Mol. Cell. Biol. 8, 1504–1508.
Ketter, J. S., Jarai, G., Fu, Y. H., and Marzluf, G. A. (1991). Nucleotide sequence, messenger RNA stability, and DNA recognition elements of cys-14, the structural gene for sulfate permease II in Neurospora crassa. Biochemistry 30, 1780–1787.
Marzluf, G. A. (1970). Genetic and biochemical studies of distinct sulfate permease species in different development stages of Neurospora crassa. Arch. Biochem. Biophys. 138, 254–263.
Marzluf, G. A. (1972). Control of the synthesis, activity and turnover of enzymes of sulfur metabolism in Neurospora crassa. Arch. Biochem. Biophys. 150, 714–724.
Marzluf, G. A. (1974). Uptake and efflux of sulfate in Neurospora crassa. Biochim. Biophys. Acta 339, 374–381.
Marzluf, G. A., and Metzenberg, R. L. (1968). Positive control by the cys-3 locus in regulation of sulfur metabolism in Neurospora. J. Mol. Biol. 33, 423–437.
Metzenberg, R. L. (1968). Repair of multiple defects of a regulatory mutant of Neurospora by high osmotic pressure and by reversion. Arch. Biochem. Biophys. 125, 532–541.
Metzenberg, R. L., and Ahlgren, S. K. (1970). Mutants of Neurospora deficient in aryl sulfatase. Genetics 64, 409–422.
Metzenberg, R. L., and Ahlgren, S. K. (1971). Structural and regulatory control of aryl sulfatase in Neurospora: The use of interspecific differences in structural genes. Genetics 68, 369–381.
Metzenberg, R. L., and Parson, J. W. (1966). Altered repression of some enzymes of sulfur utilization in a temperature-conditional lethal mutant of Neurospora. Proc. Natl. Acad. Sci. U.S.A. 55, 629–635.
Metzenberg, R. L., Kappy, M. S., and Parson, J. W. (1964). Irreparable mutations and ethionine resistance in Neurospora. Science 145, 1434–1435.
Nagai, S., and Flavin, M. (1967). Acetylhomoserine. An intermediate in the fungal biosynthesis of methionine. J. Biol. Chem. 242, 3884–3895.
Natorff, R., Balinska, M., and Paszewski, A. (1993). At least four regulatory genes control sulphur metabolite repression in Aspergillus nidulans. Mol. Gen. Genet. 238, 185–192.
Paietta, J. V. (1989). Molecular cloning and regulatory analysis of the arylsulfatase structural gene of Neurospora crassa. Mol. Cell. Biol. 9, 3630–3637.
Paietta, J. V. (1990). Molecular cloning and analysis of the scon-2 negative regulatory gene of Neurospora crassa. Mol. Cell. Biol. 10, 5207–5214.
Paietta, J. V. (1992). Production of the CYS3 regulatory, a bZIP DNA-binding protein, is sufficient to induce sulfur gene expression in Neurospora crassa. Mol. Cell. Biol. 12, 1568–1577.
Paietta, J. V., Akins, R. A., Lambowitz, A. M., and Marzluf, G. A. (1987). Molecular cloning and characterization of the cys-3 regulatory gene of Neurospora crassa. Mol. Cell. Biol. 7, 2506–2511.
Pall, M. L. (1971). Amino acid transport in Neurospora crassa II: properties and regulation of a methionine transport system. Biochim. Biophys. Acta 233, 201–214.
Paszewski, A., and Grabski, J. (1974). Regulation of S-amino acids biosynthesis in Aspergillus nidulans. Mold. Gen. Genet. 132, 307–320.
Thomas, D., Jacquemin, I, and Surdin-Kerjan, Y. H. (1992). MET4, a leucine zipper protein, and centromere-binding factor 1 are both required for transcriptional activation of sulfur metabolism in Saccharomyces cerevisiae. Mol. Cell. Biol. 12, 1719–1727.

GENETIC AND BIOCHEMICAL ANALYSIS
OF ALTERNATIVE RNA SPLICING

Dianne Hodges[1] and Sanford I. Bernstein

Biology Department and Molecular Biology Institute, San Diego State University,
San Diego, California 92182-0057

[1] Present address:: Isis Pharmaceuticals, 2280 Faraday Ave., Carlsbad, CA 92008.

ADVANCES IN GENETICS, Vol. 31

I. Introduction

The process of alternative splicing, whereby transcripts from a single gene are cut and joined to yield multiple forms of mature RNA, is still poorly defined. In order to understand alternative splicing, it is imperative to define the mechanics of constitutive splicing. Since intron removal occurs in both processes, most constitutive splicing factors are expected to be utilized in alternative splicing as well. Furthermore, it now appears that several of the trans-acting molecules involved in alternative splicing are constitutive splicing factors which regulate splice site choice by their concentration and/or activity. We therefore devote the first section of this chapter to the process of constitutive splicing, and emphasize the implications for understanding alternative splicing.

The second portion of the chapter focuses on biochemical studies and *in vitro* mutagenesis experiments directed at understanding alternative splicing. These analyses have provided insight by identifying important cis-acting sequences and a few trans-acting factors that permit selective use of splice junctions. Such analyses have shown that RNA secondary structure, steric constraints, splice junction affinity, and splice site competition play important regulatory roles.

The final portion deals with genetic approaches to understanding alternative RNA splicing. Interestingly, the best-defined genetic system for studying this (sex determination in *Drosophila*) was understood at the genetic level well before RNA splicing was defined as its key regulatory component. This system is a paradigm for how alternative RNA splicing can be genetically dissected, including the eventual molecular, reverse genetic and biochemical analyses of the trans-acting factors and their targets. We also discuss studies of several other genes in *Drosophila* and yeast which illustrate the success of applying genetics to alternative RNA splicing, and we comment on possible genetic approaches for further using *Drosophila* to define trans-acting alternative splicing factors. Finally, we focus on the observations that several human diseases result from exon skipping or the activation of cryptic splice junctions. Some of these studies provide insight into the process of alternative RNA splicing.

II. Constitutive Intron Removal

Most of our knowledge of the cis- and trans-acting elements that permit constitutive splicing, which are most likely implicated in splic-

ing of alternative exons as well, has been obtained through biochemical and genetic (or reverse genetic) manipulations in yeast and mammalian systems. Splicing of nuclear pre-mRNAs containing a single
intron is a two-step, ATP-dependent process (for reviews see Green,
1986; Padgett *et al.*, 1986; Sharp, 1987; Krainer and Maniatis, 1988;
Bindereif and Green, 1990; Guthrie, 1991; Wassarman and Steitz, 1991).
The first step involves cleavage at the 5' end of the intron (the 5' splice
site) and formation of a 2'–5' phosphodiester bond between the 5' terminal guanine of the intron and the 2' hydroxyl of a conserved adenosine located in the intron (known as the branchpoint). Intermediates
generated during the first step of splicing are free exon 1 as well as the
intron with its 5' end joined to the branchpoint and its 3' end still
joined to exon 2. The latter is known as a lariat structure due to its
configuration. In the second step of splicing, cleavage at the 3' end of
the intron (the 3' splice site) and concomitant ligation of the two exons
occurs. The spliced exons and the intron lariat are the end products
(Fig. 1). The various products and intermediates of the splicing process

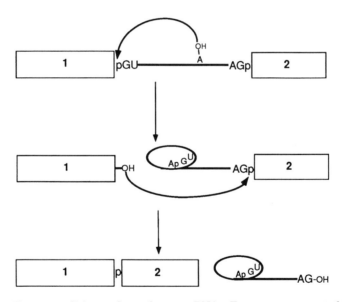

Fig. 1. Two-step splicing pathway for pre-mRNAs. Exons are represented by open
boxes and introns by lines. Step one involves joining of the branchpoint adenosine to the
5' end of the intron via a 5'-2' phosphodiester bond. This results in free exon 1 and a
lariat structure consisting of the intron attached to exon 2. The second step involves
cleavage at the 3' end of the intron and ligation of exon 1 to the 5' end of exon 2. This
results in exon 1 and 2 being joined by a phosphodiester bond and release of the intron
lariat.

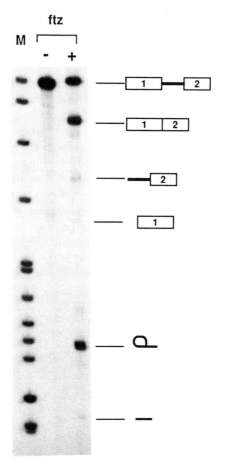

FIG. 2. *In vitro* splicing of *Drosophila ftz* transcripts as detected by gel electrophoresis and autoradiography. ^{32}P-labeled pre-mRNA was transcribed *in vitro* and added to a *Drosophila* Kc cell nuclear extract under appropriate ionic conditions, in the presence (+) or absence (−) of ATP. Following a 2-hr incubation, RNA was purified and separated by electrophoresis on a 5% polyacrylamide, 8 *M* urea gel. The size of the spliced products was determined relative to ^{32}P-labeled *Msp*I fragments of pBR322 (M). Exons are represented by open boxes and introns by lines. The RNA transcript (exons 1 and 2 separated by an intron) remains unprocessed in the absence of ATP. When ATP is present, the end products (exon 1 spliced to exon 2, and the lariat-shaped intron) are obvious, as well as some of the intermediates in the splicing reaction.

have been identified using *in vitro* systems, where enzymatically synthesized RNA is processed in nuclear extracts in the presence of added ATP and salts (see Fig. 2). In contrast to the *in vitro* systems, only mature spliced products accumulate to detectable levels *in vivo*.

A. RNA SPLICING OCCURS IN SPLICEOSOMES

Splicing reactions are catalyzed in spliceosomes, complexes containing small nuclear ribonucleoprotein particles (snRNPs), heterogeneous ribonucleoprotein particles (hnRNPs), and various splicing factors (Guthrie and Patterson, 1988; Gall, 1991; Guthrie, 1991; Wassarman and Steitz, 1991). Spliceosome assembly is initiated by formation of the H complex, which contains pre-mRNAs and hnRNPs (this complex is not specific to splicing since splice sites are not required for its assembly). This is followed by production of the E complex, which contains U1 snRNP and U2AF in addition to hnRNPs, and commits the pre-mRNA to being spliced (Michaud and Reed, 1993). Then in a series of ATP-dependent steps, the A complex containing U1 and U2 snRNPs; the B complex containing U4, U5, and U6 snRNPs in addition to U1 and U2; and the C complex, which has the same snRNP components (although U1 and U4 are only loosely associated) are produced. Both biochemical and genetic studies show that phosphorylation and dephosphorylation of spliceosome protein components are important in pre-mRNA splicing (Mermound et al., 1992; Tazi et al., 1992, 1993; Alahari et al., 1993). Following splicing, the spliced product dissociates from the complex. See Michaud and Reed (1991) for a discussion of spliceosome assembly and Bennett et al. (1992a) for details of the associated protein components.

Electron microscopy and three-dimensional reconstruction analyses indicate that the snRNP proteins are distributed in a network or reticulum that extends between the nucleoli and the nuclear membrane (Spector, 1990). These domains were identified by electron microscopy as interchromatin granules and perichromatin fibrils (Fu and Maniatis, 1990; Spector et al., 1991). Interchromatin granules are proposed to be sites of storage or assembly of splicing components whereas perichromatin fibrils contain nascent transcripts as well as splicing components (Spector et al., 1991, and references therein).

Studies performed to localize splicing components suggest that component storage, spliceosome assembly, and/or splicing occur in discrete regions within the interphase nucleus. Twenty to fifty immunofluorescent "speckles" per nucleus are detected, with antibodies directed against snRNP proteins or the m^3G cap structure of snRNAs (Spector, 1990). An essential, non-snRNA splicing factor, SC35 (discussed in more detail in Section II,E), colocalizes with snRNPs in these speckled regions (Fu and Maniatis, 1990; Spector et al., 1991). A serine-arginine-rich domain found in many RNA-binding proteins appears to act as a signal for their localization to speckles (Li and Bingham, 1991). Wu et al. (1991) showed that "B snurposomes" found in the germinal vesicles of amphibian oocytes are immunoreactive with a wide variety of snRNP and hnRNP probes as well as anti-SC35 antibodies. B snurposomes may thus be equivalent to mammalian speckles. Some evidence points to speckled domains being active RNA processing sites rather than just storage or assembly centers. Microinjected, immunofluorescently labeled pre-mRNA colocalizes with snRNP antigens and SC35 in 30–60 speckled regions (Wang et al., 1991), suggesting that pre-mRNA processing occurs at these sites. Along these lines, Huang and Spector (1991) have shown that endogenous c-fos transcripts and speckles containing snRNPs and SC35 are closely associated. Likewise, Carter et al. (1991) have colocalized immunofluorescent poly(A) RNA and snRNPs to 10–20 clusters within the nucleus. Digital imaging microscopy revealed that these sites are discrete and are generally not associated with the nuclear envelope (Carter et al., 1993). This observation does not concur with electron microscopic observations showing that an snRNP reticulum links the nucleolus to the nuclear envelope (Spector, 1990; Spector et al., 1991).

Whereas several reports suggest that transcription and splicing both occur in speckled domains (Carter et al., 1993; Jiménez-Garcia and Spector, 1993; Xing et al., 1993), Wansink et al. (1993) observed no correlation between major sites of SC35 accumulation and nascent RNA. It is also controversial whether transport of splicing substrates from the site of transcription is an active process or occurs by diffusion through nuclear channels (see Zachar et al., 1993, for discussion).

In contrast to the speckled distribution of the splicing factors discussed above, Carmo-Fonseca et al., (1991) found by in situ hybridization that the major snRNAs involved in splicing are present in only 3–4 "foci" per nucleus. In addition, common snRNP proteins and the non-snRNP protein U2AF (discussed in Section II,E) colocalize to these foci (U1 snRNA and U2AF were also detected throughout the nucleoplasm). When Huang and Spector (1992) used increased hybridization times, they also detected foci containing U1 and U2 snRNAs, but only in some cell types. They detected snRNAs in speckled domains as well. SC35 colocalized with U1 and U2 snRNA in the speckles but was not detected in the foci. Huang and Spector concluded that foci are cell-type specific and correspond to coiled bodies, RNP-containing structures of unknown function that are unlikely to be splicing centers since they are absent in many cell types. Recently Carmo-Fonseca et al. (1992) showed that U1 and U2 snRNP components could be found in both speckles and foci when SDS extraction was incorporated in their procedure. Interestingly, disrupting transcription also disrupted accumulation of snRNPs in foci, but not in speckles, indicating that foci could serve as splicing centers. Thus the site of pre-mRNA splicing in the nucleus, whether it be speckles or foci, and the colocalization of transcription and splicing remain controversial.

B. Determinants of Splice Site Use

Sequence comparisons and mutagenesis experiments have identified RNA elements necessary for splice site definition. The conserved sequences $^C/_A$ AG ↓ GURAGU or G ↓ GUAUGU, in mammals or yeast, respectively, are present at the 5' splice site, and the sequence $Py_{11}NYAG$ ↓ is generally found at the 3' splice site in mammals (where ↓ denotes the splice junction; Breathnach et al., 1978; Mount, 1982; Teem et al., 1984; Ruskin and Green, 1985). Choice of a 5' splice site depends upon its match with the consensus sequence, its surrounding sequences, and its distance from the 3' splice site (Reed and Maniatis, 1986; Lear et al., 1990; Nelson and Green, 1990). The terminal Gs of the intron may interact via non-Watson-Crick base pairing to bring the 5' and 3' exons together (Parker and Siliciano, 1993).

The branchpoint sequences UACUAAC in yeast and YNCUGAC in mammals (where the underlined adenosine is the residue that forms the 2' to 5' link) are usually found 17 to 40 nucleotides (nt) upstream of the 3' splice site (Langford et al., 1984; Ruskin et al., 1984, 1985; Zeitlin and Efstratiadis, 1984; Keller and Noon, 1984; Konarska et al., 1985; Nelson and Green, 1989). This conforms with the observation that cryptic branchpoints 22 to 37 nucleotides upstream of the 3' splice site are activated when the natural branchpoint is mutated (Reed and Maniatis, 1988). In contrast to this, branchpoints located as far as 177 nts upstream from the 3' splice junction are used efficiently in several alternatively spliced pre-mRNAs (Helfman and Ricci, 1989; Smith and Nadal-Ginard, 1989). Smith et al. (1989) proposed that the branchpoint is specified by association with a polypyrimidine tract and by its precise sequence context but not by proximity to the 3' splice site AG. They also suggested that the 3' splice site chosen is the first AG downstream of the branchpoint-polypyrimidine tract. The AG must be in a favorable context, that is, at least 12 nt 3' of the branchpoint and not sequestered in a hairpin (Smith et al., 1993). In a further refinement of these observations, Reed (1989) showed that distant branchpoints can be activated in pre-mRNAs containing a long polypyrimidine tract downstream of the branchpoint but that branchpoints followed by short polypyrimidine tracts require an adjacent AG dinucleotide. She also observed that branchpoints followed by long polypyrimidine tracts are efficiently used even if the 3' splice site is purine rich.

C. Role of snRNPs in Splicing

The snRNPs U1, U2, U4/U6, and U5 are trans-acting factors directly involved in splicing (for reviews see Maniatis and Reed, 1987; Dreyfuss *et al.*, 1988a; Guthrie and Patterson, 1988; Mattaj, 1989; Bindereif and Green, 1990; Gall, 1991; Wassarman and Steitz, 1991; Guthrie, 1991). Suppression of 5' splice site mutations by compensatory mutations near the 5' end of U1 snRNAs provided unequivocal proof that U1 snRNPs interact with the conserved sequences at the 5' splice sites via base pairing (Zhuang and Weiner, 1986; Séraphin *et al.*, 1988; Siliciano and Guthrie, 1988). U2 snRNP was first shown to bind to the branchpoint region by Black *et al.* (1985) and mutagenesis of yeast branchpoint sequences in combination with compensatory U2 snRNP mutations confirmed that this interaction was via base pairing (Parker *et al.*, 1987). Genetic manipulations in mammalian cells also confirmed that U2 snRNP base-pairs with the loosely conserved branchpoint sequence (Zhuang and Weiner, 1989; Wu and Manley, 1989). In mammalian cells, however, the binding of U2 snRNA to the branchpoint is weak and requires the assistance of U2 auxiliary factor (U2AF) (Ruskin *et al.*, 1988; Zamore and Green, 1989, 1991), which is discussed in greater detail below.

U1 snRNP also interacts with sequences at the 3' end of the intron. This was first suggested by Zillmann *et al.* (1987), who analyzed the snRNP content of spliceosomes by immunoprecipitation with anti-snRNP antibodies. Spliceosomes purified from HeLa cell extracts in which the 5' end of U1 snRNA and been inactivated by oligonucleotide-directed RNase H cleavage contained decreased amounts of all of the snRNPs, which suggests that U2, U5, and U4/U6 association with the spliceosome is dependent upon U1 snRNPs. Interestingly, low levels of complexes containing the cleaved U1 snRNA were detected and formation of these complexes was not dependent upon a 5' splice site but required sequences at the 3' end of the intron. Barabino *et al.* (1990) subsequently depleted HeLa nuclear extracts of individual snRNAs using affinity chromatography of biotinylated antisense 2' O-methyl oligonucleotides and convincingly demonstrated that U1 snRNP promoted stable binding of U2 snRNP to the branchpoint. This interaction required the branchpoint sequence but not the 5' splice site of the pre-mRNA or the 5' end of U1 snRNA.

A role for U1 snRNA in branchpoint recognition has also been observed in yeast (Ruby and Abelson, 1988). Additional evidence was provided by Séraphin and Rosbash (1989), who identified "commitment complexes" formed in nuclear extracts prepared from yeast strains conditionally expressing U1 or U2 genes. The commitment complex formed in the absence of U2 snRNP and ATP. However it required the branchpoint sequence, U1 snRNP and the 5' splice site. Upon addition of U2 snRNP and ATP, the pre-mRNA could be chased into mature spliceosomes. Subsequently the commitment complex was resolved into two separate species; formation of commitment complex 1 required only the 5' splice site whereas formation of commitment complex 2 was dependent upon the branchpoint and the 5' splice site (Séraphin and Rosbash, 1991).

Michaud and Reed (1991) used gel filtration to purify an ATP-independent, U1 snRNP-containing mammalian spliceosomal complex, the E complex, which they suggest may be the mammalian homolog of the yeast commitment complex. Goguel et al. (1991) identified yeast U1 snRNA mutations that affect 3' splice site selection in vivo, enhancing the hypothesis that U1 snRNP interacts with the 3' splice site region. The authors suggest that the yeast U1 snRNP may have two separable domains, one that interacts with the 5' splice site and another that interacts with the 3' splice site region. Reich et al. (1992) showed that mutations in the 3' AG of a Saccharomyces pombe intron which inhibit the first step of splicing can be rescued by compensatory changes in two universally conserved nucleotide residues of U1 snRNA. These U1 residues are also capable of binding just upstream of the 5' splice site and are adjacent to the sequence that binds the 5' end of the intron. Thus U1 snRNP may serve a role in bringing the two exons together during the splicing process. Interestingly, these residues of U1 appear to be important for 3' splice site binding only in S. pombe, indicating that 3' splice site recognition may occur by different means in different organisms (Séraphin and Kandels-Lewis, 1993).

Early experiments indicated that U5 snRNP interacts directly or indirectly with the 3' splice site in mammalian cells (Chabot et al., 1985) and complementation of snRNP-depleted nuclear extracts determined that it is absolutely required for the second step of splicing (Winkelmann et al., 1989). U5 snRNPs were also found to be essential for splicing yeast pre-mRNAs in vivo (Patterson and Guthrie, 1987). Analysis of splicing and spliceosome formation in yeast cell extracts genetically depleted of U5 snRNA confirmed that U5 snRNP is required for spliceosome formation as well as for both steps of splicing (Séraphin et al., 1991). Identification of yeast U5 snRNA mutants that suppress a 5' splice site mutation suggests that this snRNA is involved in specification of the 5' cleavage site (Newman and Norman, 1991). The suppression involved selection by the mutant U5 snRNA of a cryptic 5' splice site. Newman and Norman (1992) subsequently determined that the choice of cryptic 5' splice site is dependent upon the ability of U5 snRNA to base-pair with a sequence immediately upstream of the cleavage site. Wyatt *et al.* (1992) showed that U5 snRNA interacts with the 5' exon sequence at position −2 by a non-Watson-Crick base-pairing mechanism. Pre-mRNAs containing 4-thiouridine at the −2 position of the 5' exon are cross-linked to the invariant 9-nt loop of U5 snRNA. The cross-link is ATP-dependent and the RNA–RNA interaction dissolves before the first step of splicing. Newman and Norman (1992) demonstrated possible involvement of U5 snRNA in the second step of splicing by the ability of several U5 snRNA mutants to resolve dead-end lariat-3' exon intermediates generated from 5' cleavage of the pre-mRNA 5' splice site mutant. This was mediated by base-pairing of U5 snRNA with the first two nucleotides of the downstream exon. Additional cross-linking studies performed by Sontheimer and Steitz (1993) identified cross-links between the conserved U5 snRNA loop and the last and first nucleotides of the 5' and 3' exons, respectively. Thus U5 snRNA interacts with both the 5' and 3' exons to hold them together in the catalytic site of the spliceosome, serving an analogous function to specific intron sequences in self-splicing introns.

In both yeasts and mammals, U4 and U6 snRNAs are found base-paired to each other in a single snRNP (Bringmann *et al.*, 1984; Hashimoto and Steitz, 1984; Siliciano *et al.*, 1987; Brow and Guthrie, 1988). In addition, U4 and U6 form a tripartite complex with U5 snRNA (Pikielny *et al.*, 1986; Bindereif and Green, 1987; Cheng and Abelson, 1987; Konarska and Sharp, 1987, 1988). This U4/U5/U6 complex may act as a link between the 5' and 3' splice sites via interaction with U1 and U2 snRNPs (Bindereif and Green, 1987). Indeed, evidence has been obtained indicating that U2 and U6 snRNAs base-pair. Base-pairing between mammalian U2 and U6 snRNAs was originally suggested by psoralen photocross-linking experiments (Hausner *et al.*, 1990). This was later confirmed by elegant reverse genetic experiments in mammalian cells (Datta and Weiner, 1991; Wu and Manley, 1991). Second-site mutations in the 5' end of a suppressor U2 snRNA prevented rescue of splicing of pre-mRNAs containing mutant branchpoints. However, compensatory mutations in the region of U6 snRNA which had been proposed to base-pair with the 5' end of U2 snRNA permitted splicing of the mutant pre-RNAs.

The role of U6 snRNA in splicing, spliceosome assembly, and snRNP complex formation has been analyzed in a mammalian *in vitro* system by reconstitution of U4/U6 snRNP-depleted nuclear extracts (Wolf and Bindereif, 1992). The 5' end of U6 is dispensable for all three of these functions whereas sequences near or within the U4 interaction domain are required. The 3' end of U6 snRNA that base-pairs with U2 snRNA is required for splicing activity and spliceosome formation but not for snRNP complex formation.

Vankan *et al.* (1992) have analyzed the effects of deletion and substitution mutations in U4 and U6 snRNAs on splicing and the assembly of splicing complexes in *Xenopus* oocytes. Several single-stranded regions in U4 snRNA (outside of the region involved in base pairing with U6 snRNA) are required for splicing complementation after *in vivo* depletion of the endogenous U4 snRNA. A perfect correlation with the ability of the mutant U4 and U6 snRNAs to complement splicing and formation of complexes with the pre-mRNA was observed. Interestingly, in the absence of U6 snRNA, U4 snRNA was detected in complexes with pre-mRNA and U1 and U2 snRNAs; however, U6 snRNA was not present in the complex in the absence of U4 snRNA. U5 snRNA was only associated with the pre-mRNA when intact U4 and U6 were both present. This suggests that U4/U6 or U4/U5/U6 snRNP interaction with the spliceosome is mediated by U4 snRNP. However, as stated above, the 3' end of U6 snRNA base-pairs to U2 snRNA and deletion of this region of U6 snRNA inhibits its association with the splicing complex even though U4/U6 assembly is not affected. Therefore, stable association of U6 with the splicing complex may be mediated through its interaction with both U2 and U4 snRNAs.

Although it is clear that U1, U2, and U4/5/6 are in close association with the 5' and 3' splice sites, the actual catalytic entity involved in cleavage and ligation is still unclear. The unusually high evolutionary conservation of the U6 snRNA sequence suggested that this snRNA might assume the catalytic function (Brow and Guthrie, 1988; Guthrie and Patterson, 1988). After the tripartite U4/U5/U6 complex assembles onto the pre-mRNA, U4/U6 base-pairing is disrupted concomitant with the appearance of splicing intermediates and products (Pikielny *et al.*, 1986; Cheng and Abelson, 1987; Lamond *et al.*, 1988; Blencowe *et al.*, 1989). This observation led to the proposal that U4 snRNA sequesters the active site of U6 snRNA and that unwinding of the two snRNAs is required for catalysis (Brow and Guthrie, 1989).

The discovery of an intron in *S. pombe* U6 snRNA (Tani and Ohshima, 1989) in the region which base-pairs with U4 snRNA led to the hypothesis that the intron was inserted by an aberrant reverse splicing reaction followed by reverse transcription, and implied that the site of insertion was near the U6 snRNA catalytic site (Brow and Guthrie, 1989). Further evidence supporting this hypothesis was obtained by isolation of a number of lethal or temperature-sensitive *S. cerevisiae*

U6 snRNA mutants within the U4/U6 pairing region or the region corresponding to the intron insertion site of the *S. pombe* U6 gene Madhani *et al.*, 1990; Fabrizio and Abelson, 1990). It has recently been established that U6 snRNA can be cross-linked to the 5' splice site region of the pre-mRNA (Sawa and Abelson, 1992; Sawa and Shimura, 1992; Wassarman and Steitz, 1992) as well as to the branchpoint of the intron lariat (Sawa and Shimura, 1992). The above data are consistent with a catalytic role for U6 snRNA, and a model for this based, in part, on similarities between U6 and domain 5 of group II self-splicing introns is discussed by Wassarman and Steitz (1992). Further evidence for this model is provided by recent studies suggesting that the 5' and 3' exons are brought into the active site of the spliceosome by RNA–RNA interactions between the intron and U2 and U6 snRNAs, and that the exons are held there by interaction with U5 snRNA (Kandels-Lewis and Séraphin, 1993; Lesser and Guthrie, 1993; Sontheimer and Steitz, 1993; reviewed by Wise, 1993).

D. Splicing Factors and Exon Recognition

Proper choice of 5' and 3' splice site pairs in premessenger RNAs is not just dependent upon recognition of the conserved sequence elements since there are many sequences which match the consensus but are not used (Ohshima and Gotoh, 1987). Insight into how proper 5' and 3' splice site pairs are chosen in constitutively spliced transcripts has been obtained by experiments addressing communication among splicing factors. It is well established that splicing factors bound at the 5' and 3' ends of introns interact directly or indirectly to promote recognition and removal of introns (Lamond *et al.*, 1987). However, the exon definition model for early spliceosome assembly postulates that splicing factors communicate across exons as well. Robberson *et al.* (1990) proposed this model, which suggests that recognition of an exon by U1 snRNP involves interaction of the snRNP with the 3' end of the intron, followed by translocation of the U1 snRNP to the downstream 5' splice site.

As discussed earlier, there is abundant evidence that U1 snRNP interacts at both ends of an intron. U2 snRNP is also proposed to play a role in exon definition. In addition to base-pairing to the branchpoint, it appears to interact with the 5' splice site since its inactivation prevents protection of the 5' splice site from oligonucleotide-mediated cleavage (Robberson *et al.*, 1990). These observations suggest that communication across exons occurs between U2 and U1 snRNPs. The exon definition model is further supported by results showing that some substrates containing two exons separated by an intron are assembled into a splicesomal complex and are spliced more efficiently if the downstream exon contains a 5' splice site (Robberson et al., 1990). This is especially obvious if the second exon is long. Thus identification of an intron by the splicing apparatus appears to be dependent upon defining the downstream exon. This definition can apparently be aided by purine-rich "exon recognition sequences," located within the exon, which interact with U1 snRNP (Watakabe et al., 1993). Such purine-rich elements can apparently promote use of both constitutive and alternative 3' splice sites (Xu et al., 1993; Yeakley et al., 1993).

In another study, Talerico and Berget (1990) assayed in vitro splicing of three-exon transcripts containing mutations at the 5' splice site of the internal exon, and confirmed that these mutations greatly decreased the efficiency of removal of the upstream intron. The failure to remove the upstream intron, resulting in skip splicing, is also observed in vivo with known vertebrate 5' splice site mutations in internal exons (Talerico and Berget, 1990, and references therein). 5' splice site mutations that improve U1 snRNP binding would thus be expected to enhance removal of the upstream intron, and inhibit skip splicing, as has been demonstrated for preprotachykinin pre-mRNA splicing (Nasim et al., 1990; Kuo et al., 1991; Grabowski et al., 1991). These results are discussed in relation to models of alternative splicing in Section V, D.

The exon definition model also postulates that the exon at the 5' end of the pre-mRNA is recognized by the interaction of splicing factors with factors that recognize the 5' end cap structure. Such a role for the 5' cap structure in splicing in a mammalian in vitro system was suggested by experiments which demonstrated that the presence of a m⁷GpppG cap in pre-mRNA greatly increased the efficiency of removal of the 5' proximal intron but had no effect on removal of the downstream introns (Ohno et al., 1987). Inoue et al. (1989) showed this was true in vivo by analysis of m⁷GpppG versus ApppG capped pre-mRNAs injected into Xenopus oocytes.

The exon definition model suggests that splicing of the 3'-most intron involves interaction of the splicing and polyadenylation systems. The preferred order of splicing and polyadenylation in vivo is still unclear. Removal of internal introns appears to occur on nascent transcripts prior to completion of transcription and polyadenylation (Beyer and Osheim, 1988), but these studies have not addressed the relationship of polyadenylation and removal of the terminal intron. In vitro, pre-mRNA that is spliced cannot be chased into polyadenylated product, indicating that polyadenylation must precede splicing to produce functional mRNA (Niwa and Berget, 1991a). Polyadenylation is stimulated by the presence of an upstream intron (Niwa et al., 1990; Nesic et al., 1993). The reciprocal effect was also observed in vitro; polyadenylation enhances removal of the upstream intron whereas mutation of the conserved AAUAAA polyadenylation signal drastically decreases splicing (Niwa and Berget, 1991b).

Interestingly, U1 snRNP binds to sequences within polyadenylation substrates; U1 binding and polyadenylation are enhanced in RNAs containing an upstream 3' splice site, which suggests that U1 plays a role in coordinating splicing and polyadenylation (Wassarman and Steitz, 1993). In transcripts containing multiple introns, polyadenylation only enhances removal of the most 3' intron, which is consistent with the exon definition model (Niwa and Berget, 1991b). Niwa et al. (1992) showed that insertion of a 5' splice site within 300 nt downstream of the 3' splice site of a terminal exon depresses polyadenylation and decreases UV cross-linking of a polyadenylation factor. A 5' splice site inserted downstream of the poly (A) site does not affect polyadenylation efficiency. These results suggest that polyadenylation sites in introns are ignored because they are preceded by 5' splice sites, and that exons are defined either by their flanking 3' and 5' splice sites or by a 3' splice site and a polyadenylation signal.

E. Additional Constitutive Splicing Factors

A great deal of progress has recently been made in identifying and purifying proteins that influence use of 5' and 3' splice sites, and in molecularly cloning their coding sequences. ASF/SF2 (Ge and Manley, 1990; Krainer et al., 1990a,b, 1991; Ge et al., 1991) is an essential 33-kDa factor that is required for the first cleavage-ligation step of splicing. ASF/SF2 contains an amino terminal RNA binding domain consisting of an RNP consensus sequence (an 80–90 amino acid sequence present in a number of RNA binding proteins) and a carboxy terminal serine-arginine (SR)-rich domain (for review of RNA binding proteins, see Bandziulis et al., 1989). A potential Drosophila homolog termed B52 is found at transcriptionally active loci (Champlin et al., 1991). Although ASF/SF2 binds RNA nonspecifically, it has been shown to promote annealing of complementary RNAs. The finding that it stimulates use of proximal 5' splice sites in a dose-dependent manner suggests that it may mediate alternative splicing choices as well. Recent structure and function studies on ASF/SF2 shows that the SR domain is essential for the general splicing process, but that it is not required for activating proximal 5' splice site use (Cáceres and Krainer, 1993;

Zuo and Manley, 1993). These investigators found a second, poorly conserved, RNP consensus sequence located C-terminal to the first. Each RNP consensus sequence can independently bind RNA, but affinity is much better when both sequences are present. Proteins having both RNP consensus sequences function best in both general splicing and in activating proximal 5' splice sites.

In contrast with ASF/SF2, distal splicing factor (DSF) is a partially purified micrococcal-nuclease-insensitive activity which promotes use of distal 5' splice sites (Harper and Manley, 1991). DSF is not believed to be required for constitutive splicing (Harper and Manley, 1991). Mayeda and Krainer (1992) purified a similar activity termed SF5 from HeLa cells. Interestingly, the 34-kDa SF5 protein was determined to be hnRNP A1, an RNA binding protein known to interact with the 3' splice site and to promote RNA renaturation (Munroe and Dong, 1992). The hnRNP A1 N-terminus contains two copies of the RNP consensus sequence and the C-terminal has 12 repeats of a glycine-rich sequence believed to contribute to cooperative binding of the protein (Buvoli *et al.*, 1988; Biamonti *et al.*, 1989; Mayeda and Krainer, 1992). While DSF is present in relatively low quantities, hnRNP protein A1 is abundant and believed to act stoichiometrically. However, both activities counteract the action of ASF/SF2, and the relative amounts of ASF/SF2 versus DSF or hnRNP A1 determine whether a proximal or distal 5' splice site is selected. These activities are thus likely to play an important role in alternative RNA splicing (Mayeda *et al.*, 1993).

Much less is known about other proteins that interact with the 5' splice site. Cross-linking to pre-mRNAs of two partially characterized proteins, SSP-1 and SPP-2, was shown to be dependent upon a 5' splice site but not U1 snRNP or the 3' end of the intron (Stolow and Berget, 1991). SF4 is another non-snRNP factor that has been partially purified and is required for 5' cleavage and ligation but not for functional spliceosome assembly (Utans and Kramer, 1990).

A number of proteins that interact with the branchpoint-3' splice site region have been identified as well. The 65-kDa polypeptide component of U2 auxiliary factor (U2AF) is an essential mammalian splicing factor that binds to the polypyrimidine tract-3' splice site and promotes U2 snRNP binding to the branchpoint (Ruskin *et al.*, 1988; Zamore and Green, 1989, 1991). The DNA encoding the 65-kDa component of U2AF has now been cloned (Zamore *et al.*, 1992). Not surprisingly, the protein contains three RNP consensus domains, which have been shown to bind preferentially to pyrimidine-rich sequences; all three domains are required for optimal affinity and specificity of binding. Like ASF/SF2, U2AF[65] contains an SR domain that is dispensable for RNA binding but is required for U2AF function (Zamore *et al.*, 1992). Arginine side chains in other RNA binding proteins have been proposed to mediate RNA recognition by interaction with specific RNA structures (Calnan *et al.*, 1991). Zamore *et al.* (1992) speculate that U2AF bound at the polypyrimidine tract may interact through its SR domain with U2 snRNA bound at the upstream branchpoint and thus stabilize its binding. U2AF^{65} is essential for viability in *Drosophila* (Kanaar *et al.*, 1993) and it is required for splicing in *S. pombe* (Potashkin *et al.*, 1993). Analysis of a cDNA encoding U2AF^{35}, the other component of U2AF, indicates that this protein contains an SR sequence but lacks RNP consensus domains (Zhang *et al.*, 1992).

Kramer and Utans (1991) have identified two protein activities termed SF1 and SF3 that are required for the formation of a presplicing complex on pre-mRNA lacking a 5' splice site. They suggest that these factors, in addition to U2AF, promote U2 snRNP interaction with the branchpoint. A U5 snRNP-associated protein (intron binding protein or 70-kDa protein) that recognizes the AG dinucleotide at the 3' splice site and the polypyrimidine tract has been identified (Tazi *et al.*, 1986; Gerke and Steitz, 1986) but has not been further characterized.

Polypyrimidine tract binding protein (PTB) is a splicing factor of 57–62 kDa that mediates 3' splice site selection by directly binding to the upstream polypyrimidine tract and is required for efficient U2 snRNP interaction with the pre-mRNA (Garcia-Blanco *et al.*, 1989; Wang and Pederson, 1990; Gil *et al.*, 1991; Patton *et al.*, 1991). It binds in the absence of the 5' splice site, the branchpoint, and the AG at the 3' splice site; there is a good correlation between the strength of binding of PTB to a particular polypyrimidine tract and splicing efficiency (Garcia-Blanco *et al.*, 1989; Patton *et al.*, 1991), although in one case PTB binds to a nonpyrimidine-rich sequence and appears to negatively regulate exon inclusion (Mulligan *et al.*, 1992). In the fractionation scheme used by Patton *et al.* (1991), proteins of 33 kDa and 100 kDa copurify with PTB. The 33-kDa protein has been identified as hnRNP A1 (cited in Bothwell *et al.*, 1991, as personal communication from B. Nadal-Ginard). Thus, the role of hnRNP A1 in 5' splice site selection may involve association with PTB at the 3' splice site.

Unlike the ~60-kDa human PTB described above, murine PTB was identified as a 25-kDa component of a complex also containing 100-kDa and 35-kDa proteins; however, the 25-kDa proteins was determined to be a C-terminal proteolytic fragment of a larger 60-kDa subunit (Bothwell *et al.*, 1991). This C-terminal fragment binds the polypyrimidine tract with the same specificity as the whole protein

(Bothwell *et al.*, 1991). PTB is distantly related to hnRNP L protein, which is associated with the landmark giant loops of amphibian lampbrush chromosomes (Piñol-Roma *et al.*, 1989); both proteins contain nonconsensus RNA binding domains. PTB was recently shown to be hnRNP I (Bennett *et al.*, 1992b). Interestingly, PTB promotes translation of picornavirus RNA by binding to a large stem-loop structure in the 5′ untranslated region that is the internal ribosomal entry site (Hellen *et al.*, 1993). Thus PTB functions in the nucleus and cytoplasm may be related, that is, promotion of assembly of large macromolecular RNP complexes on pre-mRNA or mRNA, respectively.

cDNAs encoding the 100-kDa protein associated with PTB (PTB-associated splicing factor or PSF) have been characterized (Patton *et al.*, 1993). PSF contains two RNA-binding domains and its amino terminal is proline and glutamine rich, a motif found to mediate protein–protein interactions of transcription factors. Like PTB, PSF binds preferentially to poly (U) tracts both independently and in a complex with PTB. Studies involving antibody inhibition and reconstitution of immune-depleted nuclear extracts showed that PSF is an essential factor required for early spliceosome formation and splicing catalysis; similar experiments determined that PTB is not essential for splicing. Patton *et al.* propose that since PTB is quite abundant, it may bind to pre-mRNA early, measure the strength of the polypyrimidine tract, and then recruit other factors, such as PSF, to the spliceosome.

Fu and Maniatis (1990) identified a 35-kDa non-snRNP protein, SC35, which is required for the first step of splicing. SC35 is tightly associated with the nuclear matrix, possibly by protein–protein interactions since its immunolocalization to "speckles" is not affected by RNAse digestion (Spector *et al.*, 1991). The colocalization of SC35 and U1 and U2 snRNA as well as other snRNP antigens to the same nuclear regions indicates that SC35 may be involved in assembly and attachment of splicing components to the nuclear matrix (Spector *et al.*, 1991; Huang and Spector, 1991). SC35 has an RNP consensus sequence at its NH_2 terminus, an internal proline-glycine-rich sequence, and an SR domain at its C terminus (Fu and Maniatis, 1992b). In addition to sharing these structural features with ASF/SF2, SC35 also appears to be a nonspecific RNA-binding protein. SC35 is involved in mediating interaction of U1 snRNP with splicing factors bound at the 3′ splice site as well as aiding U2 snRNP interaction with the branchpoint; these interactions are ATP dependent (Fu and Maniatis, 1992a).

A number of splicing factors, including the U1 70-kDa protein, ASF/SF2 and SC35, share the SR motif. Some of these proteins also share functional attributes. A comparison of the activities of ASF/SF2

224   DIANNE HODGES AND SANFORD I. BERNSTEIN

and SC35 indicate that both factors can complement the splicing activity of S100 or SC35-immunodepleted nuclear extracts; in addition, both promote use of proximal 5' and 3' splice sites (Fu *et al.*, 1992). Zahler *et al.* (1992) recently described the isolation of four proteins related to SC35 and ASF/SF2 that have serine-arginine-rich carboxy terminals and are likely involved in pre-mRNA splicing. Each of these "SR" proteins is unique in sequence, but each is able to rescue ASF/SF2-deficient splicing extracts. It is important to note however, that clear differences in SR protein preferences for proximal or distal 5' splice sites exist, indicating a possible role in alternative RNA splicing; this is also suggested by quantitative and qualitative differences in their tissue-specific expression (Zahler *et al.*, 1993). The differential ability of SR proteins to commit specific mRNAs to the splicing pathway further supports this idea (Fu, 1993).

Studies on the phosphorylation state of SR proteins suggest that this type of post-translational modification may regulate their ability to mediate splicing. This is clearly the case for the U1 70-kDa protein, where inhibition of phosphate removal prevents splicing (Tazi *et al.*, 1993). Interestingly, Tazi *et al.* (1993) cite unpublished evidence that the U1 snRNP displays kinase activity which phosphorylates the SR domain of the U1 70-kDa protein. It is conceivable that phosphorylation of non-snRNP SR proteins can regulate their role in alternative RNA splicing.

Wu and Maniatis (1993) very recently showed that SC35 and ASF/SF2 can act as a bridge between the U1 70-kDa protein at the 5' splice site and U2AF at the 3' splice site. Interaction of SC35 or ASF/SF2 occurs with the 35-kDa component of U2AF, which is strongly bound to the 65-kDa component at the polypyrimidine tract. This joining of protein components at the 5' and 3' splice cites may facilitate proper splice site selection.

PRP8 is a yeast splicing factor which has been identified genetically (see Section II,G). This 260-kDa protein is associated with U5 snRNP (Lossky *et al.*, 1987) and is an integral component of spliceosomes (Whittaker *et al.*, 1990). PRP8 is required for formation of the triple snRNP, U4/U5/U6, as well as the assembly of the triple snRNP into the spliceosome (Brown and Beggs, 1992). Homologs in higher organisms have been identified and characterized by immunological and biochemical means. Anti-PRP8 antibodies cross-react with p220, a 200-kDa protein purified from mammalian U5 snRNPs (Anderson *et al.*, 1989; Bach *et al.*, 1989; Pinto and Steitz, 1989; Garcia-Blanco *et al.*, 1990). Both the mammalian and yeast proteins are in close proximity to the pre-mRNA in the spliceosome since they can be UV cross-linked to splicing-competent substrates (Garcia-Blanco et al., 1990; Whittaker and Beggs, 1991). Interestingly, p220 does not cross-link to mammalian substrates containing mutations in the 3′ splice site or polypyrimidine tract (Garcia-Blanco et al., 1990) whereas PRP8 cross-linking to yeast substrates is dependent upon the presence of an intact 5′ splice site and branchpoint but not a 3′ splice site (Whittaker and Beggs, 1991). In Drosophila, a U5 snRNP-associated protein of >200 kDa cross-reacts with the anti-PRP8 antibody (Paterson et al., 1991).

Other splicing components that have been partially characterized include an 88-kDa protein from mammalian cells that is essential for splicing (Ast et al., 1991). Nuclear extracts immunodepleted of this protein were inactive for spliceosome formation and in vitro splicing, but both activities could be restored by reconstitution with the affinity-purified 88-kDa protein. Purification of mammalian spliceosomes by affinity chromatography led to the discovery of a large number of other proteins which are likely functional splicing components; they range in size from less than 14 kDa to 200 kDa (Reed, 1990). Five proteins of molecular weights 15.5, 20, 27, 60, and 90 kDa have been found specifically associated with and required for formation of the tri-snRNP U4/U5/U6 complex (Behrens and Lührmann, 1991).

F. ROLE OF HNRNP PROTEINS IN SPLICING

Approximately 20 proteins, including a set of abundant core proteins (A1/A2, B1/B2, C1/C2), are found in heterogeneous nuclear ribonucleoprotein (hnRNP) particles prepared from newly transcibed pre-mRNA (Piñol-Roma et al., 1988). Owing to their abundance and the fact that many of them bind RNA nonspecifically, the role of the hnRNP proteins in splicing has been somewhat overlooked. However, the finding that several of these proteins have specific binding preferences, as well as the identification of hnRNP A1 as SF5 and hnRNP I as PTB (see Section II,E), has increased the interest in these proteins as potential splicing regulators. hnRNP proteins A1, C, and D bind specifically to the 3′ end of introns within the conserved polypyrimidine stretch between the branchpoint and the 3′ splice site (Dreyfuss et al., 1998a,b; Swanson and Dreyfuss, 1988a,b).

Pre-mRNA splicing is inhibited by a monoclonal antibody to the hnRNP C proteins (Choi et al., 1986) and by a polyclonal antibody to the hnRNP core proteins (Sierakowska et al., 1986), indicating a functional role for these hnRNPs in splicing. Mayrand and Pederson (1990) observed that cross-linking of A1 and C hnRNPs to pre-mRNA was eliminated by oligonucleotide-directed RNase H cleavage of nucleotides 28–42 of U2 snRNA and that cleavage of the 5' end of U1 snRNA eliminated cross-linking of hnRNP A1. It thus appears that there is interaction between these hnRNPs and U1 and U2 snRNPs. This is consistent with the identification of hnRNP A1 as a factor involved in promoting use of distal 5' splice sites. Bennett *et al.* (1992b) reported that the proteins found in mammalian hnRNP complexes are the same as those in the H complex precursors to spliceosomes. Interestingly, the ratios of the hnRNP proteins differ, depending on the particular pre-mRNAs used as assembly substrates. Thus the differential binding of hnRNP proteins (as is already clear for hnRNPs A1 and PTB) may affect the process of alternative splicing by activating or masking specific splice junctions.

G. RNA HELICASES AND ATPASES AS ESSENTIAL SPLICING COMPONENTS

Years before RNA splicing was discovered, Hartwell (1967) chemically created yeast (*S. cerevisiae*) mutants with the goal of obtaining temperature-sensitive defects in synthesis of macromolecules such as protein, RNA, and DNA. Approximately 5% of the 400 mutants identified in this screen fell into 10 complementation groups that affected RNA synthesis, and many were subsequently shown to specifically affect pre-mRNA splicing (Rosbash *et al.*, 1981). Since this screen was not saturating, Vijayraghavan *et al.* (1989) performed another mutagenesis screen and looked directly for RNA splicing mutations by assaying for the accumulation of intron-containing precursors in yeast grown at the nonpermissive temperature. They obtained 6 new complementation groups involved in various stages of pre-mRNA splicing. Strauss and Guthrie (1991) also mutagenized yeast and selected both heat- and cold-sensitive mutants defective in pre-mRNA splicing.

These various genetic approaches have identified a large number of PRP (precursor RNA processing) mutants, many of which have recently been characterized genetically and biochemically [see tables in Guthrie (1991) and Ruby and Abelson (1991)]. Nuclear extracts prepared at the nonpermissive temperature have been used to characterize yeast spliceosomes and to elucidate the steps in the splicing process disrupted by the mutations. In addition, the genetic approach has been useful for obtaining additional components of the splicing apparatus by selection for second-site suppressor mutants whose gene products interact with mutant PRP proteins (Jamieson *et al.*, 1991; Shannon and Guthrie, 1991; Strauss and Guthrie, 1991).

Molecular characterization of several of the PRP genes identified conserved sequence motifs (G-protein, zinc finger-like, RNP, and DEAD or DEAH) that provide clues to their functions (for reviews see Warner, 1987; Guthrie, 1991; Ruby and Abelson, 1991; Wassarman and Steitz, 1991). The DEAD or DEAH (acronyms for the single amino acid codes found in the conserved sequence elements) classes consist of proteins with homology to RNA-dependent ATPases or ATP-dependent RNA helicases. Biochemical analyses of two of the members of the DEAD-DEAH class, PRP2 and PRP16, confirmed that they can hydrolyze ATP in the presence of RNA (Schwer and Guthrie, 1991; Kim *et al.*, 1992). PRP2 is an RNA-dependent ATPase that is required for 5' splice site cleavage and lariat formation.

Kim and Lin (1993) showed that spliceosomes isolated from *prp2* mutants undergo the first step of splicing when supplemented with purified PRP2 and ATP, which suggests that ATP hydrolysis is required for the first catalytic event. PRP16 is required for the second step of splicing and promotes a conformational change in the spliceosome that leads to protection of the 3' splice site from oligonucleotide-directed RNAse H degradation (Schwer and Guthrie, 1992). These types of data indicate that the ATP dependence of splicing most likely involves conformational changes generated by ATP hydrolysis. Such conformational changes requiring ATPases or RNA helicases might involve unwinding of the pre-mRNA or dissolution of other RNA–RNA base-pairings (such as U4 with U6, U2 with U6, U1 with the 5' splice site, U2 with the branchpoint) that occur as catalysis proceeds. The PRP8 protein appears to counteract the effect of RNA helicases and stabilize U4-U6 association (Brown and Beggs, 1992).

The yeast PRP9 protein is required for U2 snRNP binding to the spliceosome but is not tightly associated with the snRNP (Abovich *et al.*, 1990). PRP9 contains an amino-terminal leucine zipper structure and two carboxy-terminal zinc-finger-like motifs of the Cys-His type (Legrain and Choulika, 1990). *prp 9-1* and *9-2* mutants are defective in splicing and target unspliced RNA to the cytoplasm (Legrain and Rosbash, 1989). The *spp91-1* suppressor rescues these defects (Chapon and Legrain, 1992). SPP91-1 (also known as PRP21; Arenas and Abelson, 1993) is a 33-kDa protein with limited homology to the *Drosophila su(w^a)* protein. SPP91/PRP21 contains a nuclear localization signal that is distinct from the serine-arginine-rich sequence believed to target splicing factors to sites of splicing within the nucleus. PRP9 and SPP91/PRP21 interact with PRP11 to form a multimeric complex that bridges U1 and U2 snRNPs, resulting in prespliceosome formation; PRP9 and PRP11 do not directly interact but can simultaneously bind to SPP91/PRP21 (Legrain and Chapon, 1993). PRP5 (an

RNA-dependent ATPase) and PRP11 are detected in association with
U1 snRNP in ATP-independent complexes prior to the association of
PRP9, SPP91/PRP21, and U2 snRNP to form the prespliceosome (Ruby
et al., 1993). Homologs of several of these proteins occur in mammalian
spliceosomes (Behrens *et al.*, 1993; Bennett and Reed, 1993; Brosi
et al., 1993). Based on these types of studies in yeast and the presence
of homologs to the PRP proteins in other species, it is clear that the
yeast PRP mutants are valuable tools that will continue to provide
insight into the mechanisms of constitutive as well as alternative
splicing.

III. Approaches to Defining the Mechanism of Alternative Splicing

In Fig. 3, a variety of alternative splicing patterns are illustrated.
Figures 3A and 3B show that by utilizing alternative 5' or 3' splice

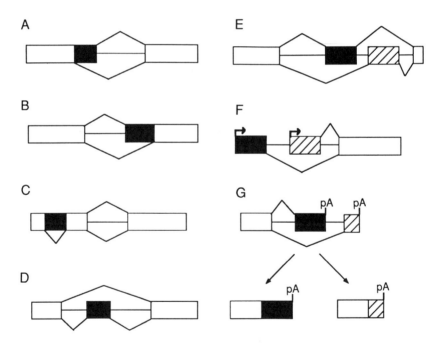

FIG. 3. Alternative splicing patterns. Constitutive exons are represented by open
boxes and alternative exons by black or cross-hatched boxes. Introns are indicated by
horizontal lines. Lines above and below each gene structure indicate alternative modes
of splicing. In (F), two potential transcriptional start sites are indicated by arrows. In
(G), alternative polyadenylation sites are indicated by pA.

sites, exons can be lengthened or shortened. Figure 3C shows that when both an alternative 5' splice site and an alternative 3' site within an exon are activated, an internal intron can be eliminated. An entire exon can be skipped during the splicing process, as shown in Fig. 3D. Exons can be mutually exclusive, in that one or the other is present in the final transcript (Fig. 3E). Alternative transcriptional promoters (Fig. 3F) or polyadenylation sites (Fig. 3G) may influence use of alternative splice junctions, simply by providing a template containing or lacking specific splice sites. In the following section we describe well-studied examples of alternatively spliced transcripts and how biochemical and reverse genetic experiments have provided insight into regulated splicing mechanisms.

A. USE OF ALTERNATIVE 5' SPLICE SITES:

SV40 early pre-mRNA is alternatively spliced to produce two mRNAs encoding large T or small t antigen by use of two different 5' splice sites (Fig. 4). Use of the more distal 5' splice site (relative to the common 3' splice site) generates large T mRNA whereas using the more proximal 5' splice site (only 66 nt upstream of the 3' splice site) produces small t mRNA. The ratios of these two products varies among different cell lines; in HeLa cells five times more large T than small t mRNA is produced but the reverse ratio is observed in embryonic kidney 293 cells. Ge and Manley (1990) purified a splicing factor, ASF/

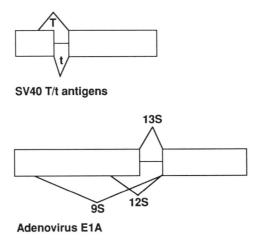

SV40 T/t antigens

Adenovirus E1A

FIG. 4. SV40 large T and small t antigens and adenovirus E1A transcripts (9S, 12S, and 13S) are produced by use of alternative 5' splice sites.

SF2, from 293 cells and demonstrated that it promotes use of the proximal small t 5′ splice site and decreases use of the large T 5′ splice site. The properties of this splicing factor are discussed in Section II,E.

A number of characteristics of the SV40 precursor RNA make the small t 5′ splice site an unfavorable choice (Fu and Manley, 1987; Noble *et al.*, 1987, 1988; Zhuang *et al.*, 1987; Fu *et al.*, 1988). The size of the small t intron sterically hinders the simultaneous interaction of splicing factors with the 5′ and 3′ splice sites. The small intron size also constrains the small t splice to using the branchpoints at positions −18 and −19 relative to the 3′ splice site, whereas large T splicing utilizes these as well as numerous upstream branchpoints. The polypyrimidine tract just downstream of these branchpoints is interrupted by several purines, which may not allow efficient binding of splicing factors to the 3′ splice site. Mutations that increase the strength of the polypyrimidine tract increase the T/t mRNA ratio, which suggests that increased interaction of splicing factors at the 3′ end of the intron inhibits the binding of splicing factors to the nearby small t 5′ splice site. Both small t and large T 5′ splice sites are suboptimal, and improvement of the large T 5′ splice site permits it to out-compete the small t 5′ splice site *in vivo*. Therefore, loose and transient binding of splicing factors at both splice sites may be required to overcome the steric hindrance imposed by the small intron size on small t splicing.

Adenovirus E1A RNA is spliced to generate 9S, 12S, and 13S mRNA by ligation of any of three alternative 5′ splice sites to a common 3′ splice site (Fig. 4). Use of the more proximal 12S and 13S 5′ splice sites predominates early in infection but utilization of the distal 9S site increases late in infection. Replacement of the common E1A branchpoint-3′ splice site region with a heterologous branchpoint and 3′ splice site increased use of the 12S and 9S 5′ splice sites, with concomitant decreased use of the 13S site (Ulfendahl *et al.*, 1989). Thus there is a cooperative interaction of particular 5′–3′ splice site pairs. The authors proposed that late in infection the virus modifies splicing factors that bind to the E1A 3′ splice site, resulting in selection of different 5′ splice sites. Schmitt *et al.* (1987) also induced a switch from predominant use of the proximal 5′ splice sites to use of the distal 5′ splice sites by altering ion concentrations *in vitro*. This *in vitro* effect is most likely due to pre-mRNA structural changes. Whether such changes occur during the course of infection is unknown. Nuclear extracts prepared from late infected HeLa cells were also able to modulate a similar switch as a result of a micrococcal nuclease-sensitive activity identified as high-molecular-weight nucleic acid and proposed to be viral RNA (Gattoni *et al.*, 1991). The authors suggest that the high-molecular-weight RNA sequesters splicing factors required for proximal 5' splice site utilization which are present in limiting quantities in the cell.

Eperon *et al.* (1993) attempted to determine if there is a correlation between U1 snRNP occupancy and the choice of alternative 5' splice sites of varying strengths. They assayed protection of 5' splice sites from RNase H-mediated cleavage. Binding of U1 snRNP to duplicated consensus sites is apparently random. snRNP binding at both sites frequently occurs simultaneously. Eperon *et al.* propose that this results in selection of the 5' splice site closest to the downstream 3' splice site. Their model predicts that competition between weak 5' splice sites depends on the dissociation rates of U1 snRNP with each site. Addition of purified ASF/SF2 to the protection assays increased binding of U1 snRNP to both 5' splice sites; reversal of the inhibitory effect of hnRNP A1 on U1 snRNP binding is a possible mechanism for this action. Equalization of U1 snRNP binding by ASF/SF2 would then lead to selection of the 5' splice site most proximal to the 3' splice site.

B. EXON INCLUSION AND EXCLUSION

A particular alternative exon may be included in one tissue type and excluded in another, or both patterns may occur in a single cell. For example, exons 4 through 8 of troponin T gene transcripts in differentiated fast skeletal muscles are utilized in a combinatorial fashion, allowing up to 32 different combinations to be expressed from a single gene. Breitbart and Nadal-Ginard (1987) demonstrated that unlike differentiated muscles, nonmuscle cells or myoblasts are unable to recognize and include any of these muscle-specific exons in mature transcripts. They suggested that differences in cell architecture or trans-acting factor expression and activity between muscle and nonmuscle cells affect recognition of these exons.

Mutagenesis experiments have identified some cis-acting sequences that mediate exon inclusion and exclusion. Elements within calcitonin/calcitonin gene-related peptide (CT/CGRP) pre-mRNAs regulate recognition of the 3' splice site of alternative exon 4 (see Section III,D for a gene map). The CT/CGRP primary transcript is alternatively spliced by exclusion of this exon in neuronal cells to produce CGRP and inclusion in most other cells to produce calcitonin. Expression studies of the CT/CGRP gene in transgenic animals suggested that inclusion of exon 4 is the default choice whereas the skip splicing observed in neuronal cells represents the regulated pathway (Crenshaw *et al.*, 1987). Emeson *et al.* (1989) identified an element upstream of the exon 4 branchpoint that is required for exon 4 exclusion in neuronal cells, which suggests that a neuronal-specific inhibitor might interact with this region. The unusual uridine branchpoint of exon 4 was also shown to play a role in exon 4 skipping, since mutating the branchpoint to an adenosine resulted in the predominant inclusion of exon 4 in neuronal cells (Adema and Baas, 1991). Detailed mutational analysis of the intron region in front of exon 4 by Yeakley *et al.* (1993) failed to identify a single cis-acting element involved in exon 4 alternative splicing. They showed, however, that the suboptimal 3' splice site and branchpoint are important in exclusion of this exon. Cote *et al.* (1992) mapped an element required for recognition of exon 4 to the first 30 nts of this exon. They used UV cross-linking to identify a possible trans-acting factor from HeLa cell nuclei that binds to this site.

Inclusion and exclusion of exon 18 in mature neural cell adhesion molecule (NCAM) mRNA is dependent upon the state of differentiation of the nerve cell; exon 18 is skipped in mitotic cells but included in postmitotic cells. Regulatory elements that mediate alternative splicing of NCAM transcripts have been identified at the 5' splice site of the alternative exon (Tacke and Goridis, 1991). Replacement of the 5' splice site of exon 18 with an α-globin 5' splice site resulted in exclusive use of exon 18 regardless of the differentiation state. Interestingly, large deletions within exon 18 of NCAM RNA increased levels of exon 18 skipping, but even larger deletions resulted in increased exon 18 inclusion. This suggested that overall structure rather than a particular sequence element within exon 18 is required for its recognition.

Mutation of the 5' splice site of an internal exon of preprotachykinin pre-mRNA in order to improve its potential for base-pairing to U1 snRNA promotes inclusion of this exon in cells that normally preferentially exclude it (Nasim *et al.*, 1990; Grabowski *et al.*, 1991; Kuo *et al.*, 1991). In addition to enhancing binding of U1 snRNP to the 5' splice site, this mutation promotes preferential removal of the upstream intron which contains a weak 3' splice site. This appears to be mediated by increased binding of U2AF (Hoffman and Grabowski, 1992). The results from splicing of NCAM and preprotachykinin transcripts lend support to the exon definition model (see Sections II,D and V,D) since improved binding of splicing factors to the 5' splice site of the alternative exon would be expected to enhance interaction across the exon with splicing factors bound at the 3' splice site. These inherently poor 5' splice sites may normally be used in a stage- or tissue-specific manner by alterations in the concentrations or activities of constitutive splicing factors or unidentified tissue-specific factors. These factors may act to increase the recognition of the 5′ splice site by directly assisting in U1 snRNP binding or by increasing its accessibility to these sites by altering the pre-mRNA structure. In some cases alternative exons themselves have been identified as elements essential for correct regulation of alternative splicing. An exon of just 18 nts, exon N1 of c-src pre-mRNA, is only included in mature mRNA in neuronal cells. Steric hindrance between splicing factors bound at both ends of the exon may be responsible for skip splicing since this neuron-specific exon was included in non-neuronal cells when its size was incrased (Black, 1991). This led to the hypothesis that neuron-specific factors alter the strength of one of the splice sites of exon N1. If binding of splicing factors to one end of the exon was delayed until the intron at the other end was removed, the steric hindrance would be eliminated, suggesting that sequential removal of the introns rather than simultaneous removal is required for efficient inclusion of exon N1. Black (1992) recently showed that a positive-acting sequence just downstream of the N1 exon is required for its regulated splicing.

Exon 5 of cardiac troponin T pre-mRNA is another small exon (30 nts) that is alternatively included or excluded. Improving the poor 5′ splice site of exon 5 results in increased recognition of this exon. Cooper and Ordahl (1989) identified a region within exon 5 that when mutated eliminates or reduces use of this exon by disrupting 3′ splice site recognition (Cooper, 1992). Recently this region has been shown to be a purine-rich positive element that is important for activating the 3′ splice site of exon 5 (Xu et al., 1993). Although this sequence is apparently important for recognition of the small exon, it is not key to its developmentally regulated alternative splicing (Xu et al., 1993).

Leukocyte-common antigen pre-mRNAs are also differentially spliced in a cell-specific manner. Exons 4, 5, and 6 are spliced into mature mRNA in B cells but all three exons are skipped in thymocytes and other nonhematopoietic cells. Three sequence elements within exon 4 were found to be required for proper regulation, as mutation of these elements resulted in inclusion of exon 4 in thymocytes (Streuli and Saito, 1989). Two models were suggested to explain these results: (1) the constitutive splicing machinery cannot recognize exon 4 due to these negative regulatory elements and B cell-specific factors override this negative effect, or (2) thymocytes contain an inhibitor that prevents recognition of exon 4 but it is either not present or is inactive in B cells.

DIANNE HODGES AND SANFORD I. BERNSTEIN

C. MUTUALLY EXCLUSIVE EXONS

The best examples of mutually exclusive exon use are found in the α- and β tropomyosin genes. Figure 5 illustrates the splicing patterns of these transcripts. We review rat α-tropomyosin splicing first and then contrast it to the splicing patterns of rat and chicken β-tropomyosin transcripts.

Exons 2 and 3 of rat α-tropomyosin pre-mRNAs are spliced in a strict, mutually exclusive manner to generate mRNAs containing exons 1, 2, and 4 in smooth muscle or 1, 3, and 4 in all other tissues. In α-tropomyosin transcripts, exon 2 is never spliced to exon 3. Smith and Nadal-Ginard (1989) showed that the mutually exclusive nature of these exons is enforced by the close proximity of the branchpoint of exon 3 to the 5' splice site of exon 2. The location of the branchpoint, just 41 nt downstream of the 5' splice site bordering exon 2, sterically prevents simultaneous interaction of splicing factors at these two sites.

Mullen *et al.* (1991) found that the preferential inclusion of exon 3

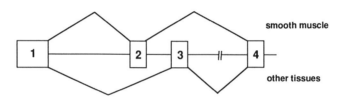

Alpha-tropomyosin (exons 1-4 only)

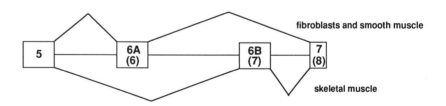

Beta-tropomyosin (exons 5-8 only)

FIG. 5. Vertebrate tropomyosin transcripts contain mutually exclusive exons. Rat α-tropomyosin mRNAs include either exon 2 (smooth muscle) or exon 3 (other tissues). Chicken β-tropomyosin uses either exon 6A (fibroblasts and smooth muscle) or 6B (skeletal muscle). Rat β-tropomyosin transcripts show the same type of splicing pattern, but the mutually exclusive exons are labeled 6 and 7. Note that only the relevant portions of the genes are depicted.

in mature α-tropomyosin transcripts is due to the very long polypyrimidine tract (50 nt) that follows the consensus branchpoint sequence 172 nt upstream of the 3' splice site of exon 3. The default choice of exon 3 clearly results from cis competition between the branchpoint-polypyrimidine tracts of exons 2 and 3, since substitution of the exon 2 branchpoint and polypyrimidine tract with the branchpoint and polypyrimidine tract of exon 3 resulted in a total switch from 1-3-4 to 1-2-4 splicing (Mullen *et al.*, 1991). In an attempt to identify factors mediating the preferential selection of exon 3, Patton *et al.* (1991) purified a 57-kDa protein named PTB that binds specifically to the polypyrimidine tract. PTB was previously identified as a ubiquitous splicing factor by Garcia-Blanco *et al.* (1989). See Section II,E for a further description of this protein.

The selection of α-tropomyosin exon 2 in smooth muscle cells results from negative regulation of exon 3 and is dependent on elements within and flanking that exon (Mullen *et al.*, 1991; Nadal-Ginard *et al.*, 1991). The mechanisms by which these elements act have not yet been determined but they do not appear to form a secondary structure that would mask recognition of exon 3. Unlike what is observed for 3' splice site choice, 5' splice site selection does not appear to be a primary determinant in choosing exon 2 or exon 3; improvement of the 5' splice site of exon 2 did not affect the predominant 1-3-4 splicing observed *in vivo* in nonmuscle cells (Mullen *et al.*, 1991).

The rat β-tropomyosin gene encodes skeletal muscle β-tropomyosin and fibroblast tropomyosin by mutually exclusive exon selection. Exon 7 is used in skeletal muscles while exon 6 is selected in fibroblasts and smooth muscle cells. Muscle-specific use of exon 7 is mediated by branchpoints located far upstream of the 3' splice site (positions -144, -147, and -153; Helfman and Ricci, 1989). The distant branchpoints in β-tropomyosin transcripts, however, do not account for the mutually exclusive nature of exons 6 and 7. These branchpoints are located more than 130 nt downstream of the 5' slice site, and steric constraints would therefore not be expected to prevent splicing of exon 6 to exon 7. Several mutations that lead to inclusion of muscle-specific exon 7 in nonmuscle cells have been identified (Helman *et al.*, 1990; Guo *et al.*, 1991). These included changes between the branchpoint and 3' splice site (but not at the 3' splice site region) as well as within exon 7. Overexpression of minigene transcripts *in vivo* also promoted inclusion of exon 7 in nonmuscle cells, suggesting that large amounts of RNA titrate out an inhibitor (Guo *et al.*, 1991).

A cellular factor that binds to both the upstream polypyrimidine tract and the intron regulatory element preceding exon 7 has been identified as PTB (Mulligan *et al.*, 1992). PTB binds poorly to mutant intron regulatory elements that allow inclusion of exon 7 in nonmuscle cells. In addition to possible regulation by PTB, a potential secondary structure that could sequester the 5' and 3' splice sites of exon 7 has been predicted by phylogenetic comparisons of the sequences of rat and chicken β-tropomyosins (Helfman *et al.*, 1990; Libri *et al.*, 1990, 1991). Thus a negative factor as well as the secondary structure may modulate exon 7 exclusion in nonmuscle cells (Guo *et al.*, 1991). This model implies that the negative regulator is absent or inactive in skeletal muscle and/or the secondary structure is disrupted. Exon 7 would then be recognized as the default choice owing to the presence of the long polypyrimidine tract and its positive interaction with PTB. The fact that the negative regulator appears to be PTB is difficult to explain in light of its presence in all cell types tested. It is possible that subsequent recruitment of additional tissue-specific regulatory factors to the negative regulatory site occurs, or that PTB activity is modulated in a tissue-specific manner by post-transcriptional or post-translational processing.

Transcripts produced from the chicken homolog of the rat β-tropomyosin gene are also alternatively spliced in a similar tissue-specific manner. Chicken β-tropomyosin exons 6A and 6B correspond to rat β-tropomyosin exons 6 and 7 respectively (Libri *et al.*, 1989a). Libri *et al.* (1989b) demonstrated correct developmental and tissue-specific splicing of a chicken β-tropomyosin minigene (containing exons 5 through 7 and associated introns) by stable transfection into quail myoblasts, but this regulation was lost in transient transfections. They proposed that loss of regulation might be due to titration of muscle-specific factors that mediate exon 6B use. Goux-Pelletan *et al.* (1990) identified a long polypyrimidine tract upstream of exon 6B which when partially deleted resulted in a switch from exon 6A to exon 6B utilization *in vitro*. The deletion appears to have disrupted a secondary structure that masks exon 6B recognition in nonmuscle cells.

As mentioned above, both rat and chicken β-tropomyosin pre-mRNAs are predicted to form similar secondary structures surrounding their respective skeletal muscle exons (Libri *et al.*, 1989a; Helfman *et al.*, 1990). Mutations in exon 6B and the upstream intron predicted to disrupt this secondary structure were analyzed *in vivo* (Libri *et al.*, 1990). These analyses identified two negative regulatory elements: the 3' end of intron 6A and the beginning of exon 6B (proposed to form stem I), and a region upstream of exon 6B proposed to base-pair with a purine-rich region in intron 6B and form stem III. The intron element upstream of exon 6B is in a location similar to the negative element in rat β-tropomyosin that is proposed to bind a negative factor in nonmuscle cells. However, Libri *et al.* (1990) favor a model in which secondary structure precludes recognition of exon 6B in nonmuscle cells, whereas secondary structure is altered in muscle cells, permitting exon 6B recognition. This was confirmed by compensatory mutations that restored base pairing as well as exon 6B exclusion in nonmuscle cells (Libri *et al.*, 1991).

In vitro splicing of transcripts containing an artificial stem III (created by hybridization of two transcripts in trans) also showed that removal of intron 6B is inhibited by such a secondary structure (Clouet D'Orval *et al.*, 1991a). In addition, analysis of the susceptibility of the region surrounding exon 6B to cleavage by various chemical and enzymatic probes (Clouet D'Orval *et al.*, 1991b) determined that mutations in stem I and in stem III induced structural changes around the 5' splice site of exon 6B and the exon 7 branchpoint, even though these sequence elements are not included in these stems. The authors proposed that the 5' splice site of exon 6B and the downstream branchpoint are unmasked by a general unfolding of the pre-mRNA in the derepressed state (as would be seen in skeletal muscle), resulting in the increased ability of spliceosomal components to interact with these sites. Results from studies of rat and chicken β-tropomyosin transcripts thus confirm a role for secondary structure in preventing utilization of the muscle-specific exons 7 and 6B in nonmuscle cells. However, the negative elements in stem I individually inhibit splicing when placed in a heterologous gene, indicating that secondary structure is not the lone regulatory mechanism for exclusion of exon 6B (Gallego *et al.*, 1992). Further, a positive regulatory element that enhances use of the 5' splice site of exon 6A is located just downstream of this exon (Gallego *et al.*, 1992). It is therefore likely that multiple mechanisms interact to maintain the strict regulation of β-tropomyosin expression. These include splice site blocking due to secondary structure formation, branchpoint and splice site competition (Libri *et al.*, 1992), cellular factors that stabilize or destabilize secondary structures, as well as trans-acting factors that enhance or block splice site or branchpoint recognition (Gallego *et al.*, 1992).

In contrast to the trans regulation exemplified by the above examples, alternative splicing of rat skeletal myosin light-chain 1/3 (MLC 1/3) transcripts is most likely dependent only upon cis-acting elements (Gallego and Nadal-Ginard, 1990). Two different pre-mRNAs are transcribed from the MLC 1/3 gene by use of alternative promoters for exons 1 and 2, but both transcripts are produced simultaneously in adult

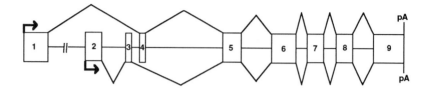

Myosin light chain 1/3

Fig. 6. Alternative promoters (indicated by arrows) and alternative splicing are used in vertebrate myosin light chain genes to produce myosin light chain 1 (upper splicing pattern) and myosin light chain 3 (lower splicing pattern) mRNAs.

muscle (Fig. 6). In MLC1 pre-mRNAs, exon 1 is spliced to exon 4 and exon 3 is skipped (exon 2 cannot be ligated to exon 1 due to the lack of a 3' splice site preceding it). However, exon 2 is spliced to exon 3 to generate MLC3 when transcription initiates within exon 2. MLC exons 3 and 4 are not intrinsically mutually exclusive, since Gallego and Nadal-Ginard (1990) demonstrated that the intron between the two exons is removed, both *in vitro* and *in vivo*, in the absence of competing splice sites. The failure to splice together the two exons is due to a low compatibility of the two splice sites within the intron separating exons 3 and 4 relative to the compatibility with the flanking splice sites. This preference for specific splice site pairing reflects cooperative interactions that do not appear to be based upon proximity or homology to the conserved sequence elements found at the 5' and 3' splice sites.

D. ROLE OF POLYADENYLATION IN ALTERNATIVE SPLICING

When pre-mRNAs contain multiple polyadenylation site choices, differential selection of such sites may correlate with alternative exon use. This is observed for calcitonin/CGRP RNA processing. In nonneuronal cells, the first four exons of the calcitonin/CGRP primary transcript are spliced and polyadenylated at the 3' end of exon 4, generating calcitonin mRNA (Fig. 7). However, in neuronal cells, cleavage and polyadenylation at the terminus of exon 6 occurs during splicing of exon 3 to exon 5, producing the neuropeptide CGRP. Although it is tempting to speculate that polyadenylation site choice regulates alternative exon choice, the regulated event appears to be the commitment to a particular splicing pathway. Leff *et al.* (1987) showed that both cell types are capable of using both poly(A) sites, regardless of the linear order of the sites in the primary transcript, when polyadenylation is assayed in the absence of the competing splicing reactions. When

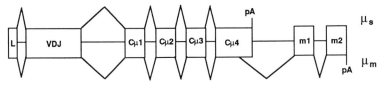

Immunoglobulin μ heavy chain

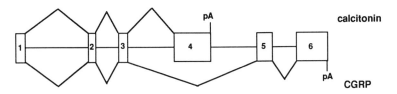

Calcitonin/CGRP

FIG. 7. Alternative polyadenylation sites in immunoglobulin μ heavy chain and cal-
citonin/CGRP transcripts are correlated with alternative splicing patterns. Polyadenylation at the end of the Cμ4 exon of immunoglobulin μ heavy chain RNA results in
a transcript encoding the secreted form of the protein, μ_s. Polyadenylation at the end of
exon m2 and use of a 5' splice site in exon Cμ4 results in inclusion of exons m1 and m2
in the mRNA, which encodes the membrane-bound form of the protein, μ_m. In calcitonin/
CGRP transcripts, inclusion of exon 4 and polyadenylation at its end results in mRNA
encoding calcitonin. Exclusion of exon 4 and inclusion of exons 5 and 6, with polyadenylation at the end of exon 6, results in mRNA coding for calcitonin gene-related peptide
(CGRP).

the calcitonin-specific poly(A) site flanking exon 4 was inactivated
and cells that express calcitonin were transfected with this construct,
the mutated precursor RNA was polyadenylated at the downstream
poly(A) site. However, these RNAs were incapable of undergoing either splice choice. The failure to splice in the calcitonin mode suggests
that polyadenylation at the calcitonin-specific poly(A) site removes
sequences that inhibit splicing of exon 3 to 4. Two regions flanking
exon 4 were found to be essential for proper splicing regulation. The
3' end of intron 3 contains an element that inhibits CGRP splicing in
non-neuronal cells whereas sequences flanking the 3' end of exon 4
were found to be necessary for CGRP splicing in neuronal cells (see
further discussion in Section III,B). The authors proposed that neuronal cells contain tissue-specific factor(s) or an increased amount or
activity of a constitutive splicing factor that promotes an RNA structure required for efficient splicing of exon 3 to exon 5. This structure
may simultaneously mask the calcitonin poly(A) site.

Another example of linkage between polyadenylation site choice and alternative splicing is the immunoglobulin μ heavy chain transcription unit which contains two potential polyadenylation sites that are separated by approximately 1900 nt (Fig. 7). The more 5' polyadenylation site is within the large intron (approximately 1860 nt) that separates the Cμ4 exon from the M1 exon. Therefore, cleavage and polyadenylation within the intron eliminate the possibility of splicing Cμ4 to M1 and generate the secreted form of the immunoglobulin, μ_s. Use of the more 3' polyadenylation site and splicing together of the Cμ4, M1, and M2 exons generates the membrane-bound form of the immunoglobulin, μ_m. In pre-B and mature B cells, about equal amounts of μ_s and μ_m are produced but upon differentiation into plasma cells, the ratio of μ_s to μ_m increases to 6–8:1. The determination of whether poly(A) site choice or splice site choice is the regulated event is critical to understanding this change in expression.

Experiments by Galli *et al.* (1987) demonstrated that the most 3' poly(A) site is intrinsically stronger than the upstream site and that transcription termination prior to the μ_m poly(A) site contributes to the preferential use of the μ_s poly(A) site in plasma cells. In addition, they proposed that plasma cells contain an increased amount of a poly(A) site factor that increases the use of the weaker μ_s poly(A) site. Peterson and Perry (1989) determined that the weak 5' splice site of exon Cμ4 contributes to this regulation because a mutation that converted it to an optimal 5' splice site resulted in a shift from predominantly μ_s to μ_m poly(A) site use in plasma cells. In addition, the 3' splice site of the M1 exon was found to be sequestered in a stable stem-loop structure that inhibited its use *in vitro* (Watakabe *et al.*, 1989). Therefore, the weak 5' splice site of Cμ4, the masked 3' splice site of M1, the different strengths of the poly(A) sites, and the large distance between the two poly(A) sites all play a role in ensuring a proper balance between cleavage-polyadenylation at the suboptimal μ_s poly(A) site and removal of the intron that contains this site.

In order to determine if there is an increase in μ_s cleavage-polyadenylation or a decrease in the efficiency of removal of the μ_m intron upon B cell differentiation, the efficiency of splicing and polyadenylation in B cells and plasma cells was compared using transcripts in which these two events were uncoupled (Peterson *et al.*, 1991). No differences in splicing efficiencies were observed whereas there was a 2-fold increase in use of the 5' poly(A) site in plasma cells compared to that in mature B cells. Thus the increased ability to utilize the most 5' poly(A) site as well as premature transcription termination probably accounts for the increased use of the μ_s poly(A) site in plasma cells. Although competition between splicing and polyadenylation plays a role in alternative production of secreted versus membrane-bound immunoglobulin μ, the event that is upregulated during B cell differentiation appears to be polyadenylation at the weaker poly(A) site.

E. ROLE OF SECONDARY STRUCTURE IN ALTERNATIVE SPLICING

Since nascent transcripts are bound by hnRNPs soon after transcription, the general perception has been that pre-mRNA is unable to form a large amount of secondary structure *in vivo* (Choi *et al.*, 1986; Dreyfuss, 1986; Beyer and Osheim, 1988). However, as discussed earlier, secondary structure has frequently been suggested to play a role in alternative splicing, and *in vitro* splicing evidence is quite compelling. Pre-mRNA utilized in splicing assays *in vitro* is transcribed in the absence of cellular RNA-binding proteins and most likely folds into secondary structures whose complexity is dependent upon the sequence context. Once the RNA is added to the splicing reaction, the stability of the structure in the presence of the nuclear extract components will determine if splicing is affected. *In vitro*, the secondary structure stability is dependent upon the degree and nature of the complementarity, the ionic strength of the buffer components, as well as the quantity and activity of helix-destabilizing components (hnRNPs, helicases, and other RNA unwinding activities) present in the nuclear extract.

A number of studies in addition to those previously discussed have addressed the influence of secondary structure on alternative splicing, both *in vitro* and *in vivo*. Solnick (1985) demonstrated that skipping of an internal exon was greatly enhanced by insertions of long inverted repeats of 105 nts into the flanking introns in an *in vitro* assay. The inverted repeats presumably sequestered the internal exon in the loop of a long stem-loop structure and masked its recognition by the splicing machinery. When similar transcripts were assayed by transient expression in HeLa cells, skip splicing was observed only at very low levels. There were also striking differences in the amount of secondary structure required to promote skip splicing between the in vitro and in vivo systems (Solnick and Lee, 1987). Pre-mRNAs with stems containing only 25 nts demonstrated skip splicing behavior *in vitro* but *in vivo* the stem had to exceed 50 nt to produce a discernible effect, suggesting that structures formed *in vivo* do not necessarily correspond to those formed *in vitro*. In contrast to these results, Goguel *et al.* (1993) recently showed that short hairpins that sequester the 5' splice site and branchpoint of yeast transcripts inhibit splicing both *in vitro* and *in vivo*. These authors showed that formation of the commitment complex in the spliceosome assembly pathway is blocked *in vitro* by the presence of such hairpin structures.

hnRNPs or other splicing factors may bind to the upstream portion of a stem structure prior to completion of transcription of the down-stream half of the stem and thus prevent secondary structure forma-tion *in vivo*. This hypothesis was tested by Eperon *et al.* (1986, 1988) by analysis of the relative use of test 5′ splice sites located 25 nt up-stream of a reference 5′ splice site. When analyzed by transient trans-fection into HeLa cells, the upstream test site was preferentially used. The test site was then placed within inverted repeats separated by variable distances to determine the effect of loop size on utilization of the test site. As expected, sequestration of the test site in a stem-loop structure *in vitro* resulted in almost exclusive use of the reference site regardless of the loop size. *In vivo* experiments revealed that there is a threshold loop size of 50–56 nt below which use of the test site was eliminated. Above this value, there was elevated use of the test site. The increased loop size may increase the window of time during which splicing factors can bind and prevent secondary-structure formation due to the delay in transcription of the downstream complementary sequences.

It is thus clear that artificially engineered secondary structures can affect splicing of pre-mRNAs both *in vitro* and *in vivo*. However, analy-sis of the secondary structure of naturally occurring transcripts, as dis-cussed in Sections III,C and III,D, is required in order to definitively establish a mechanistic role for secondary structure in alternative splicing. Some additional examples of this approach are discussed here. Chebli *et al.* (1989) demonstrated a positive role for secondary structure in splicing out of a small intron in adenovirus E1A pre-mRNAs *in vitro*. Mutagenesis of a short stem–loop structure in the intron between the distant branchpoints and the 3′ splice site de-creased removal of this intron but compensatory mutations that regen-erated the stem restored splicing. This suggests that the stem–loop decreased the operational distance between the branchpoints and the 3′ splice site to within the range determined to be optimal for efficient splicing in the absence of long polypyrimidine tracts.

Similarly, base pairing of sequences upstream of the 5′ splice site with sequences just 5′ of the branchpoint is proposed to play a positive role in the splicing of yeast RP15 pre-mRNA by decreasing the effec-tive distance between the 5′ splice site and the branchpoint (Goguel and Rosbash, 1993). Domenjoud *et al.* (1991) determined that a long stem–loop structure present in the exon just upstream of intron 1 of early adenovirus transcripts promotes use of the natural 5′ splice site over a cryptic 5′ splice site located 74 nt downstream. Watakabe *et al.* (1989) demonstrated a negative role for secondary structure using an *in vitro* system. They found that a stem–loop structure which sequestered the 3' splice site of an immunoglobulin exon was generated under *in vitro* splicing conditions and that mutations that disrupt the stem increase the removal of the upstream intron. Secondary structure affects splicing of yeast ribosomal RPL32 transcripts; splicing regulation is also dependent upon the protein product (L32) of these same transcripts (Dabeva *et al.*, 1986). Phylogenetic sequence comparisons and mutagenesis of conserved sequence elements identified a weak secondary structure formed between the 5' splice site and upstream sequences that is essential for regulated splicing (Eng and Warner, 1991). The L32 protein may inhibit splicing by stabilizing this weak secondary structure and thereby blocking access of U1 snRNA to the 5' splice site. It is likely that similar mechanisms operate in higher organisms and will be discovered in the near future. Use of both genetic and reverse genetic approaches should speed such analyses.

IV. Use of Drosophila to Study Alternative Splicing

As detailed below, alternative splicing can be studied in *Drosophila* by standard genetic approaches, by transforming the organism with *in vitro* mutagenzied genes and by using nuclear extracts. *Drosphila* and mammals share features of the constitutive and alternative splicing processes. Spikes and Bingham (1992) showed that spliceosome assembly has similar kinetics and components in *Drosophila* nuclear extracts and in vertebrate *in vitro* splicing systems. They also found that in *Drosophila,* as in mammals, introns that precede alternative exons can have remote branchpoints. In contrast to these similarities, Guo *et al.* (1993) found differences between HeLa and *Drosophila* nuclear extracts in their abilities to recognize a short *Drosophila* intron. While the *Drosophila* extract efficiently removed a 74-nt intron, the HeLa extract required that the intron be lengthened for efficient splicing. These authors further showed that the presence of a polypyrimidine tract is much more important for stimulation of intron removal in HeLa cell extracts. Overall, important generalities may be drawn from using *Drosophila* to study the process of alternative splicing; however, it is important to keep in mind that all aspects of the splicing process are not shared among the phyla.

A.   SEX DETERMINATION GENE TRANSCRIPTS

With the discovery that the *Drosophila melanogaster transformer* (*tra*) gene generates a female-specific product by alternative splicing, it became apparent that alternative splicing regulates the *Drosophila* sex determination pathway (Boggs *et al.*, 1987). This correlation was possible because the identity of the sex determination genes and their order of action had already been defined genetically (for reviews see Baker and Belote, 1983; Cline, 1985; Baker *et al.*, 1987; Baker, 1989; Steinmann-Zwicky *et al.*, 1990; Maniatis, 1991). In *Drosophila*, sex is determined by the ratio of the number of X chromosomes to autosomes (A); an X/A ratio of 1 results in female differentiation and an X/A ratio of 0.5 produces males. The gene *Sex lethal* (*Sxl*) is activated by the female X/A ratio, but the lower 0.5 ratio in males does not permit *Sxl* expression in the early embryo. The gene products of the *transformer* (*tra*) and *transformer-2* (*tra-2*) loci act downstream of *Sxl* but upstream of the *doublesex* (*dsx*) and *intersex* (*ix*) genes. *Sxl*, *tra*, *tra-2*, and *ix* gene products are all required for female but not male somatic sexual differentiation, indicating that maleness is the default state. In contrast, the *dsx* gene is required for both male and female sexual differentiation and it is expressed in alternative modes in the two sexes. The two *dsx* products have opposing effects; female *dsx* along with *ix* represses male differentiation whereas male *dsx* inhibits female differentiation. Molecular analysis of *tra* and *dsx* expression in various mutant backgrounds confirmed the above genetically defined pathway (Nagoshi *et al.*, 1988). The genetic and molecular actions of the sex determination genes are summarized in Fig. 8, and are discussed in more detail below.

1. Sex lethal

Molecular studies suggest that Sxl gene regulates sex-specific splicing of its own transcripts by binding to them and inhibiting use of a male-specific 3' splice site, resulting in utilization of an alternative female-specific 3' splice site (discussed in detail later). The absence of Sxl results in default use of the male-specific 3' splice site (Fig. 8). The *Sxl* gene has been cloned and the complex temporal and sex-specific expression pattern determined (Maine *et al.*, 1985; Salz *et al.*, 1989). Two sets of transcripts are produced; one set is expressed transiently during the blastoderm stage and a second set appears slightly later and continues to be expressed throughout development. Multiple late transcripts are detected in both males and females, with the only sex-specific differences being the inclusion of exon 3 in all male mRNAs. Bell *et al.* (1988) obtained and characterized male- and female-specific *Sxl* cDNAs. The female cDNA encodes two RNP consensus motifs, which is consistent with the hypothesis that Sxl is an RNA binding protein involved in directly regulating splicing of its own and other transcripts in the sex determination cascade. The male cDNA contains an extra exon which introduces two termination codons into the open reading frame. The truncated protein predicted from this cDNA lacks the RNA binding domains and would presumably be inactive in regulating splicing. Sequencing and expression studies of additional *Sxl* cDNAs confirmed that alternative splicing of exon 3 is the only sex-specific splicing event detected (Samuels *et al.*, 1991).

The female-specific *Sxl* transcripts have the potential to encode up to six different polypeptides but all would contain the RNP motifs (Samuels *et al.*, 1991). Abundant 36- and 38-kDa polypeptides as well as lower levels of Sxl proteins ranging in size from 36 to 42 kDa were identified on immunoblots of female fly proteins (Bopp *et al.*, 1991). The female Sxl proteins are localized in the nucleus consistent with their predicted splicing function and as expected, no full-length proteins were detected in males at any stage (Bopp *et al.*, 1991). Interestingly, when a female *Sxl* cDNA transgene encoding the 38-kDa protein was ectopically expressed in males, the 38-kDa protein as well as a 36-kDa protein were synthesized. Since the transgene only encoded the 38-kDa protein, the 36-kDa protein must have been produced from the endogenous *Sxl* gene (Bell *et al.*, 1991). This shows that the Sxl protein regulates the *Sxl* gene. Such autoregulation, which stably maintains the female mode of expression, was predicted on the basis of earlier genetic studies (Cline, 1984).

The ectopic expression of the Sxl protein in males induced a rapid (within 15 mins) switch from male-specific to female-specific splicing of *Sxl-lacZ* reporter transcripts (Bell *et al.*, 1991). The rapidity of this event suggests that the Sxl protein directly interacts with the splicing substrates. Sakamoto *et al.* (1992) transfected cultured *Drosophila* Kc cells with various *Sxl* constructs and showed that U-rich sequences surrounding the male-specific exon are important elements regulating exclusion of this exon. They further demonstrated that partially purified Sxl protein binds to these regions *in vitro*. These studies and *in vivo* analysis of heterologus transcripts containing the male-specific exon and surrounding intron sequences (Horabin and Schedl, 1993) suggest that the Sxl protein blocks use of the male-specific exon of *Sxl* transcripts.

Although the sexual state of the fly is maintained by alternative splicing of *Sxl*, *tra*, and *dsx* transcripts, the initial activation of the pathway appears to be regulated by sex-specific transcription of *Sxl* from its early promoter (Keyes *et al.*, 1992). The early transcripts are expressed only in 2–4-hr female embryos and differ from the late transcripts in that transcription initiates at a location analogous to intron 1 of the late pre-mRNAs (Fig. 8). The early transcripts splice early

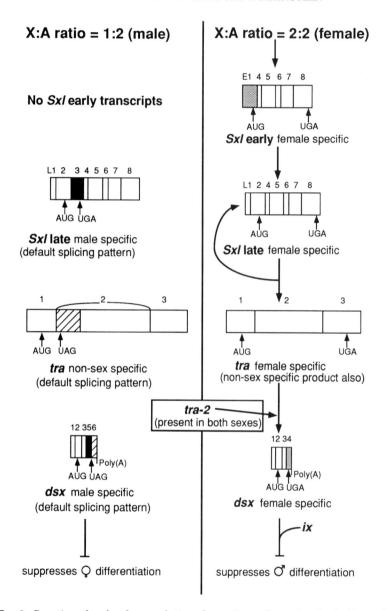

FIG. 8. Genetic and molecular regulation of somatic sex determination in *Drosophila*. Exons are designated as boxes. Male-specific exons are black or cross-hatched while female-specific exons are stippled. The female X to autosome ratio (2:2) results in *Sxl* transcription early in development, while no early transcripts are present in male embryos. Female-specific Sxl affects the alternative splicing pattern of late *Sxl* tran-

exon 1 directly to exon 4, thereby bypassing male-specific exon 3. Proteins generated from the early and late transcripts thus differ at their amino terminals but both contain the RNA binding motifs. When an early *Sxl* cDNA transgene was ectopically expressed in males, the endogenous *Sxl* gene was activated, confirming the initiation function of the early transcripts. Female-specific transcriptional activation of the early promoter is dependent upon the maternal *daughterless* (*da*) gene product as well as the dosage of *sisterless-a* (*sis-a*), *sisterless-b* (*sis-b*), and *runt* (Cline, 1986, 1988; Parkhurst *et al.*, 1990; Duffy and Gergen, 1991; Erickson and Cline, 1991). *sis-a* and *sis-b* are X-linked genes which are involved in communicating the X/A ratio to *Sxl;* a single X chromosome dosage of these gene products is not sufficient for transcriptional activation of *Sxl* in males. In the absence of the early transcript products, the late *Sxl* transcripts cannot be spliced to remove the translational terminating exon and thus only nonfunctional protein is produced in males.

2. *transformer*

As previously mentioned, *tra* is required throughout the *Drosophila* life cycle for normal female, but not male, development. Loss-of-function *tra* mutations transform XX females into sterile pseudomales and *tra* is thought to act in conjunction with *tra-2* to regulate the sex-specific splicing of *dsx* gene transcripts (see Fig. 8). Cloning of the *tra* gene (Butler *et al.*, 1986; McKeown *et al.*, 1987) and cDNA analysis (Boggs *et al.*, 1987; McKeown *et al.*, 1987) indicate that alternative splicing gives rise to a 1-kb female-specific RNA and a 1.2-kb non-specific RNA. The female-specific RNA encodes a protein of approximately 22-kDa containing a serine-arginine (SR) domain. The non-

---

scripts, resulting in exclusion of male-specific exon 3, possibly by blocking its 3' splice site. Female-specific late Sxl maintains this splicing pattern. The absence of the early transcripts in males results in default splicing of *Sxl* transcripts, yielding *Sxl* RNAs encoding a nonfunctional protein. Note that *Sxl* transcripts in both sexes can be polyadenylated at either of two sites within exon 8 or can be spliced to downstream exons 9 and 10 (not shown; see Keyes *et al.*, 1992). In both males and females, *tra* transcripts are spliced in the default mode, resulting in the production of a nonfunctional protein. However, there is also a female-specific pathway that results in a functional protein. These *tra* transcripts are spliced at an internal 3' splice site in exon 2, as a result of blocking of the nonsex-specific 3' splice site by Sxl. Female-specific tra acts in concert with tra-2 on *dsx* transcripts to activate inclusion of female-specific exon 4 and perhaps to enhance polyadenylation at the end of this exon. Female-specific dsx acts with ix to repress male somatic differentiation. Male-specific dsx prevents female somatic differentiation.

specific RNA contains translation termination signals and generates truncated, nonfunctional protein (McKeown et al., 1988). As predicted by the requirement for a functional tra gene for normal female development, P element-mediated expression of a female tra cDNA under heat shock control permits tra-null individuals with an XX background to develop as females instead of males; in addition, expression of this construct feminizes tra males that would otherwise be phenotypically wild type (McKeown et al., 1988).

At the molecular level, generation of the two tra transcripts involves splicing of exon 1 to one of two different 3' splice sites. The most downstream 3' splice site is used only in females. Both males and females use the upstream 3' splice site. Deletion of this nonspecific 3' splice site no longer permits sex-dependent splicing; both males and females now utilize the female 3' splice site but unspliced transcripts accumulate, indicating the poor efficiency of this splicing reaction (Sosnowski et al., 1989). In defining the cis-acting elements required for sex-specific splicing of tra RNAs, Sosnowski et al. (1989) showed that the intron between exon 1 and the female-specific 3' splice site confers sex-specific splicing to thymidine kinase pre-mRNAs in vivo, indicating that this stretch of DNA contains the necessary sequences for regulated splicing.

Since tra function is not required for regulated tra splicing (McKeown et al., 1988), it was suggested that alternative splicing of tra transcripts is directly regulated by Sxl, perhaps by a mechanism analogous to that used by Sxl to regulate splicing of its own RNA (Boggs et al., 1987). Comparison of the DNA sequence upstream of the nonsex-specific 3' splice site of tra with the male-specific 3' splice site of Sxl identified a conserved element containing a tract of 8 U residues (Bell et al., 1988; Sosnowski et al., 1989). Point mutations within the U tract of tra resulted in almost exclusive utilization of the nonspecific 3' splice site in females, implying that the mutations prevented Sxl binding. These results support the model in which Sxl blocks use of the nonsex-specific splice site and thereby promotes use of the downstream, less efficient, female 3' splice site. Inoue et al. (1990) independently arrived at this conclusion by coexpressing the tra gene and Sxl cDNAs in cultured Kc cells. When an expression vector containing a male Sxl cDNA or a female Sxl cDNA with a frame shift mutation was cotransfected with the tra gene, the nonsex-specific 3' splice site of tra was used almost exclusively. In contrast, expression of the wild-type female Sxl cDNA induced production of high levels of female-specific tra mRNA.

Inoue et al. (1990) demonstrated a direct interaction of Sxl with the conserved poly(U)-containing intron sequences by binding of *tra* RNA to bacterially expressed Sxl protein. *tra* RNA with U to C substitutions in this region, similar to those analyzed *in vivo* by Sosnowski *et al.* (1989), failed to bind the Sxl protein. Recently, Valcárcel *et al.* (1993) used an *in vitro* splicing system to show that Sxl prevents binding of U2AF to the poly(U) sequence in front of the nonsex-specific splice site of *tra* transcripts. Sxl does not bind efficiently to other types of polypyrimidine tracts, whereas U2AF does. Furthermore Sxl lacks a Ser-Arg-rich "effector domain" that is required to activate the 3′ splice site. A chimeric protein containing the effector domain of U2AF and the RNA binding sequence of Sxl does not block use of the nonsex-specific splice site. Valcárcel *et al.* conclude that Sxl most likely prevents use of the nonsex-specific splice site by out-competing U2AF, thereby permitting activation of the female-specific 3′ splice site.

3. transformer-2

Loss-of-function mutations of *tra-2* cause XX flies to develop as pseudomales. *tra-2*, like *tra*, is required for production of female *dsx* transcripts (Nagoshi *et al.*, 1988). The *tra-2* gene encodes polypeptides with an RNP consensus sequence as well as a Ser-Arg-rich region, motifs present in other proteins implicated in RNA processing (Amrein *et al.*, 1988; Goralski *et al.*, 1989). The importance of the RNA binding domain was demonstrated by the sexual transformation observed in XX flies carrying a *tra-2* allele with a single amino acid substitution within the RNP motif (Amrein *et al.*, 1990).

Although multiple *tra-2* transcripts are produced by use of different promoters and alternative splicing events, there are no sex-specific differences of expression in somatic tissues (Amrein *et al.*, 1990; Mattox *et al.*, 1990). Protein produced from a *tra-2/lac-Z* fusion gene was detected in both males and females and localized to the nucleus, indicating a lack of sex-specific translational control (Mattox *et al.*, 1990). One proposal to explain the female specificity of tra-2 action on *dsx* splicing is that it is mediated by its interaction with the female-specific tra protein (Amrein *et al.*, 1988). tra and tra-2 may form protein–protein associations required for RNA binding, or one protein may directly modify the other protein to its functional state.

4. doublesex

The *dsx* gene, like the *Sxl* and *tra* genes, generates different-sized transcripts in males and females (Baker and Wolfner, 1988; see also Fig. 8). These transcripts differ at their 3′ ends as a result of alternative splicing and polyadenylation (Burtis and Baker, 1989). In XX flies the first three common exons are spliced to a female-specific exon and the transcript is polyadenylated at the end of this exon; in XY flies the female exon is skipped and exon 3 is spliced to two downstream male-specific exons. Comparisons of the *dsx* 3′ splice sites indicates that the sequence upstream of the female-specific exon is purine, rather than pyrimidine, rich and thus is likely to compete poorly with the stronger male-specific 3′ splice site. Conservation of this weak female 3′ splice site between *Drosophila melanogaster* and *Drosophila virilis* suggests that it represents an important regulatory element (Burtis and Baker, 1989). Six copies of the 13-mer sequence TC$^T/_A$$^T/_A$CAATCAACA are found in the 5′ half of the female-specific exon and these appear to be regulatory elements as well (Burtis and Baker, 1989). The repeats are located in a region in which four dominant *dsx* mutations map (Nagoshi and Baker, 1990). The mutations result in expression of the male *dsx* transcript in females, confirming that these elements are essential for female-specific splicing. Since female-specific splicing of *dsx* is dependent upon both tra and tra-2 (Nagoshi *et al.*, 1988), these 13-mer sequences may represent sites of interaction with these proteins.

Ryner and Baker (1991) analyzed *dsx* pre-mRNA splicing more directly by transfection into *Drosophila* Schneider line 2 cells of a *dsx* minigene along with tra- and tra-2 producing plasmids. Expression of tra and tra-2 was shown to promote a switch from predominant use of the *dsx* male 3′ splice site to exclusive use of the female 3′ splice site. This is most likely due to activation of the female 3′ splice site rather than blockage of the male 3′ splice site since removal of the male 3′ splice site did not promote utilization of the female 3′ splice site in the absence of tra and tra-2. A large deletion that removed all six 13-mer repeats eliminated the tra/tra-2 mediated activation of the female 3′ splice site. In addition, these six repeats conferred tra and tra-2-dependent splicing to a heterologous pre-mRNA when inserted in its 3′ exon. Hoshijima *et al.* (1991) performed similar cotransfection experiments in Kc cells. In the absence of tra and tra-2, they found that *dsx* pre-mRNA was processed into the male-specific mRNA almost exclusively. While either a female-specific *tra* cDNA or a *tra-2* cDNA alone activated production of significant amounts of female *dsx* transcripts, coexpression of tra and tra-2 induced a complete switch to use of the female-specific *dsx* exon. The requirement for tra and tra-2, as well as for the 13-mer sequences in the exon, is eliminated by increasing the pyrimidine content of the female-specific 3′ splice site, indicating that binding of tra and tra-2 may increase recognition of the female-specific 3′ splice site by the splicing machinery. Hoshijima *et al.* also determined that deletion of three of the 13-mer sequences in the female-specific exon did not affect use of this exon, but deletion of all six repeats drastically reduced female-specific splicing.

Hedley and Maniatis (1991) demonstrated direct binding of bacterially produced tra-2 protein to *dsx* RNA containing six copies of the 13-mer repeat. The efficiency of the binding was dependent upon the number of repeats. By cotransfection experiments, they determined that deletion of the tra-2 binding sites (13-mer repeats) eliminated female-specific splicing. Results from all three laboratories are thus consistent with the hypothesis that tra and tra-2 binding to the female-specific exon enhances recognition of the weak upstream 3' splice site by the constitutive splicing machinery.

The possibility that tra and tra-2 activation of the female-specific 3' splice site of *dsx* results from activation of the female-specific 3' poly(A) site is somewhat controversial. Ryner and Baker (1991) concluded that increased use of the female poly(A) site only occurred in the presence of splicing and required tra and tra-2, indicating that 3' splice site activation was the regulated event. However, Hedley and Maniatis (1991) concluded that both splicing and polyadenylation were regulated by tra-2 since use of the female poly(A) site was affected by deletion of the tra-2 binding site.

Splicing of *dsx* exon 3 to female-specific exon 4 in an *in vitro* system is dependent upon the addition of tra and tra-2 proteins as well as the presence of the 13-mer repeats in the female exon (Tian and Maniatis, 1992). This study also determined that tra and tra-2 activate splicing of hybrid *dsx*/human β-globin transcripts containing the 13-mer repeats in the downstream exon; these transcripts contained suboptimal branchpoint or polypyrimidine tracts and were defective for splicing in the absence of tra and tra-2 and the repeat sequence.

Tian and Maniatis (1992) showed by UV cross-linking that tra binds nonspecifically to RNA in the absence of nuclear extract. However, cross-linked proteins of 29 and 38 kDa, most likely tra and tra-2 respectively, specifically interact with the *dsx* RNA 13-mer repeat region in the presence of nuclear extract. In addition, nuclear proteins of 24, 33, and 65 kDa also bind to the repeat sequence. This suggests that cooperative interactions between tra and these nuclear factors mediate the specificity of tra interaction with the female-specific exon. The SR domain of tra was proposed to mediate contact with these nuclear factors since this domain functions in both protein-protein and protein–RNA interactions. Tian and Maniatis (1993) recently showed that several SR proteins are among those recruited to the female-specific *dsx* exon by tra and tra-2, and that commitment to the female splicing pattern requires specific members of the SR family. The mechanism by

which the female-specific 3' splice site is activated by the binding of
tra, tra-2, and other nuclear factors is not clear, but may involve recruitment of U2AF and other splicing factors that promote recognition
of the suboptimal 3' splice site of this exon. Wu and Maniatis (1993)
have shown that this recruitment may occur by direct binding of tra
and tra-2 to U2AF[35]. This would permit U2AF[65] (which is tightly
bound to U2AF[35], but normally could not recognize the poor polypyrimidine tract preceding this exon) to initiate U2 snRNP interaction at
the 3' splice site.

5. Role of Alternative Splicing in the Germ Line

Although alternative splicing of tra-2 transcripts does not play a
role in sex determination of somatic tissues, alternative splicing occurs
in tra-2 pre-mRNAs in the male germ line (Mattox and Baker, 1991).
Two transcripts specific to the male germ line originate from a downstream promoter in exon 3 and differ only in the presence or absence
of the intron between exons 3 and 4 (M1 intron). M1-containing transcripts make up 60–70% of the tra-2 mRNA. Both transcripts have the
potential to encode polypeptides with RNP consensus sequences. One
of these polypeptides may be involved in promoting retention of intron
M1 in mature mRNA, since in tra-2 nulls there is a 2- to 3-fold increase
in removal of intron M1, and unspliced transcripts are undetectable.
This hypothesis was confirmed by the demonstration that retention of
the M1 intron in tra-2/lac Z reporter transcripts in vivo was dependent
on the presence of functional tra-2 protein. The polypeptide encoded by
M1-containing transcripts is not sufficient for regulation of its own
synthesis, suggesting that the fully-spliced male germ line specific
transcripts encode this function. Thus, increased splicing of the M1
intron leads to increased levels of the products mediating repression of
M1 intron splicing. As described in Sections III,E and IV,B splicing of
yeast ribosomal protein gene L32 transcripts and Drosophila su(wᵃ)
transcripts is regulated by similar negative feedback mechanisms (Dabeva et al., 1986; Zachar et al., 1987a,b).

B. SUPPRESSOR OF WHITE APRICOT

Mutations in the suppressor of white apricot, su(wᵃ)⁺, gene of Drosophila increase eye pigmentation in whiteᵃᵖʳⁱᶜᵒᵗ,wᵃ, flies. The wᵃ mutation arises from insertion of a copia transposable element into an
intron of the white gene (Bingham et al., 1981; Goldberg et al., 1982;
Levis et al., 1984; Pirrotta and Brockl, 1984; Zachar et al., 1985). Since
white is required for normal red pigmentation of the eye, insertion of
the copia element and truncation of the transcript within the element

by polyadenylation drastically reduces pigment deposition, resulting in an apricot eye color. The minor amount of pigment deposition appears to arise from occasional transcriptional read-through and splicing out of the intron containing the copia element (Levis *et al.*, 1984; Pirrotta and Brockl, 1984; Zachar *et al.*, 1985). Mutation of the $su(w^a)^+$ gene apparently increases this read-through and splicing process by suppressing polyadenylation and/or increasing splicing efficiency.

Molecular cloning of the $su(w^a)^+$ gene led to the discovery that its transcripts are alternatively spliced (Zachar *et al.*, 1987b). Early in embryogenesis, all introns are removed; however, following the cellular blastoderm stage, decreased levels of completely spliced RNAs accumulate, with transcripts that retain the first intron or the first two introns dominating during larval, pupal, and adult stages (Chou *et al.*, 1987). The partially spliced transcripts apparently do not encode functional protein products, while the completely spliced RNA codes for a protein with a high serine-arginine content at its C-terminal (Chou *et al.*, 1987; Zachar *et al.*, 1987a). As described above, this is a motif common to several RNA binding proteins, and appears to serve as a signal for localizing them to nuclear domains containing splicing components (Li and Bingham, 1991).

By producing transgenic flies with frame-shift mutations in the $su(w^a)^+$ gene, Zachar *et al.* (1987a) elegantly demonstrated that the open reading frames of the completely spliced $su(w^a)^+$ transcript are required for the gene to function. Mutations within the coding regions increased pigmentation of w^a eyes, indicating that the $su(w^a)^+$ product normally functions to allow polyadenylation within the transcribed copia element insert of w^a and/or to prevent splicing together of the adjacent exons. Perhaps most interestingly, accumulation of partially spliced $su(w^a)$ transcripts is dependent upon the production of the functional protein product from the completely spliced transcripts. Thus the $su(w^a)^+$ protein appears to act upon transcripts from its own gene by negatively regulating splicing out of the first, and perhaps the second, intron of $su(w^a)^+$ premessenger RNAs. The cis-acting signals on the $su(w^a)^+$ pre-mRNA and the molecular details of how the trans-acting factor blocks splicing are still to be determined, although the presence of a distant branchpoint in intron 1 has been suggested to play a role in regulated splicing (Chou *et al.*, 1987).

C. SUPPRESSOR OF SABLE

In *Drosophila melanogaster,* a *vermilion (v)* eye color mutation arising from insertion of a transposable element can be suppressed by mutations at the *suppressor sable* [*su(s)*] locus. The *su(s)* mutations affect splicing patterns of the RNA produced from the mutant v allele by activating any of four cryptic 5' splice sites and one 3' splice site within the retrotransposon, resulting in removal of most of the transcribed element from the mature RNA (Fridell et al., 1990). A reversion mutation in this v allele results from insertion of a second transposable element into the first; this permits efficient splicing out of the transposable elements using a 5' splice site in the second element and the aforementioned 3' splice site (Pret and Searles, 1991). $su(s)$ mutations cause a further 2-fold increase in use of this splicing pathway. Since cryptic splice sites are activated in $su(s)$ mutants, the $su(s)$ wild-type gene product may be involved in suppressing their use. The 150-kDa suppressor of sable protein is likely to bind RNA since it contains an RNP consensus motif and a highly charged region similar to those found in other RNA binding proteins, including the 70-kDa snRNP protein, the suppressor of white apricot protein, and the transformer and transformer-2 proteins (Voelker et al., 1991). The suppressor of sable protein is found in the nucleus, further implying a role in RNA processing (Voelker et al., 1991). It is possible that the effect of $su(s)$ mutations on splicing may be indirect; the absence of suppressor of sable protein could increase the stability of mutant v pre-mRNAs, allowing them to be processed instead of degraded (Voelker et al., 1991).

D. P ELEMENT TRANSCRIPT SPLICING

Transposition of the P element in Drosophila melanogaster is responsible for the phenomenon of hybrid dysgenesis (see Rio, 1990, for review), in which offspring of a P strain male and an M strain female display high levels of infertility and germ line mutations (Kidwell et al., 1977). The P element is 2.9 kb in length with terminal inverted repeats, and contains four exons that encode a transposase which is responsible for movement of both intact and internally deleted P elements (Karess and Rubin, 1984; O'Hare and Rubin, 1983). P element mobility is normally limited to the germ line (Laski et al., 1986; McElwain, 1986). Somatic cell transcripts retain the 3'-most intron of the P element. Mutation of the splice junctions flanking this intron prevents germ line transposition, leading to the conclusion that germ line-specific removal of this intron is critical for transposase activity (Laski et al., 1986). These experiments were performed using a convenient genetic assay for P element mobility that depends on observation of a singed bristle phenotype. In this assay, transposition of defective P elements at the singed^{weak} locus results in wild-type or singed^{extreme} phenotypes.

High-level transcription of a P element (under control of the heat shock 70 promoter) does not permit somatic cell transposition, presumably owing to the failure to splice the two terminal exons together (Laski *et al.*, 1986). P element transposition can occur somatically when the 3'-most intron is deleted from the gene prior to injection of the construct (Laski *et al.*, 1986). Depending on the genotypes of the offspring, somatic transposition resulted in organisms with either mosaic bristle phenotypes (due to the instability at the *singed*[weak] locus in somatic cells) or mottled eye color (resulting from excision of a P element containing an active *white*[+] gene or transposition of an element containing a barely-active *white*[+] gene to a location where it is more highly expressed).

1. In Vivo Mapping of cis-Regulatory Regions

To map the cis-acting sequences required for germ line-specific splicing, Laski and Rubin (1989) created a series of deletions within the 3'-most intron and assayed each construct's ability to induce germ line or somatic transposition using the singed bristle and white eye phenotypic assays described earlier. They showed that most of the intron sequences are not required for regulated splicing. Furthermore, the suboptimal flanking splice sites are not key regulatory elements in this process since mutating them to the consensus sequences did not affect tissue-specific splicing. The branchpoint sequence has also been ruled out as regulating alternative splicing of the P element intron (Chain *et al.*, 1991).

To further delineate the important cis-regulatory elements for germ line-specific splicing of P element transcripts, Laski and Rubin (1989) constructed a fusion gene by joining the two 3' terminal exons and their associated intron to the *Escherichia coli lacZ* gene. The heat shock 70 promoter was used to drive expression of this construct, and production of a functional fusion protein (displaying β-galactosidase activity) was dependent upon splicing out of the regulated P element intron. Germ line and somatic cell expression were subsequently assayed by incubation of dissected organisms from transformant lines in the chromogenic substrate X-gal. Analysis of these hybrid genes and a series of additional constructs reported by Chain *et al.* (1991) mapped the cis-acting domain to the exon preceding the regulated intron, between 12 and 31 bases upstream of the 5' splice site. Interestingly, this region contains one of several nearby "pseudo-5' splice sites" which may compete with the true 5' splice site to regulate splicing in somatic tissues (for further discussions see Siebel and Rio, 1990; Rio, 1991).

2. Biochemical Analysis of trans-Acting Factor Involvement

In vitro splicing studies have begun to yield information about trans-acting factors involved in regulating P element transcript splicing. Siebel and Rio (1990) found that Drosophila Kc cell nuclear extracts were not able to remove the 3'-most intron from P element transcripts, but HeLa cell extracts produced accurately spliced product. Preincubation of the transcript with increasing amounts of Kc cell extracts prior to addition to the HeLa splicing system progressively inhibited splicing. Since no intermediate products accumulated, inhibition appears to occur prior to 5' splice site cleavage. Inhibition was prevented by heat treatment of the Kc cell extract, suggesting that protein components are involved. RNA–protein cross-linking experiments showed that a 97-kDa protein from Kc cell extracts (but not HeLa extracts) bind to the exon region upstream of the 5' splice site. Competition experiments indicated that short transcripts containing this region of the RNA prevent the binding of an inhibitory factor in the Kc cell extract to a longer P element transcript, permitting splicing to occur in the HeLa extract that was supplemented with Kc cell extract. This may be due to removal of the 97-kDa protein, although several other proteins that are somewhat less specific in their binding abilities interact with this region of the transcript or the intron (two of the proteins are 40 kDa and one is 65 kDa). Siebel and Rio (1990) hypothesize that the 97 kDa protein acts as a blocking factor to prevent 5' splice site recognition in somatic cells or perhaps acts to channel splicing factors to "pseudo-5' splice sites" located near the true 5' splice site. When Chain et al. (1991) used UV cross-linking to examine Drosophila Kc cell proteins binding to the 5' splice site and surrounding regions, they detected proteins of 95, 45, and 40 kDa. These proteins bind preferentially to wild-type transcripts compared with transcripts containing mutations in the cis-acting regulatory domain upstream of the 5' splice site.

In agreement with the results of Siebel and Rio (1990), Tseng et al. (1991) failed to observe splicing out of the 3'-most intron of P element transcripts in Kc cell extracts. In HeLa extracts, Tseng et al. (1991) found that mutations in the cis-acting regulatory sequence of the penultimate exon greatly increased the efficiency of removal of the downstream intron. This suggested that a negative-acting factor in HeLa extracts inhibited regulated splicing of wild-type transcripts. Evidence for this was obtained by showing that introduction of short competitor transcripts containing the cis-acting regulatory sequence into the splicing reaction permitted much higher levels of splicing of wild-type transcripts. In cross-linking experiments using HeLa cell extracts, proteins of 32–35 kDa and 43 kDa were shown to bind preferentially to wild-type transcripts rather than to transcripts having mutations in the cis-acting regulatory region. Tseng *et al.* (1991) suggest that these proteins may serve to block splicing, or perhaps permit transport from nucleus to cytoplasm of the unspliced RNA.

The absence of splicing of the 3′-most intron in *Drosophila* Kc extracts and the presence of proteins that bind the cis-acting regulatory sequence suggest that negative regulation occurs in somatic tissues to inhibit intron removal. It is currently not clear how splicing occurs in germ-line cells. Perhaps the absence of splicing inhibitory factors allows intron removal in this tissue. However, since overexpression of P element transcripts in somatic cells *in vivo* (Laski *et al.*, 1986) fails to titrate out the negative factor and permit removal of the regulated intron, a positive factor may be required to permit P element transcript splicing in *Drosophila*.

E. Contractile Protein Gene Transcripts

1. Myosin Heavy Chain

Muscle myosin heavy chain (MHC) in *Drosophila* is encoded by a single gene, but as many as 480 isoforms of the protein may result from alternative RNA processing (Bernstein *et al.*, 1986; Rozek and Davidson, 1986; George *et al.*, 1989; Collier *et al.*, 1990; Kronert *et al.*, 1991). An interesting mutation which prevents accumulation of MHC in the indirect flight muscles arises as a result of a GU to GA transversion at the 5′ splice site of alternative exon 9a (Kronert *et al.*, 1991). Exon 9a, rather than exons 9b or 9c, is normally utilized in this muscle type, and the mutation results in the production of aberrant transcripts which contain intron 8, exon 9a, intron 9a, exon 9b, intron 9b, exon 9c and a portion of intron 9c joined at a cryptic 5′ splice site to exon 10. As in the wild-type indirect flight muscles, the small cells of the jump muscles normally use exon 9a. Remarkably, the jump muscles of the mutant produce fully processed transcripts, all containing exon 9b.

Unlike the flight muscles, the jump muscles have the plasticity to switch to utilizing exon 9b when exon 9a is inactivated. This indicates there is a fundamental difference between the two muscle types in how the choice for alternative exon 9 is made. In the flight muscles there may be a negative factor preventing inclusion of exons 9b and 9c, since these exons are not utilized, even when exon 9a is inactivated. In the jump muscles, exon 9b is accessible to the splicing apparatus, even though exon 9a is normally used. Therefore splice site competition, rather than negative factors, may regulate choice between alternative exons in this muscle type. It is of course possible that factors required for the activation of exon 9a or 9b splice sites are restricted to specific muscle types. This possibility is reasonable since the 5' splice sites of the exon 9 family differ significantly from the consensus for U1 snRNA binding.

Alternative splicing of exon 18 of the MHC gene has been studied by reverse genetic analysis *in vivo* (Hess and Bernstein, 1991) and in a *Drosophila in vitro* splicing system (Hodges and Bernstein, 1992). Exon 18 is excluded from all larval and some adult muscles, but is included in the bulk of the adult musculature (Bernstein *et al.*, 1986; Kazzaz and Rozek, 1989). Transcripts from a minigene containing exons 17, 18, and 19 along with associated introns and polyadenylation signals are correctly processed after the minigene is introduced into organisms by germ-line transformation (Hess and Bernstein, 1991). Thus this substrate contains the necessary cis regulatory signals for stage-specific splicing. Exon 18 is skipped when a similar substrate is spliced in nuclear extracts from undifferentiated Kc cells (Hodges and Bernstein, 1992). Since these cells are of embryonic origin, it is perhaps not surprising that exon 18 is skipped; however, since Kc cells do not express the MHC gene, skipping may represent the constitutive splicing pattern. In the *in vitro* system, exon 17 is unable to splice to exon 18, even when exon 19 is not in the primary transcript. Likewise, exon 18 splices poorly to exon 19, even in the absence of the competing exon 17. Poor recognition of exon 18 splice sites appears to be due to the nonconsensus splice sites themselves, since mutation toward consensus allows *in vitro* splicing (Hodges and Bernstein, 1992). Interestingly, both splice junctions must be mutated to consensus for exon 18 to be included in the final spliced product. If only one splice junction is changed to consensus, exon 18 is skipped *in vitro*. These results are borne out *in vivo* as well, where inclusion of exon 18 at the larval stage occurs at significant levels only when both splice cites of exon 18 are changed to the consensus sequences (Hess and Bernstein, 1991; D. Hodges, R. M. Cripps, and S. I. Bernstein, unpublished observations).

2. α-Actinin

Roulier *et al.* (1992) have found that the *fliA³* mutation disrupts *Drosophila* α-actinin mRNA splicing by changing the alternative splice site AG↓GUUGGAA to AA↓GUUGGAA. Although this mutation does not alter the coding sequence of the exon, it prevents the use of this splice junction, which is normally activated in adult muscle. Instead, a downstream 5' splice site normally used only in larval muscles appears to be activated. Thus, splice site competition in the adult muscle environment favors the larval 5′ splice site over the mutated adult splice site, or a positive factor required for activation of the adult splice site cannot recognize the mutated sequence.

F. Genetic Approaches for Identifying trans-Acting Splicing Factors in *Drosophila*

As illustrated above, genetic analysis in *Drosophila* has provided important insights into mechanisms of alternative RNA splicing. Interestingly, the sex determination pathway, which to date provides our best understanding of trans-acting alternative splicing factors, was defined at the genetic level well before alternative splicing was implicated in this process. Likewise, *su(wᵃ)* was isolated to study a genetic phenomenon and later found to encode an alternative splicing factor by molecular analysis.

Drosophila genetics should also be able to be used to identify trans-acting factors for other genes known to display alternative splicing. For example, selection for revertants of the *fliA³* α-actinin mutation should permit the isolation of cis-acting mutations that allow the mutant splice site to function. More important, mutations at other loci which permit splicing at the mutant splice junction should delineate trans-acting factors which regulate alternative splice junction function. If a phenotype (like flight ability) is not available for screening purposes, it may be possible to employ a selectable marker. For instance, in-frame addition of a neomycin-resistance gene to the terminal exon of the P element or to the adult-specific exon of a MHC minigene could permit selection for mutants displaying resistance to the drug G418. Since the RNA splices in question are not normally made in somatic tissue or larval muscle, respectively, resistance would indicate either a cis-acting mutation or a trans-acting factor muta tion that permits splicing. These genetic approaches are fraught with possible pitfalls since it is unclear whether a readily isolated domi nant mutation will be elicited or whether such a mutation would be lethal because it would affect alternative splicing events in transcripts of other genes. However, such genetic approaches have proved useful in delineating yeast RNA splicing factors (see Section II,G), and therefore merit attention for alternative splicing studies in *Drosophila*.

V. Mutations at Vertebrate Splice Sites and Their Effects on RNA Splicing

Defective RNA splicing is at the root of a number of human genetic diseases. The molecular lesions are often splice site mutations which result in aberrant RNA processing and yield a nonfunctional transcript. Such mutations illustrate the effects that might be expected when a particular alternative splice site is inactivated by RNA secondary structure or masked by negative splicing regulators. Results from such studies are therefore instructive for studying the mechanism of alternative RNA splicing. Three classes of splice site mutation effects are known: failure to remove an intron, cryptic splice site activation, and exon skipping. These phenomena and their relationships to models for alternative RNA splicing are discussed in this section.

A. Intron Retention

Intron retention has occasionally been reported to result from a splice junction mutation. Ohno and Suzuki (1988) showed that intron 13 can be retained in transcripts of the β-hexosaminidase gene due to a G to C mutation at site $+1$ of the intron. This mutation results in Tay-Sachs disease. Another example of intron retention arises from a mutation at a 5' splice site of the p53 gene (Sameshima et al., 1990). The resulting absence of the p53 protein apparently can cause small cell lung carcinoma.

It is important to note that in vivo instability of unspliced transcripts may prevent observation of intron retention. Unspliced transcripts may not accumulate due to a coupling between splicing and transport into the cytoplasm or perhaps because of the observed instability of transcripts containing a premature stop codon (Baserga and Benz, 1988; Cheng et al., 1990). If an unspliced transcript were stable (as is certainly the case for some alternatively spliced RNAs), then blockage of a splice junction leading to intron retention clearly would be a reasonable mode of alternative RNA splicing.

B. Cryptic Splice Site Activation

Many of the splice junction mutations first studied were within human globin genes and such mutations result in thalassemias (see review in Treisman et al., 1983b). Mutation of splice junctions in these genes sometimes leads to activation of cryptic splice sites. This is true for mutations of a 3' splice site which results in activation of a cryptic 3' splice site within the adjacent exon and some mutations of a 5' splice site which results in activation of several cryptic 5' splice sites (Aebi et al., 1986). Another more complicated example occurs when a mutation creates a 5' splice junction in the intron of the β-globin gene (Treisman et al., 1983a). This new splice junction is used as well as a cryptic 3' splice site located upstream in the intron, thus creating a new exon. These examples demonstrate that blockage of a particular splice site, or indeed availability of a new splice site, can activate a cryptic splice junction, resulting in an alternative splicing pattern.

C. Exon Skipping

A number of mutations resulting in human disease induce exon skipping, and most of these are within the 5' splice site of the skipped exon (see Talerico and Berget, 1990, for a review). In particular, several mutations in human collagen genes arise from exon skipping due to 5' splice site mutations (see Kuivaniemi et al., 1991, for review). Since mutation of a 5' splice site can cause exon skipping, an attractive hypothesis to explain exon skipping during alternative splicing involves blockage of a 5' splice site by RNA secondary structure or factor binding (Mitchell et al., 1986; Robberson et al., 1990).

Exon skipping may be favored when cryptic branchpoints or splice sites are absent in the pre-mRNA. Exon skipping may also appear to be favored over intron retention in some mutations, but only because of the instability of unspliced transcripts (see Section V,A). On the other hand, exon skipping may not be an option for processing some mutant transcripts. For instance, if the splice junctions upstream and downstream of the exon with the mutated splice site are incompatible (Gallego and Nadal-Ginard, 1990), they may not be able to be spliced together. Furthermore, if the mutated exon is not flanked by exons on both sides, that is, it is the first or last exon of the gene, then it obviously will not be skipped.

D. Exon Definition Model

As discussed in Section II,D, Robberson et al. (1990) have devised an interesting model which explains exon skipping. This model is important in considering the mechanics of alternative RNA splicing. The exon definition model states that both 5' and 3' splice sites of an exon must be recognized by the splicing apparatus in order for the exon to be "defined" and included in the mature transcript. Certainly the abundant examples of exon skipping induced by naturally occurring 5' splice site mutations (see review in Talerico and Berget, 1990) support this model since the failure to recognize the 5' splice site of an exon usually prevents inclusion of that exon in the mRNA. Experimental support for this hypothesis is provided by studies which demonstrate that assembly of spliceosome components upon a splicing substrate lacking an upstream 5' splice site is greatly enhanced by the presence of a downstream 5' splice site (Robberson et al., 1990). In studies directed at analyzing the mechanism of exon skipping in vitro, Talerico and Berget (1990) have shown that splicing of an internal exon to its flanking exons is strongly inhibited by mutations at the 5' splice site of the internal exon. Kuo et al. (1991) and Grabowski et al. (1991) demonstrated that enhanced U1 binding to the 5' splice site of an internal alternative exon of rat preprotachykinin transcripts dramatically increases inclusion of that exon in the processed transcript. In agreement with the exon definition model, removal of both the upstream and downstream introns is dependent upon strong U1 binding at the 5' splice site of this exon. Strong binding of U1 and the 70-kDa U1 protein might allow interactions with SR proteins and $U2AF^{35}$. This could bring the associated $U2AF^{65}$ subunit into contact with the upstream 3' splice site and initiate splice site recognition by U2 followed by upstream intron removal (Wu and Maniatis, 1993).

Overall, the exon definition model is a useful conceptual framework for further exploration of exon skipping. Masking of a single splice junction of an exon can clearly result in its exclusion from the mature mRNA. Thus secondary structures that inhibit recognition of a splice site (Libri et al., 1991), trans-acting factors that block a particular splice site, or perhaps even post-transcriptional modifications of splicing signals [by such processes as RNA editing (Feagin et al., 1987), inosine conversion of base-paired adenosine residues (Bass and Weintraub, 1988), or other base conversions (Powell et al., 1987)] may be implicated in exon skipping during alternative RNA splicing.

VI. Conclusions and Perspectives

Since the discovery of pre-mRNA splicing nearly 20 years ago, a tremendous amount has been learned regarding the mechanics of splice site definition, and the factors involved in RNA cutting and ligation. As we have discussed, both biochemical and genetic approaches have been key elements in these advances. Unlike self-splicing RNA ribozymes, which can be cut and joined without additional protein or RNA components (Cech and Bass, 1986), pre-mRNAs are processed in a spliceosome complex containing snRNAs and various proteins. These components enter and leave the complex during the splicing process and many of their roles are yet to be elucidated. In addition, new splicing factors are still being discovered. As with self-splicing RNAs, it is likely that RNA acts to catalyze the transesterification reactions involved in cutting and joining of the pre-RNA, but protein components are likely to be involved as well. This is not unlike the situation for the ribosome, a protein–RNA complex that performs protein synthesis in which the RNA component has recently been implicated in catalyzing the formation of peptide bonds (Noller *et al.*, 1992). Genetic analyses of RNA splicing in yeast, which involve isolating mutations that disable splicing as well as suppressors of these mutations, have defined a series of proteins involved in the splicing process. Continued mutant analysis in concert with biochemical assays of protein properties will elucidate their functions.

A number of the components involved in constitutive splicing also play a role in alternative splicing. It is becoming clear that different regulators or combinations of regulators can affect alternative splicing of different transcripts. Brute force biochemistry coupled with *in vitro* splicing systems have been useful in defining a few of these. Perhaps most instructive has been the discovery of several genes in *Drosophila* which encode alternative splicing regulators. Biochemical systems are being developed that should allow determination of precisely how these *Drosophila* factors control alternative splicing. Expansion of the genetic approach to include direct selection for alternative splicing factors should permit identification of additional splicing components and definition of how they function. The genetic approach may prove a useful future direction for eventually understanding the basis of alternative RNA splicing in a wider variety of pre-mRNAs.

ACKNOWLEDGMENTS

We thank Richard Cripps, Phillip Singer, Linda Wells, and Michelle Mardahl for critical comments on the manuscript, and numerous colleagues in the field who provided relevant reprints and manuscripts. We also thank Ted Wright for soliciting this paper. Our research has been supported by grants from the National Institutes of Health (GM32443) and the Muscular Dystrophy Association. S. I. Bernstein is an established investigator for the American Heart Association.

REFERENCES

Abovich, N., Legrain, P., and Rosbash, M. (1990). The yeast PRP6 gene encodes a U4/U6 small nuclear ribonucleoprotein particle (snRNP) protein, and the PRP9 gene encodes a protein required for U2 snRNP binding. *Mol. Cell. Biol.* **10**, 6417–6425.
Adema, G. J., and Baas, P. D. (1991). Deregulation of alternative processing of calcitonin/CGRP-I pre-mRNA by a single point mutation. *Biochem. Biophys. Res. Commun.* **178**, 985–992.
Aebi, M., Hornig, H., Padgett, R. A., Reiser, J., and Weissmann, C. (1986). Sequence requirements for splicing of higher eukaryotic nuclear pre-mRNA. *Cell* **47**, 555–565.

Alahari, S. K., Schmidt, H., and Käufer, N. F. (1993). The fission yeast prp4⁺ gene involved in pre-mRNA splicing codes for a predicted serine/threonine kinase and is essential for growth. *Nucleic Acids Res.* **21,** 4079–4083.

Amrein, H., Gorman, M., and Nöthiger, R. (1988). The sex-determining gene *tra-2* of *Drosophila* encodes a putative RNA binding protein. *Cell* **55,** 1025–1035.

Amrein, H., Maniatis, T., and Nöthiger, R. (1990). Alternatively spliced transcripts of the sex-determining gene *tra-2* of *Drosophila* encode functional proteins of different size. *EMBO J.* **9,** 3619–3629.

Anderson, G. J., Bach, M., Lührmann, R., and Beggs, J. D. (1989). Conservation between yeast and man of a protein associated with U5 small nuclear ribonucleoprotein. *Nature (London)* **342,** 819–821.

Arenas, J., and Abelson, J. (1993). The *Saccharomyces cerevisiae PRP21* gene product is required for pre-spliceosome assembly. *Proc. Natl. Acad. Sci. U.S.A.* **90,** 6771–6775.

Ast, G., Goldblatt, D., Offen, D., Sperling, J., and Sperling, R. (1991). A novel splicing factor is an integral component of 200S large nuclear ribonucleoprotein (lnRNP) particles. *EMBO J.* **10,** 425–432.

Bach, M., Winkelmann, G., and Lührmann, R. (1989). 20S small nuclear ribonucleoprotein U5 shows a surprisingly complex protein composition. *Proc. Natl. Acad. Sci. U.S.A.* **86,** 6038–6042.

Baker, B. S. (1989). Sex in flies: the splice of life. *Nature (London)* **340,** 521–524.

Baker, B. S., and Belote, J. M. (1983) Sex determination and dosage compensation in *Drosophila melanogaster. Annu. Rev. Genet.* **17,** 345–393.

Baker, B. S., and Wolfner, M. F. (1988). A molecular analysis of *doublesex,* a bifunctional gene that controls both male and female sexual differentiation in *Drosophila melanogaster. Genes Dev.* **2,** 477–489.

Baker, B. S., Nagoshi, R. N., and Burtis, K. C. (1987). Molecular genetic aspects of sex determination in *Drosophila. BioEssays* **6,** 66–70.

Bandziulis, R. J., Swanson, M. S., and Dreyfuss, G. (1989). RNA-binding proteins as developmental regulators. *Genes Dev.* **3,** 431–437.

Barabino, S. M. L., Blencowe, B. J., Ryder, U., Sproat, B. S., and Lamond, A. I. (1990). Targeted snRNP depletion reveals an additional role for mammalian U1 snRNP in spliceosome assembly. *Cell* **63,** 293–302.

Baserga, S., and Benz, E. J., Jr. (1988). Nonsense mutation in the human β-globin gene affect mRNA metabolism. *Proc. Natl. Acad. Sci. U.S.A.* **85,** 2056–2060.

Bass, B. L., and Weintraub, H. (1988). An unwinding activity that covalently modifies its double-stranded RNA substrate. *Cell* **55,** 1089–1098.

Behrens, S.-E., and Lührmann, R. (1991). Immunoaffinity purification of a [U4/U6.U5] tri-snRNP from human cells. *Genes Dev.* **5,** 1439–1452.

Behrens, S.-E., Galisson, F., Legrain, P., and Lührmann, R. (1993). Evidence that the 60-kDa protein of 17S U2 small nuclear ribonucleoprotein is immunologically and functionally related to the yeast PRP9 splicing factor and is required for the efficient formation of prespliceosomes. *Proc. Natl. Acad. Sci. U.S.A.* **90,** 8229–8233.

Bell, L. R., Maine, E. M., Schedl, P., and Cline, T. W. (1988). *Sex-lethal,* a *Drosophila* sex determination switch gene, exhibits sex-specific RNA splicing and sequence similarity to RNA binding proteins. *Cell* **55,** 1037–1046.

Bell, L. R., Horabin, J. I., Schedl, P., and Cline, T. W. (1991). Positive autoregulation of *Sex-lethal* by alternative splicing maintains the female determined state in *Drosophila. Cell* **65,** 229–239.

Bennett, M., and Reed, R. (1993). Correspondence between a mammalian spliceosome component and an essential yeast splicing factor. *Science* **262,** 105–107.

Bennett, M., Michaud, S., Kingston, J., and Reed, R. (1992a). Protein components specifically associated with prespliceosome and spliceosome complexes. *Genes Dev.* **6**, 1986–2000.

Bennett, M., Piñol-Roma, S., Staknis, D., Dreyfuss, G., and Reed, R. (1992b). Differential binding of heterogeneous nuclear ribonucleoproteins to mRNA precursors prior to spliceosome assembly *in vitro*. *Mol. Cell. Biol.* **12**, 3165–3175.

Bernstein, S. I., Hansen, C. J., Becker, K. D., Wassenberg, D. R., II, Roche, E. S., Donady, J. J., and Emerson, C. P., Jr. (1986). Alternative RNA splicing generates transcripts encoding a thorax-specific isoform of *Drosophila melanogaster* myosin heavy chain. *Mol. Cell. Biol.* **6**, 2511–2519.

Beyer, A. L., and Osheim, Y. N. (1988). Splice site selection, rate of splicing, and alternative splicing on nascent transcripts. *Genes Dev.* **2**, 754–765.

Biamonti, G., Buvoli, M., Bassi, M. T., Morandi, C., Cobianchi, F., and Riva, S. (1989). Isolation of an active gene encoding human hnRNP protein A1; evidence for alternative splicing. *J. Mol. Biol.* **207**, 491–503.

Bindereif, A., and Green, M. R. (1987). An ordered pathway of snRNP binding during mammalian pre-mRNA splicing complex assembly. *EMBO J.* **6**, 2415–2424.

Bindereif, A., and Green, M. R. (1990). Identification and functional analysis of mammalian splicing factors. *Genet. Eng.* **12**, 201–224.

Bingham, P. M., Levis, R., and Rubin, G. M. (1981). Cloning of DNA sequences from the *white* locus of *D. melanogaster* by a novel and general method. *Cell* **25**, 693–704.

Black, D. L. (1991). Does steric interference between splice sites block the splicing of a short c-*src* neuron-specific exon in non-neuronal cells? *Genes Dev.* **5**, 389–402.

Black, D. L. (1992). Activation of c-*src* neuron-specific splicing by an unusual RNA element *in vivo* and *in vitro*. *Cell* **69**, 795–807.

Black, D. L., Chabot, B., and Steitz, J. A. (1985). U2 as well as U1 small nuclear ribonucleoproteins are involved in pre-mRNA splicing *in vitro*. *Cell* **42**, 737–750.

Blencowe, B. J., Sproat, B. S., Ryder, U., Barabino, S., and Lamond, A. I. (1989). Antisense probing of the human U4/U6 snRNP with biotinylated 2′-OMe RNA oligonucleotides. *Cell* **59**, 531–539.

Boggs, R. T., Gregor, P., Idriss, S., Belote, J. M., and McKeown, M. (1987). Regulation of sexual differentiation in *D. melanogaster* via alternative splicing of RNA from the *transformer* gene. *Cell* **50**, 739–747.

Bopp, D., Bell, L. R., Cline, T. W., and Schedl, P. (1991). Developmental distribution of female-specific *Sex-lethal* proteins in *Drosophila melanogaster*. *Genes Dev.* **5**, 403–415.

Bothwell, A. L. M., Ballard, D. W., Philbrick, W. M., Lindwall, G., Maher, S. E., Bridgett, M. M., Jamison, S. F., and Garcia-Blanco, M. A. (1991). Murine polypyrimidine tract binding protein; purification, cloning, and mapping of the RNA binding domain. *J. Biol. Chem.* **266**, 24657–24663.

Breathnach, R., Benoist, C., O'Hare, K., Gannon, F., and Chambon, P. (1978). Ovalbumin gene: evidence for a leader sequence in mRNA and DNA sequences at the exon–intron boundaries. *Proc. Natl. Acad. Sci. U.S.A.* **75**, 4853–4857.

Breibart, R. E., and Nadal-Ginard, B. (1987). Developmentally induced, muscle-specific trans factors control the differential splicing of alternative and constitutive troponin T exons. *Cell* **49**, 793–803.

Bringmann, P., Appel, B., Rinke, J., Reuter, R., Theissen, H., and Lührmann, R. (1984). Evidence of the existence of snRNAs U4 and U6 in a single ribonucleoprotein complex and for their association by intermolecular base pairing. *EMBO J.* **3**, 1357–1363.

Brosi, R., Gröning, K., Behrens, S.-E., Lührmann, R., and Krämer, A. (1993). Interaction of mammalian splicing factor SF3a with U2 snRNP and relation of its 60-kd subunit to yeast PRP9. *Science* **262**, 102–105.

Brow, D. A., and Guthrie, C. (1988). Spliceosomal RNA U6 is remarkably conserved from yeast to mammals. *Nature (London)* **334**, 213–218.

Brow, D. A., and Guthrie, C. (1989). Splicing a spliceosomal RNA. *Nature (London)* **337**, 14–15.

Brown, J. D., and Beggs, J. D. (1992). Role of PRP8 protein in the assembly of splicing complexes. *EMBO J.* **11**, 3721–3729.

Burtis, K. C., and Baker, B. S. (1989). *Drosophila doublesex* gene controls somatic sexual differentiation by producing alternatively spliced mRNAs encoding related sex-specific polypeptides. *Cell* **56**, 997–1010.

Butler, B., Pirrotta, V., Irminger-Finger, I., and Nöthiger, R. (1986). The sex-determining gene *tra* of *Drosophila:* molecular cloning and transformation studies. *EMBO J.* **5**, 3607–3613.

Buvoli, M., Biamonti, G., Tsoulfas, P., Bassi, M. T., Ghetti, A., Riva, S., and Morandi, C. (1988). cDNA cloning of human hnRNP protein A1 reveals the existence of multiple mRNA isoforms. *Nucleic Acids Res.* **16**, 3751–3770.

Cáceres, J. F., and Krainer, A. R. (1993). Functional analysis of pre-mRNA splicing factor SF2/ASF structural domains. *EMBO J.* **12**, 4715–4726.

Calnan, B. J., Tidor, B., Biancalana, S., Hudson, D., and Frankel, A. D. (1991). Arginine-mediated RNA recognition: the arginine fork. *Science* **252**, 1167–1171.

Carmo-Fonseca, M., Tollervey, D., Pepperkok, R., Barabino, S. M. L., Merdes, A., Brunner, C., Zamore, P. D., Green, M. R., Hurt, E., and Lamond, A. I. (1991). Mammalian nuclei contain foci which are highly enriched in components of the pre-mRNA splicing machinery. *EMBO J.* **10**, 195–206.

Carmo-Fonseca, M., Pepperkok, R., Carvalho, M. T., and Lamond, A. I. (1992). Transcription-dependent colocalization of the U1, U2, U4/U6, and U5 snRNPs in coiled bodies. *J. Cell Biol.* **117**, 1–14.

Carter, K. C., Taneja, K. L., and Lawrence, J. B. (1991). Discrete nuclear domains of Poly (A) RNA and their relationship to the functional organization of the nucleus. *J. Cell Biol.* **115**, 1191–1202.

Carter, K. C., Bowman, D., Carrington, W., Fogarty, K., McNeil, J. A., Fay, F. S., and Lawrence, J. B. (1993). A three-dimension view of precursor messenger RNA metabolism within the mammalian nucleus. *Science* **259**, 1330–1335.

Cech, T. R., and Bass, B. L. (1986). Biological catalysis by RNA. *Annu. Rev. Biochem.* **55**, 599–629.

Chabot, B., Black, D. L., LeMaster, D. M., and Steitz, J. A. (1985). The 3′ splice site of premessenger RNA is recognized by a small nuclear ribonuceloprotein. *Science* **230**, 1344–1349.

Chain, A. C., Zollman, S., Tseng, J. C., and Laski, F. A. (1991). Identification of a cis-acting sequence required for germ line-specific splicing of the P element ORF2-ORF3 intron. *Mol. Cell. Biol.* **11**, 1538–1546.

Champlin, D. T., Frasch, M., Saumweber, H., and Lis, J. T. (1991). Characterization of a *Drosophila* protein associated with boundaries of transcriptionally active chromatin. *Genes Dev.* **5**, 1611–1621.

Chapon, C., and Legrain, P. (1992). A novel gene, *spp91-1,* suppresses the splicing defect and the pre-mRNA nuclear export in the *prp9-1* mutant. *EMBO J.* **11**, 3279–3288.

Chebli, K., Gattoni, R., Schmitt, P., Hildwein, G., and Stévenin, J. (1989). The 216-

nucleotide intron of the E1A pre-mRNA contains a hairpin structure that permits utilization of unusually distant branch acceptors. *Mol. Cell. Biol.* **9**, 4852–4861.

Cheng, J., Mirjana, F., and Maquat, L. E. (1990). Translation to near the distal end of the penultimate exon is required for normal levels of spliced triosephosphate isomerase mRNA. *Mol. Cell. Biol.* **10**, 5215–5225.

Cheng, S.-C., and Abelson, J. (1987). Spliceosome assembly in yeast. *Genes Dev.* **1**, 1014–1027.

Choi, Y. D., Grabowski, P. J., Sharp, P. A., and Dreyfuss, G. (1986). Heterogeneous nuclear ribonuceloproteins: role in RNA splicing. *Science* **231**, 1534–1539.

Chou, T.-B., Zachar, Z., and Bingham, P. M. (1987). Developmental expression of a regulatory gene is programmed at the level of splicing. *EMBO J.* **6**, 4095–4104.

Cline, T. W. (1984). Autoregulatory functioning of a *Drosophila* gene product that establishes and maintains the sexually determined state. *Genetics* **107**, 231–277.

Cline, T. W. (1985). Primary events in the determination of sex in *Drosophila melanogaster*. In "Origin and Evolution of Sex" (H. O. Halvorson and A. Monroy, eds.), pp. 301–327. Alan R. Liss, New York.

Cline, T. W. (1986). A female-specific lethal lesion in an X-linked positive regulator of the *Drosophila* sex determination gene, *Sex-lethal*. *Genetics* **113**, 641–663.

Cline, T. W. (1988). Evidence that *sisterless-a* and *sisterless-b* are two of several discrete "numerator elements" of the X/A sex determination signal in *Drosophila* that switch *Sxl* between two alternative stable expression states. *Genetics* **119**, 829–862.

Clouet D'Orval, B., D'Aubenton-Carafa, Y., Sirand-Pugnet, P., Gallego, M., Brody, E., and Marie, J. (1991a). RNA secondary structure repression of a muscle-specific exon in HeLa cell nuclear extracts. *Science* **252**, 1823–1828.

Clouet D'Orval, B., d'Aubenton-Carafa, Y., Marie, J., and Brody, E. (1991b). Determination of an RNA structure involved in splicing inhibition of a muscle-specific exon. *J. Mol. Biol.* **221**, 837–856.

Collier, V. L., Kronert, W. A., O'Donnell, P. T., Edwards, K. A., and Bernstein, S. I. (1990). Alternative myosin hinge regions are utilized in a tissue-specific fashion that correlates with muscle contraction speed. *Genes Dev.* **4**, 885–895.

Cooper, T. A. (1992). *In vitro* splicing of cardiac troponin-T precursors- exon mutations disrupt splicing of the upstream intron. *J. Biol. Chem.* **267**, 5330–5338.

Cooper, T. A., and Ordahl, C. P. (1989). Nucleotide substitutions within the cardiac troponin T alternative exon disrupt pre-mRNA alternative splicing. *Nucleic Acids Res.* **17**, 7905–7921.

Cote, G. J., Stolow, D. T., Peleg, S., Berget, S. M., and Gagel, R. F. (1992). Identification of exon sequences and an exon binding protein involved in alternative RNA splicing of calcitonin/CGRP. *Nucleic Acids Res.* **20**, 2361–2366.

Crenshaw, E. B., III, Russo, A. F., Swanson, L. W., and Rosenfeld, M. G. (1987). Neuron-specific alternative RNA processing in transgenic mice expressing a metallothionein-calcitonin fusion gene. *Cell* **49**, 389–398.

Dabeva, M. D., Post-Beittenmiller, M. A., and Warner, J. R. (1986). Autogeneous regulation of splicing of the transcript of a yeast ribosomal protein gene. *Proc. Natl. Acad. Sci. U.S.A.* **83**, 5854–5857.

Datta, B., and Weiner, A. M. (1991). Genetic evidence for base pairing between U2 and U6 snRNA in mammalian mRNA splicing. *Nature (London)* **352**, 821–824.

Domenjoud, L., Gallinaro, H., Kister, L., Meyer, S., and Jacob, M. (1991). Identification of a specific exon sequence that is a major determinant in the selection between a natural and a cryptic 5' splice site. *Mol. Cell. Biol.* **11**, 4581–4590.

Dreyfuss, G. (1986). Structure and function of nuclear and cytoplasmic ribonucleoprotein particles. *Annu. Rev. Cell Biol.* **2**, 459–498.

Dreyfuss, G., Philipson, L., and Mattaj, I. W. (1988a). Ribonucleoprotein particles in cellular processes. *J. Cell Biol.* **106**, 1419–1425.

Dreyfuss, G., Swanson, M. S., and Piñol-Roma, S. (1988b). Heterogenous nuclear ribonucleoprotein particles and the pathway of mRNA formation. *Trends Biochem. Sci.* **13**, 86–91.

Duffy, J. B., and Gergen, J. P. (1991). The *Drosophila* segmentation gene *runt* acts as a position-specific numerator element necessary for the uniform expression of the sex-determining gene *Sex-lethal*. *Genes Dev.* **5**, 2176–2187.

Emeson, R. B., Hedjran, F., Yeakley, J. M., Guise, J. W., and Rosenfeld, M. G. (1989). Alternative production of calcitonin and CGRP mRNA is regulated at the calcitonin-specific splice acceptor. *Nature (London)* **341**, 76–80.

Eng, F. J., and Warner, J. R. (1991). Structural basis for the regulation of splicing of a yeast messenger RNA. *Cell* **65**, 797–804.

Eperon, I. C., Ireland, D. C., Smith, R. A., Mayeda, A., and Krainer, A. R. (1993). Pathways for selection of 5′ splice sites by U1 snRNPs and SF2/ASF. *EMBO J.* **12**, 3607–3617.

Eperon, L. P., Estibeiro, J. P., and Eperon, I. C. (1986). The role of nucleotide sequences in splice site selection in eukaryotic pre-messenger RNA. *Nature (London)* **324**, 280–282.

Eperon, L. P., Graham, I. R., Griffiths, A. D., and Eperon, I. C. (1988). Effects of RNA secondary structure on alternative splicing of pre-mRNA: is folding limited to a region behind the transcribing RNA polymerase? *Cell* **54**, 393–401.

Erickson, J. W., and Cline, T. W. (1991). Molecular nature of the *Drosophila* sex determination signal and its link to neurogenesis. *Science* **251**, 1071–1074.

Fabrizio, P., and Abelson, J. (1990). Two domains of yeast U6 small nuclear RNA required for both steps of nuclear precursor messenger RNA splicing. *Science* **250**, 404–409.

Feagin, J. E., Jasmer, D. P., and Stuart, K. (1987). Developmentally regulated addition of nucleotides within apocytochrome b transcripts in *Trypanosoma brucei*. *Cell* **49**, 337–345.

Fridell, R. A., Pret, A., and Searles, L. L. (1990). A retrotransposon 412 insertion within an exon of the *Drosophila melanogaster vermilion* gene is spliced from the precursor RNA. *Genes Dev.* **4**, 559–566.

Fu, X.-D. (1993). Specific commitment of different pre-mRNAs to splicing by single SR proteins. *Nature (London)* **365**, 82–85.

Fu, X.-D., and Maniatis, T. (1990). Factor required for mammalian spliceosome assembly is localized to discrete regions in the nucleus. *Nature (London)* **343**, 437–441.

Fu, X.-D., and Maniatis, T. (1992a). The 35-kDa mammalian splicing factor SC35 mediates specific interaction between U1 and U2 small nuclear ribonucleoprotein particles at the 3′ splice site. *Proc. Natl. Acad. Sci. U.S.A.* **89**, 1725–1729.

Fu, X.-D., and Maniatis, T. (1992b). Isolation of a complementary DNA that encodes the mammalian splicing factor SC35. *Science* **256**, 535–538.

Fu, X.-D., Mayeda, A., Maniatis, T., and Krainer, A. R. (1992). General splicing factors SF2 and SC35 have equivalent activities *in vitro*, and both affect alternative 5′ and 3′ splice site selection. *Proc. Natl. Acad. Sci. U.S.A.* **89**, 11224–11228.

Fu, X.-Y., and Manley, J. L. (1987). Factors influencing alternative splice site utilization *in vivo*. *Mol. Cell. Biol.* **7**, 738–748.

Fu, X.-Y., Ge, H., and Manley, J. L. (1988). The role of the polypyrimidine stretch at the SV40 early pre-mRNA 3' splice site in alternative splicing. *EMBO J.* **7,** 809–817.

Gall, J. G. (1991). Spliceosomes and snurposomes. *Science* **252,** 1499–1500.

Gallego, M. E., and Nadal-Ginard, B. (1990). Myosin light-chain 1/3 gene alternative splicing: cis regulation is based upon a hierarchical compatibility between splice sites. *Mol. Cell. Biol.* **10,** 2133–2144.

Gallego, M. E., Balvay, L., and Brody, E. (1992). Cis-acting sequences involved in exon selection in the chicken β-tropomyosin gene. *Mol. Cell. Biol.* **12,** 5415–5425.

Galli, G., Guise, J. W., McDevitt, M. A., Tucker, P. W., and Nevins, J. R. (1987). Relative position and strengths of poly(A) sites as well as transcription termination are critical to membrane versus secreted μ-chain expression during B-cell development. *Genes Dev.* **1,** 471–481.

Garcia-Blanco, M. A., Jamison, S. F., and Sharp, P. A. (1989). Identification and purification of a 62,000-dalton protein that binds specifically to the polypyrimidine tract of introns. *Genes Dev.* **3,** 1874–1886.

Garcia-Blanco, M., Anderson, G. J., Beggs, J., and Sharp, P. A. (1990). A mammalian protein of 220 kDa binds pre-mRNAs in the spliceosome: a potential homologue of the yeast PRP8 protein. *Proc. Natl. Acad. Sci. U.S.A.* **87,** 3082–3086.

Gattoni, R., Chebli, K., Himmelspach, M., and Stévenin, J. (1991). Modulation of alternative splicing of adenoviral E1A transcripts: factors involved in the early-to-late transition. *Genes Dev.* **5,** 1847–1858.

Ge, H., and Manley, J. L. (1990). A protein factor, ASF, controls cell-specific alternative splicing of SV40 early pre-mRNA *in vitro*. *Cell* **62,** 25–34.

Ge, H., Zuo, P., and Manley, J. L. (1991). Primary structure of the human splicing factor ASF reveals similarities with *Drosophila* regulators. *Cell* **66,** 373–382.

George, E. L., Ober, M. B., and Emerson, C. P., Jr. (1989). Functional domains of the *Drosophila melanogaster* muscle myosin heavy-chain gene are encoded by alternatively spliced exons. *Mol. Cell. Biol.* **9,** 2957–2974.

Gerke, V., and Steitz, J. A. (1986). A protein associated with small nuclear ribonucleoprotein particles recognizes the 3' splice site of premessenger RNA. *Cell* **47,** 973–984.

Gil, A., Sharp, P. A., Jamison, S. F., and Garcia-Blanco, M. A. (1991). Characterization of cDNAs encoding the polypyrimidine tract-binding protein. *Genes Dev.* **5,** 1224–1236.

Goguel, V., and Rosbash, M. (1993). Splice site choice and splicing efficiency are positively influenced by pre-mRNA intramolecular base pairing in yeast. *Cell* **72,** 893–901.

Goguel, V., Liao, X., Rymond, B. C., and Rosbash, M. (1991). U1 snRNP can influence 3' splice site selection as well as 5' splice site selection. *Genes Dev.* **5,** 1430–1438.

Goguel, V., Wang, Y., and Rosbash, M. (1993). Short artificial hairpins sequester splicing signals and inhibit yeast pre-mRNA splicing. *Mol. Cell. Biol.* **13,** 6841–6848.

Goldberg, M. L., Paro, R., and Gehring, W. J. (1982). Molecular cloning of the white locus region of *Drosophila melanogaster* using a large transposable element. *EMBO J.* **1,** 93–98.

Goralski, T. J., Edstrom, J.-E., and Baker, B. S. (1989). The sex determination locus *transformer-2* of *Drosophila* encodes a polypeptide with similarity to RNA binding proteins. *Cell* **56,** 1011–1018.

Goux-Pelletan, M., Libri, D., d'Aubenton-Carafa, Y., Fiszman, M., Brody, E., and Marie, J. (1990). *In vitro* splicing of mutually exclusive exons from the chicken β-tropomyosin gene: role of the branch point location and very long pyrimidine stretch. *EMBO J.* **9**, 241–249.

Grabowski, P. J., Nasim, F.-U. H., Kuo, H.-C., and Burch, R. (1991). Combinatorial splicing of exon pairs by two-site binding of U1 small nuclear ribonucleoprotein particle. *Mol. Cell. Biol.* **11**, 5919–5928.

Green, M. R. (1986). Pre-mRNA splicing. *Annu. Rev. Genet.* **20**, 671–708.

Guo, M., Lo., P. C. H., and Mount, S. M. (1993). Species-specific signals for the splicing of a short *Drosophila* intron *in vitro*. *Mol. Cell. Biol.* **13**, 1104–1118.

Guo, W., Mulligan, G. J., Wormsley, S., and Helfman, D. M. (1991). Alternative splicing of β-tropomyosin pre-mRNA: *cis*-acting elements and cellular factors that block the use of a skeletal muscle exon in nonmuscle cells. *Genes Dev.* **5**, 2096–2107.

Guthrie, C. (1991). Messenger RNA splicing in yeast: Clues to why the spliceosome is a ribonucleoprotein. *Science* **253**, 157–163.

Guthrie, C., and Patterson, B. (1988). Spliceosomal snRNAs. *Annu. Rev. Genet.* **23**, 387–419.

Harper, J.E., and Manley, J. L. (1991). A novel protein factor is required for use of distal alternative 5′ splices sites *in vitro*. *Mol. Cell. Biol.* **11**, 5945–5953.

Hartwell, L. H. (1967). Macromolecule synthesis in temperature-sensitive mutants of yeast. *J. Bacteriol.* **93**, 1662–1670.

Hashimoto, C., and Steitz, J. A. (1984). U4 and U6 RNAs coexist in a single small nuclear ribonucleoprotein particle. *Nucleic Acids Res.* **12**, 3283–3293.

Hausner, T.-P., Giglio, L. M., and Weiner, A. M. (1990). Evidence for base-pairing between mammalian U2 and U6 small nuclear ribonucleoprotein particles. *Genes Dev.* **4**, 2146–2156.

Hedley, M. L., and Maniatis, T. (1991). Sex-specific splicing and polyadenylation of *dsx* pre-mRNA requires a sequence that binds specifically to tra-2 protein *in vitro*. *Cell* **65**, 579–586.

Helfman, D. M., and Ricci, W. M. (1989). Branch point selection in alternative splicing of tropomyosin pre-mRNAs. *Nucleic Acids Res.* **17**, 5633–5650.

Helfman, D. M., Roscigno, R. F., Mulligan, G. J., Finn, L. A., and Weber, K. S. (1990). Identification of two distinct intron elements involved in alternative splicing of β-tropomyosin pre-mRNA. *Genes Dev.* **4**, 98–110.

Hellen, C. U. T., Witherell, G. W., Schmid, M., Shin, S. H., Pestova, T. V., Gil, A., and Wimmer, E. (1993). A cytoplasmic 57-kDa protein that is required for translation of picornavirus RNA by internal ribosomal entry is identical to the nuclear pyrimidine tract-binding protein. *Proc. Natl. Acad. Sci. U.S.A.* **90**, 7642–7646.

Hess, N., and Bernstein, S. I. (1991). Developmentally regulated alternative splicing of *Drosophila* myosin heavy chain transcripts: *in vivo* analysis of an usual 3′ splice site. *Dev. Biol.* **146**, 339–344.

Hodges, D., and Bernstein, S. I. (1992). Suboptimal 5′ and 3′ splice sites regulate alternative splicing of *Drosophila melanogaster* myosin heavy chain transcripts *in vitro*. *Mech. Dev.* **37**, 127–140.

Hoffman, B. E., and Grabowski, P. J. (1992). U1 snRNP targets an essential splicing factor, U2AF65, to the 3′ splice site by a network of interactions spanning the exon. *Genes Dev.* **6**, 2554–2568.

Horabin, J. I., and Schedl, P. (1993). Regulated splicing of the *Drosophila Sex-lethal* male exon involves a blockage mechanism. *Mol. Cell. Biol.* **13**, 1408–1414.

Hoshijima, K., Inoue, K., Higuchi, I., Sakamoto, H., and Shimura, Y. (1991). Control of *doublesex* alternative splicing by *transformer* and *transformer-2* in *Drosophila*. *Science* **252**, 833–836.

Huang, S., and Spector, D. L. (1991). Nascent pre-mRNA transcripts are associated with nuclear regions enriched in splicing factors. *Genes Dev.* **5**, 2288–2302.

Huang, S., and Spector, D. L. (1992). U1 and U2 small nuclear RNAs are present in nuclear speckles. *Proc. Natl. Acad. Sci. U.S.A.* **89**, 305–308.

Inoue, K., Ohno, M., Sakamoto, H., and Shimura, Y. (1989). Effect of the cap structure on pre-mRNA splicing in *Xenopus* oocyte nuclei. *Genes Dev.* **3**, 1472–1479.

Inoue, K., Hoshijima, K., Sakamoto, H., and Shimura, Y. (1990). Binding of the *Drosophila Sex-lethal* gene product to the alternative splice site of *transformer* primary transcript. *Nature (London)* **344**, 461–463.

Jamieson, D. J., Rahe, B., Pringle, J., and Beggs, J. D. (1991). A suppressor of a yeast splicing mutation (*prp8-1*) encodes a putative ATP-dependent RNA helicase. *Nature (London)* **349**, 715–717.

Jiménez-Garcia, L. F., and Spector, D. L. (1993). *In vivo* evidence that transcription and splicing are coordinated by a recruiting mechanism. *Cell* **73**, 47–59.

Kanaar, R., Roche, S. E., Beall, E. L., Green, M. R., and Rio, D. C. (1993). The conserved pre-mRNA splicing factor U2AF from *Drosophila:* requirement for viability. *Science* **262**, 569–573.

Kandels-Lewis, S., and Séraphin, B. (1993). Role of U6 snRNA in 5′ splice site selection. *Science* **262**, 2035–2039.

Karess, R. E., and Rubin, G. M. (1984). Analysis of P transposable element functions in *Drosophila*. *Cell* **38**, 135–146.

Kazzaz, J. A., and Rozek, C. E. (1989). Tissue-specific expression of the alternately processed *Drosophila* myosin heavy-chain messenger RNAs. *Dev. Biol.* **133**, 550–561.

Keller, E. B., and Noon, W. A. (1984). Intron splicing: a conserved internal signal in introns of animal pre-mRNAs. *Proc. Natl. Acad. Sci. U.S.A.* **81**, 7417–7420.

Keyes, L. N., Cline, T. W., and Schedl, P. (1992). The primary sex-determination signal of *Drosophila* acts at the level of transcription. *Cell* **68**, 933–943.

Kidwell, M. G., Kidwell, J. F., and Sved, J. A. (1977). Hybrid dysgenesis in *Drosophila melanogaster:* a syndrome of aberrant traits including mutation, sterility and male recombination. *Genetics* **86**, 813–833.

Kim, S.-H., and Lin, R.-J. (1993). Pre-mRNA splicing within an assembled yeast spliceosome requires an RNA-dependent ATPase and ATP hydrolysis. *Proc. Natl. Acad. Sci. U.S.A.* **90**, 888–892.

Kim, S.-H., Smith, J., Claude, A., and Lin, R.-J. (1992). The purified yeast pre-mRNA splicing factor PRP2 is an RNA-dependent NTPase. *EMBO J.* **11**, 2319–2326.

Konarska, M. M., and Sharp, P. A. (1987). Interactions between small nuclear ribonucleoprotein particles in formation of spliceosomes. *Cell* **49**, 763–774.

Konarska, M. M., and Sharp, P. A. (1988). Association of U2, U4, U5 and U6 small nuclear ribonucleoproteins in a spliceosome-type complex in absence of precursor RNA. *Proc. Natl. Acad. Sci. U.S.A.* **85**, 5459–5462.

Konarska, M. M., Grabowski, P. J., Padgett, R. A., and Sharp, P. A. (1985). Characterization of the branch site in lariat RNAs produced by splicing of mRNA precursors. *Nature (London)* **313**, 552–557.

Krainer, A. R., and Maniatis, T. (1988). RNA splicing *In* "Transcription and Splicing" (B. D. Hames, and D. M. Glover, eds.), pp. 131–206. IRL Press, Oxford.

Krainer, A. R., Conway, G. C., and Kozak, D. (1990a) Purification and characterization of pre-mRNA splicing factor SF2 from HeLa cells. *Genes Dev.* **4**, 1158–1171.

Krainer, A. R., Conway, G. C., and Kozak, D. (1990b). The essential pre-mRNA splicing factor SF2 influences 5′ splice site selection by activating proximal sites. *Cell* **62**, 35–42.

Krainer, A. R., Mayeda, A., Kozak, D., and Binns, G. (1991). Functional expression of cloned human splicing factor SF2: homology to RNA-binding proteins, U1 70K, and *Drosophila* splicing regulators. *Cell* **66**, 383–394.

Kramer, A., and Utans, U. (1991). Three protein factors (SF1, SF3, and U2AF) function in pre-splicing complex formation in addition to snRNPs. *EMBO J.* **10**, 1503–1509.

Kronert, W. A., Edwards, K. A., Roche, E. S., Wells, L., and Bernstein, S. I. (1991). Muscle-specific accumulation of *Drosophila* myosin heavy chains: a splicing mutation in an alternative exon results in an isoform substitution. *EMBO J.* **10**, 2479–2488.

Kuivaniemi, H., Tromp, G., and Prockop, D. J. (1991). Mutations in collagen genes: causes of rare and some common diseases in humans. *FASEB J.* **5**, 2052–2060.

Kuo, H.-C., Nasim, F.-U., and Grabowski, P. J. (1991). Control of alternative splicing by the differential binding of U1 small nuclear ribonucleoprotein particle. *Science* **251**, 1045–1050.

Lamond, A. I., Konarska, M. M., and Sharp, P. A. (1987). A mutational analysis of spliceosome assembly: evidence for splice site collaboration during spliceosome formation. *Genes Dev.* **1**, 532–543.

Lamond, A. I., Konarska, M. M., Grabowski, P. J., and Sharp, P. A. (1988). Spliceosome assembly involves the binding and release of U4 small nuclear ribonucleoprotein. *Proc. Natl. Acad. Sci. U.S.A.* **85**, 411–415.

Langford, C. J., Klinz, F. J., Donath, C., and Gallwitz, D. (1984). Point mutations identify the conserved, intron-contained TACTAAC box as an essential splicing sequence in yeast. *Cell* **36**, 645–653.

Laski, F. A., and Rubin, G. M. (1989). Analysis of the *cis*-acting requirements for germ-line-specific splicing of the P-element ORF2-ORF3 intron. *Genes Dev.* **3**, 720–728.

Laski, F. A., Rio, D. C., and Rubin, G. M. (1986). Tissue specificity of *Drosophila* P element transposition is regulated at the level of mRNA splicing. *Cell* **44**, 7–19.

Lear, A. L., Eperon, L. P., Wheatley, I. M., and Eperon, I. C. (1990). Hierarchy of 5′ splice site preference determined *in vivo*. *J. Mol. Biol.* **211**, 103–115.

Leff, S. E., Evans, R. M., and Rosenfeld, M. G. (1987). Splice commitment dictates neuron-specific alternative RNA processing in calcitonin/CGRP gene expression. *Cell* **48**, 517–524.

Legrain, P., and Chapon, C. (1993). Interaction between PRP11 and SPP91 yeast splicing factors and characterization of a PRP9-PRP11-SPP91 complex. *Science* **262**, 108–110.

Legrain, P., and Choulika, A. (1990). The molecular characterization of PRP6 and PRP9 yeast genes reveals a new cysteine/histidine motif common to several splicing factors. *EMBO J.* **9**, 2775–2781.

Legrain, P., and Rosbash, M. (1989). Some cis- and trans-acting mutants for splicing target pre-mRNA to the cytoplasm. *Cell* **57**, 573–583.

Lesser, C. F., and Guthrie, C. (1993). Mutations in U6 snRNA that alter splice site specificity: Implications for the active site. *Science* **262**, 1982–1988.

Levis, R., O'Hare, K., and Rubin, G. M. (1984). Effects of transposable element insertions on RNA encoded by the *white* gene of *Drosophila*. *Cell* **36**, 471–481.

Li, H., and Bingham, P. M. (1991). Arginine/serine-rich domains of the *su(w^a)* and *tra* RNA processing regulators target proteins to a subnuclear compartment implicated in splicing. *Cell* **67**, 335–342.

Libri, D., Lemonnier, M., Meinnel, T., and Fiszman, M. Y. (1989a). A single gene codes for the β subunit of smooth and skeletal muscle tropomyosin in the chicken. *J. Biol. Chem.* **264**, 2935–2944.

Libri, D., Marie, J., Brody, E., and Fiszman, M. Y. (1989b). A subfragment of the β-tropomyosin gene is alternatively spliced when transfected into differentiating muscle cells. *Nucleic Acids Res.* **17**, 6449–6462.

Libri, D., Goux-Pelletan, M., Brody, E., and Fiszman, M. Y. (1990). Exon as well as intron sequences are cis-regulating elements for the mutually exclusive alternative splicing of the β-tropomyosin gene. *Mol. Cell. Biol.* **10**, 5036–5046.

Libri, D., Piseri, A., and Fiszman, M. Y. (1991). Tissue-specific splicing *in vivo* of the β-tropomyosin gene: dependence on an RNA secondary structure. *Science* **252**, 1842–1845.

Libri, D., Balvay, L., and Fiszman, M. Y. (1992). *In vivo* splicing of the β-tropomyosin pre-mRNA: a role for branch point and donor site competition. *Mol. Cell. Biol.* **12**, 3204–3215.

Lossky, M., Anderson, G. J., Jackson, S. P., and Beggs, J. (1987). Identification of a yeast snRNP protein and detection of snRNP-snRNP interactions. *Cell* **51**, 1019–1026.

Madhani, H. D., Bordonne, R., and Guthrie, C. (1990). Multiple roles for U6 snRNA in the splicing pathway. *Genes Dev.* **4**, 2264–2277.

Maine, E. M., Salz, H. K., Cline, T. W., and Schedl, P. (1985). The *Sex-lethal* gene of *Drosophila:* DNA alterations associated with sex-specific lethal mutations. *Cell* **43**, 521–529.

Maniatis, T. (1991). Mechanisms of alternative pre-mRNA splicing. *Science* **251**, 33–34.

Maniatis, T., and Reed, R. (1987). The role of small nuclear ribonucleoprotein particles in pre-mRNA splicing. *Nature (London)* **325**, 673–678.

Mattaj, I. W. (1989). A binding consensus: RNA-protein interactions in splicing, snRNPs, and sex. *Cell* **57**, 1–3.

Mattox, W., and Baker, B. S. (1991). Autoregulation of the splicing of transcripts from the *transformer-2* gene of *Drosophila*. *Genes Dev.* **5**, 786–796.

Mattox, W., Palmer, M. J., and Baker, B. S. (1990). Alternative splicing of the sex determination gene *transformer-2* is sex-specific in the germ line but not in the soma. *Genes Dev.* **4**, 789–805.

Mayeda, A., and Krainer, A. R. (1992). Regulation of alternative pre-mRNA splicing by hnRNP A1 and splicing factor SF2. *Cell* **68**, 365–375.

Mayeda, A., Helfman, D. M., and Krainer, A. R. (1993). Modulation of exon skipping and inclusion by heterogeneous nuclear ribonucleoprotein A1 and pre-mRNA splicing factor SF2/ASF. *Mol. Cell. Biol.* **13**, 2993–3001.

Mayrand, S. H., and Pederson, T. (1990). Crosslinking of hnRNP proteins to pre-mRNA requires U1 and U2 snRNPs. *Nucleic Acids Res.* **18**, 3307–3318.

McElwain, M. C. (1986). The absence of somatic effects of hybrid dysgenesis in *Drosophila melanogaster*. *Genetics* **113**, 897–918.

McKeown, M., Belote, J. M., and Baker, B. S. (1987). A molecular analysis of *transformer*, a gene in *Drosophila melanogaster* that controls female sexual differentiation. *Cell* **48**, 489–499.

McKeown, M., Belote, J. M., and Boggs, R. T. (1988). Ectopic expression of the female *transformer* gene product leads to female differentiation of chromosomally male *Drosophila*. *Cell* **53**, 887–895.

Mermound, J. E., Cohen, P., and Lamond, A. I. (1992). Ser/Thr-specific protein phosphatases are required for both catalytic steps of pre-mRNA splicing. *Nucleic Acids Res.* **20**, 5263–5269.

Michaud, S., and Reed, R. (1991). An ATP-independent complex commits pre-mRNA to the mammalian spliceosome assembly pathway. *Genes Dev.* 5, 2534–2546.

Michaud, S., and Reed, R. (1993). A functional association between the 5' and 3' splice sites is established in the earliest prespliceosome complex (E) in mammals. *Genes Dev.* 7, 1008–1020.

Mitchell, P. J., Urlaub, G., and Chasin, L. (1986). Spontaneous splicing mutations at the dihydrofolate reductase locus in Chinese hamster ovary cells. *Mol. Cell. Biol.* 6, 1926–1935.

Mount, S. M. (1982). A catalogue of splice junction sequences. *Nucleic Acids Res.* 10, 459–472.

Mullen, M. P., Smith, C. W. J., Patton, J. G., and Nadal-Ginard, B. (1991). Alpha-Tropomyosin mutually exclusive exon selection: competition between branchpoint/polypyrimidine tracts determines default exon choice. *Genes Dev.* 5, 642–655.

Mulligan, G. J., Guo, W., Wormsley, S., and Helfman, D. M. (1992). Polypyrimidine tract binding protein interacts with sequences involved in alternative splicing of β-tropomyosin pre-mRNA. *J. Biol. Chem.* 267, 25480–25487.

Munroe, S. H., and Dong, X. (1992). Heterogenous nuclear ribonucleoprotein A1 catalyzes RNA-RNA annealing. *Proc. Natl. Acad. Sci. U.S.A.* 89, 895–899.

Nadal-Ginard, B., Smith, C. W., Patton, J. G., and Breitbart, R. E. (1991). Alternative splicing is an efficient mechanism for the generation of protein diversity: contractile protein genes as a model system. *Adv. Enzyme Regul.* 31, 261–286.

Nagoshi, R. N., and Baker, B. S. (1990). Regulation of sex-specific RNA splicing at the *Drosophila* doublesex gene: Cis-acting mutations in exon sequences alter sex-specific RNA splicing patterns. *Genes Dev.* 4, 89–97.

Nagoshi, R. N., McKeown, M., Burtis, K. C., Belote, J. M., and Baker, B. S. (1988). The control of alternative splicing at genes regulating sexual differentiation in *D. melanogaster. Cell* 53, 229–236.

Nasim, F.-U. H., Spears, P.A., Hoffmann, H. M., Kuo, H.-C., and Grabowski, P. J. (1990). A sequential splicing mechanism promotes selection of an optional exon by repositioning a downstream 5' splice site in preprotachykinin pre-mRNA. *Genes Dev.* 4, 1172–1184.

Nelson, K. K., and Green, M. R. (1989). Mammalian U2 snRNP has a sequence-specific RNA-binding activity. *Genes Dev.* 3, 1562–1571.

Nelson, K. K., and Green, M. R. (1990). Mechanism for cryptic splice site activation during pre-mRNA splicing. *Proc. Natl. Acad. Sci. U.S.A.* 87, 6253–6257.

Nesic, D., Cheng, J., and Maquat, L. E. (1993). Sequences within the last intron function in RNA 3'-end formation in cultured cells. *Mol. Cell. Biol.* 13, 3359–3369.

Newman, A., and Norman, C. (1991). Mutations in yeast U5 snRNA alter the specificity of 5' splice-site cleavage. *Cell* 65, 115–123.

Newman, A., and Norman, C. (1992). U5 snRNA interacts with exon sequences at 5' and 3' splice sites. *Cell* 68, 743–754.

Niwa, M., and Berget, S. M. (1991a). Polyadenylation precedes splicing *in vitro. Gene Expression* 1, 5–14.

Niwa, M., and Berget, S. M. (1991b). Mutation of the AAUAAA polyadenylation signal depresses *in vitro* splicing of proximal but not distal introns. *Genes Dev.* 5, 2086–2095.

Niwa, M., Rose, S. D., and Berget, S. M. (1990). *In vitro* polyadenylation is stimulated by the presence of an upstream intron. *Genes Dev.* 4, 1552–1559.

Niwa, M., MacDonald, C. C., and Berget, S. M. (1992). Are vertebrate exons scanned during splice-site selection? *Nature (London)* 360, 277–280.

Noble, J. C. S., Pan, Z.-Q., Prives, C., and Manley, J. L. (1987). Splicing of SV40 early pre-mRNA to large T and small t mRNAs utilizes different patterns of lariat branch sites. *Cell* **50**, 227–236.

Noble, J. C. S., Prives, C., and Manley, J. L. (1988). Alternative splicing of SV40 early pre-mRNA is determined by branch site selection. *Genes Dev.* **2**, 1460–1475.

Noller, H. F., Hoffarth, V., and Zimniak, L. (1992). Unusual resistance of peptidyl transferase to protein extraction procedures. *Science* **256**, 1416–1419.

O'Hare, K., and Rubin, G. M. (1983). Structures of P transposable elements and their sites of insertion and excision in the *Drosophila melanogaster* genome. *Cell* **34**, 25–35.

Ohno, K., and Suzuki, K. (1988). Multiple abnormal β-hexosaminidase μ chain mRNAs in a compound-heterozygous Ashkenazi Jewish patient with Tay–Sachs disease. *J. Biol. Chem.* **263**, 18563–18567.

Ohno, M., Sakamoto, H., and Shimura, Y. (1987). Preferential excision of the 5' proximal intron from mRNA precursors with two introns as mediated by the cap structure. *Proc. Natl. Acad. Sci. U.S.A.* **84**, 5187–5191.

Ohshima, Y., and Gotoh, Y. (1987). Signals for the selection of a splice site in pre-mRNA. Computer analysis of splice junction sequences and like sequences. *J. Mol. Biol.* **195**, 247–259.

Padgett, R. A., Grabowski, P. J., Konarska, M. M., Seiler, S., and Sharp, P. A. (1986). Splicing of messenger RNA precursors. *Annu. Rev. Biochem.* **55**, 1119–1150.

Parker, R., and Siliciano, P. G. (1993). Evidence for an essential non-Watson–Crick interaction between the first and last nucleotides of a nuclear pre-mRNA intron. *Nature (London)* **361**, 660–662.

Parker, R., Siliciano, P. G., and Guthrie, C. (1987). Recognition of the TACTAAC box during mRNA splicing in yeast involves base pairing to the U2-like snRNA. *Cell* **49**, 229–239.

Parkhurst, S. M., Bopp, D., and Ish-Horowicz, D. (1990). X:A ratio, the primary sex-determining signal in *Drosophila,* is transduced by helix–loop–helix proteins. *Cell* **63**, 1179–1191.

Paterson, T., Beggs, J. D., Finnegan, D. J., and Lührmann, R. (1991). Polypeptide components of *Drosophila* small nuclear ribonucleoprotein particles. *Nucleic Acids Res.* **19**, 5877–5882.

Patterson, B., and Guthrie, C. (1987). An essential yeast snRNA with a U5-like domain is required for splicing *in vivo. Cell* **49**, 613–624.

Patton, J. G., Mayer, S. A., Tempst, P., and Nadal-Ginard, B. (1991). Characterization and molecular cloning of polypyrimidine tract-binding protein: a component of a complex necessary for pre-mRNA splicing. *Genes Dev.* **5**, 1237–1251.

Patton, J. G., Porro, E. B., Galceran, J., Tempst, P., and Nadal-Ginard, B. (1993). Cloning and characterization of PSF, a novel pre-mRNA splicing factor. *Genes Dev.* **7**, 393–406.

Peterson, M. L., and Perry, R. P. (1989). The regulated production of μm and μs mRNA is dependent on the relative efficiencies of μs poly(A) site usage and the $C\mu 4$-to-M1 splice. *Mol. Cell. Biol.* **9**, 726–738.

Peterson, M. L., Gimmi, E. R., and Perry, R. P. (1991). The developmentally regulated shift from membrane to secreted μ mRNA production is accompanied by an increase in cleavage-polyadenylation efficiency but no measurable change in splicing efficiency. *Mol. Cell. Biol.* **11**, 2324–2327.

Pikielny, C. W., Rymond, B. C., and Rosbash, M. (1986). Electrophoresis of ribonucleoproteins reveals an ordered assembly pathway of yeast splicing complexes. *Nature (London)* **324**, 341–345.

Piñol-Roma, S., Choi, Y. D., Matunis, M. J., and Dreyfuss, G. (1988). Immunopurification of heterogeneous nuclear ribonucleoprotein particles reveals an assortment of RNA-binding proteins. *Genes Dev.* **2**, 215–227.

Piñol-Roma, S., Swanson, M. S., Gall, J. G., and Dreyfuss, G. (1989). A novel heterogeneous nuclear RNP protein with a unique distribution on nascent transcripts. *J. Cell Biol.* **109**, 2575–2587.

Pinto, A. L., and Steitz, J. A. (1989). The mammalian analogue of the yeast PRP8 splicing protein is present in the U4/5/6 small nuclear ribonucleoprotein particle and the spliceosome. *Proc. Natl. Acad. Sci. U.S.A.* **86**, 8742–8746.

Pirrotta, V., and Brockl, C. (1984). Transcription of the *Drosophila white* locus and some of its mutants. *EMBO J.* **3**, 563–568.

Potashkin, J., Naik, K., and Wentz-Hunter, K. (1993). U2AF homolog required for splicing *in vivo*. *Science* **262**, 573–575.

Powell, L. M., Wallis, S. C., Pease, R. J., Edwards, Y. H., Knott, T. J., and Scott, J. (1987). A novel form of tissue-specific RNA processing produces apolipoprotein-B48 in intestine. *Cell* **50**, 831–840.

Pret, A., and Searles, L. L. (1991). Splicing of retrotransposon insertions from transcripts of the *Drosophila melanogaster vermilion* gene in a revertant. *Genetics* **129**, 1137–1145.

Reed, R. (1989). The organization of 3' splice-site sequences in mammalian introns. *Genes Dev.* **3**, 2113–2123.

Reed, R. (1990). Protein composition of mammalian spliceosomes assembled *in vitro*. *Proc. Natl. Acad. Sci. U.S.A.* **87**, 8031–8035.

Reed, R., and Maniatis, T. (1986). A role for exon sequences and splice-site proximity in splice site selection. *Cell* **46**, 681–690.

Reed, R., and Maniatis, T. (1988). The role of the mammalian branchpoint sequence in pre-mRNA splicing. *Genes Dev.* **2**, 1268–1276.

Reich, C. L., VanHoy, R. W., Porter, G. L., and Wise, J. A. (1992). Mutations at the 3' splice site can be suppressed by compensatory base changes in U1 snRNA in fission yeast. *Cell* **69**, 1159–1169.

Rio, D. C. (1990). Molecular mechanisms regulating *Drosophila* P element transposition. *Annu. Rev. Genet.* **24**, 543–578.

Rio, D. C. (1991). Regulation of *Drosophila* P element transposition. *Trends Genet.* **7**, 282–287.

Robberson, B. L., Cote, G. J., and Berget, S. M. (1990). Exon definition may facilitate splice site selection in RNAs with multiple exons. *Mol. Cell. Biol.* **10**, 84–94.

Rosbash, M., Harris, P. K. W., Woolford, J. L., and Teem, J. L. (1981). The effect of temperature sensitive *rna* mutants on the transcription products from cloned ribosomal protein genes of yeast. *Cell* **24**, 679–686.

Roulier, E. M., Fyrberg, C., and Fyrberg, E. A. (1992). Perturbations of *Drosophila* α-actinin cause muscle paralysis, weakness and atrophy but do not confer obvious nonmuscle phenotypes. *J. Cell Biol.* **116**, 911–922.

Rozek, C. E., and Davidson, N. (1986). Differential processing of RNA transcribed from the single-copy *Drosophila* myosin heavy chain gene produces four mRNAs that encode two polypeptides. *Proc. Natl. Acad. Sci. U.S.A.* **83**, 2128–2132.

Ruby, S. W., and Abelson, J. (1988). An early hierarchical role of U1 small nuclear ribonucleoprotein in spliceosome assembly. *Science* **242**, 1028–1035.

Ruby, S. W., and Abelson, J. (1991). Pre-mRNA splicing in yeast. *Trends Genet.* **7**, 79–85.

Ruby, S. W., Chang, T.-H., and Abelson, J. (1993). Four yeast spliceosomal proteins

(PRP5, PRP9, PRP11, and PRP21) interact to promote U2 snRNP binding to pre-mRNA. *Genes Dev.* **7**, 1909–1925.

Ruskin, B., and Green, M. R. (1985). Role of the 3' splice site consensus sequence in mammalian pre-mRNA splicing. *Nature (London)* **317**, 732–734.

Ruskin, B., Krainer, A. R., Maniatis, T., and Green, M. R. (1984). Excision of an intact intron as a novel lariat structure during pre-mRNA splicing *in vitro*. *Cell* **38**, 317–331.

Ruskin, B., Greene, J. M., and Green, M. R. (1985). Cryptic branch point activation allows accurate *in vitro* splicing of human β-globin intron mutants. *Cell* **41**, 833–844.

Ruskin, B., Zamore, P. D., and Green, M. R. (1988). A factor, U2AF, is required for U2 snRNP binding and splicing complex assembly. *Cell* **52**, 207–219.

Ryner, L. C., and Baker, B. S. (1991). Regulation of *doublesex* pre-mRNA processing occurs by 3' splice site activation. *Genes Dev.* **5**, 2071–2085.

Sakamoto, H., Inoue, K., Higuchi, I., Ono, Y., and Shimura, Y. (1992). Control of *Drosophila Sex-lethal* pre-mRNA splicing by its own female-specific product. *Nucleic Acids Res.* **20**, 5533–5540.

Salz, H. K., Maine, E. M., Keyes, L. N., Samuels, M. E., Cline, T. W., and Schedl, P. (1989). The *Drosophila* female-specific sex-determination gene, *Sex-lethal*, has stage-, tissue-, and sex-specific RNAs suggesting multiple modes of regulation. *Genes Dev.* **3**, 708–719.

Sameshima, Y., Akiyama, T., Mori, N., Mizoguchi, H., Toyoshima, K., Sugimura, T., Terada, M., and Yokota, J. (1990). Point mutation of the p53 gene resulting in splicing inhibition in small cell lung carcinoma. *Biochem. Biophys. Res. Commun.* **173**, 697–703.

Samuels, M. E., Schedl, P., and Cline, T. W. (1991). The complex set of late transcripts from the *Drosophila* sex determination gene *Sex-lethal* encodes multiple related polypeptides. *Mol. Cell. Biol.* **11**, 3584–3602.

Sawa, H., and Abelson, J. (1992). Evidence for a base-pairing interaction between U6 small nuclear RNA and the 5' splice site during the splicing reaction in yeast. *Proc. Natl. Acad. Sci. U.S.A.* **89**, 11269–11273.

Sawa, H., and Shimura, Y. (1992). Association of U6 snRNA with the 5'-splice site region of pre-mRNA in the spliceosome. *Genes Dev.* **6**, 244–254.

Schmitt, P., Gattoni, R., Keohavong, P., and Stévenin, J. (1987). Alternative splicing of E1A transcripts of adenovirus requires appropriate ionic conditions *in vitro*. *Cell* **50**, 31–39.

Schwer, B., and Guthrie, C. (1991). PRP16 is an RNA-dependent ATPase that interacts transiently with the spliceosome. *Nature (London)* **349**, 494–499.

Schwer, B., and Guthrie, C. (1992). A conformation rearrangement in the spliceosome is dependent on PRP 16 and ATP hydrolysis. *EMBO J.* **11**, 5033–5039.

Séraphin, B., and Kandels-Lewis, S. (1993). 3' splice site recognition in *S. cerevisiae* does not require base pairing with U1 snRNA. *Cell* **73**, 803–812.

Séraphin, B., and Rosbash, M. (1989). Identification of functional U1 snRNA-pre-mRNA complexes committed to spliceosome assembly and splicing. *Cell* **59**, 349–358.

Séraphin, B., and Rosbash, M. (1991). The yeast branchpoint sequence is not required for the formation of a stable U1 snRNA-pre-mRNA complex and is recognized in the absence of U2 snRNA. *EMBO J.* **10**, 1209–1216.

Séraphin, B., Kretzner, L., and Rosbash, M. (1988). A U1 snRNA: pre-mRNA base pairing interaction is required early in yeast spliceosome assembly but does not uniquely define the 5' cleavage site. *EMBO J.* **7**, 2533–2538.

Séraphin, B., Abovich, N., and Rosbash, M. (1991). Genetic depletion indicates a late role for U5 snRNP during *in vitro* spliceosome assembly. *Nucleic Acids Res.* **19**, 3857–3860.

Shannon, K. W., and Guthrie, C. (1991). Suppressors of a U4 snRNA mutation define a novel U6 snRNP protein with RNA-binding motifs. *Genes Dev.* **5**, 773–785.

Sharp, P. (1987). Splicing of messenger RNA precursors. *Science* **235**, 766–771.

Siebel, C. W., and Rio, D. C. (1990). Regulated splicing of the *Drosophila* P transposable element third intron *in vitro:* somatic repression. *Science* **248**, 1200–1208.

Sierakowska, H., Szer, W., Furdon, P. J., and Kole, R. (1986). Antibodies to hnRNP core proteins inhibit *in vitro* splicing of human beta-globin pre-mRNA. *Nucleic Acids Res.* **14**, 5241–5254.

Siliciano, P. G., and Guthrie, C. (1988). 5'-splice site selection in yeast: genetic alterations in base-pairing with U1 reveal additional requirements. *Genes Dev.* **2**, 1258–1267.

Siliciano, P. G., Brow, D. A., Roiha, H., and Guthrie, C. (1987). An essential snRNA from *S. cerevisiae* has properties predicted for U4, including interaction with a U6-like snRNA. *Cell* **50**, 585–592.

Smith, C. W. J., and Nadal-Ginard, B. (1989). Mutually exclusive splicing of alpha-tropomyosin exons enforced by an unusual lariat branch point location: implications for constitutive splicing. *Cell* **56**, 749–758.

Smith, C. W. J., Porro, E. B., Patton, J. G., and Nadal-Ginard, B. (1989). Scanning from an independently specified branch point defines the 3' splice site of mammalian introns. *Nature (London)* **342**, 243–247.

Smith, C. W. J., Chu, T. T., and Nadal-Ginard, B. (1993). Scanning and competition between AGs are involved in 3' splice site selection in mammalian introns. *Mol. Cell. Biol.* **13**, 4939–4952.

Solnick, D. (1985). Alternative splicing caused by RNA secondary structure. *Cell* **43**, 667–676.

Solnick, D., and Lee, S. I. (1987). Amount of RNA secondary structure required to induce an alternative splice. *Mol. Cell. Biol.* **7**, 3194–3198.

Sontheimer, E. J., and Steitz, J. A. (1993). The U5 and U6 small nuclear RNAs as active site components of the spliceosome. *Science* **262**, 1989–1996.

Sosnowski, B. A., Belote, J. M., and McKeown, M. (1989). Sex-specific alternative splicing of RNA from the *transformer* gene results from sequence-dependent splice site blockage. *Cell* **58**, 449–459.

Spector, D. L. (1990). Higher order nuclear organization: three-dimensional distribution of small nuclear ribonucleoprotein particles. *Proc. Natl. Acad. Sci. U.S.A.* **87**, 147–151.

Spector, D. L., Fu, X.-D., and Maniatis, T. (1991). Associations between distinct pre-mRNA splicing components and the cell nucleus. *EMBO J.* **10**, 3467–3481.

Spikes, D., and Bingham, P. M. (1992). Analysis of spliceosome assembly and the structure of a regulated intron in *Drosophila in vitro* splicing extracts. *Nucleic Acids Res.* **20**, 5719–5727.

Steinmann-Zwicky, M., Amrein, H., and Nöthiger, R. (1990). Genetic control of sex determination in *Drosophila*. *Adv. Genet.* **27**, 189–237.

Stolow, D. T., and Berget, S. M. (1991). Identification of nuclear proteins that specifically bind to RNAs containing 5' splice sites. *Proc. Natl. Acad. Sci. U.S.A.* **88**, 320–324.

Strauss, E. J., and Guthrie, C. (1991). A cold-sensitive mRNA splicing mutant is a member of the RNA helicase gene family. *Genes Dev.* **5**, 629–641.

Streuli, M., and Saito, H. (1989). Regulation of tissue-specific alternative splicing: Exon-specific cis-elements govern the splicing of leukocyte common antigen pre-mRNA. *EMBO J.* **8,** 787–796.

Swanson, M. S., and Dreyfuss, G. (1988a) Classification and purification of proteins of heterogeneous nuclear ribonucleoprotein particles by RNA-binding specificities. *Mol. Cell. Biol.* **8,** 2237–2241.

Swanson, M. S., and Dreyfuss, G. (1988b). RNA binding specificity of hnRNP proteins: a subset bind to the 3' end of introns. *EMBO J.* **11,** 3519–3529.

Tacke, R., and Goridis, C. (1991). Alternative splicing in the neural cell adhesion molecule pre-mRNA: regulation of exon 18 skipping depends on the 5'-splice site. *Genes Dev.* **5,** 1416–1429.

Talerico, M., and Berget, S. M. (1990). Effect of 5' splice site mutations on splicing of the preceding intron. *Mol. Cell. Biol.* **10,** 6299–6305.

Tani, T., and Ohshima, Y. (1989). The gene for the U6 small nuclear RNA in fission yeast has an intron. *Nature (London)* **337,** 87–90.

Tazi, J., Alibert, C., Temsamani, J., Reveillaud, I., Cathala, G., Brunel, C., and Jeanteur, P. (1986). A protein that specifically recognizes the 3' splice site of mammalian pre-mRNA introns is associated with a small nuclear ribonucleoprotein. *Cell* **47,** 755–766.

Tazi, J., Daugeron, M.-C., Cathala, G., Brunel, C., and Jeanteur, P. (1992). Adenosine phosphorothioates (ATP αS and ATP τS) differentially affect the two steps of mammalian pre-mRNA splicing. *J. Biol. Chem.* **267,** 4322–4326.

Tazi, J., Kornstädt, U., Rossi, F., Jeanteur, P., Cathala, G., Brunel, C., and Lührmann, R. (1993). Thiophosphorylation of U1-70K protein inhibits pre-mRNA splicing. *Nature (London)* **363,** 283–286.

Teem, J. L., Abovich, J., Kaufer, N. F., Schwindinger, W. F., Warner, J. R., Levy, A., Woolford, J., Leer, R. J., van Raamsdonk-Duin, M. M. C., Mager, W. H., Planta, R. J., Schultz, L., Friesen, J. D., Fried, H., and Rosbash, M. (1984). A comparison of yeast ribosomal protein gene DNA sequences. *Nucleic Acids Res.* **12,** 8295–8312.

Tian, M., and Maniatis, T. (1992). Positive control of pre-mRNA splicing in vitro. *Science* **256,** 237–240.

Tian, M., and Maniatis, T. (1993). A splicing enhancer complex controls alternative splicing of *doublesex* pre-mRNA. *Cell* **74,** 105–114.

Treisman, R., Orkin, S. H., and Maniatis, T. (1983a). Specific transcription and RNA splicing defects in five cloned β-thalassaemia genes. *Nature (London)* **302,** 591–596.

Treisman, R., Orkin, S. H., and Maniatis, T. (1983b). Structural and functional defects in β-thalassemia. *In* "Globin Gene Expression and Hematopoietic Differentiation" (G. Stamatoyannopoulos and A. W. Nienhuis, eds.), pp. 99–121. Alan R. Liss, New York.

Tseng, J. C., Zollman, S., Chain, A. C., and Laski, F. A. (1991). Splicing of the *Drosophila* P element ORF2-ORF3 intron is inhibited in a human cell extract. *Mech. Dev.* **35,** 65–72.

Ulfendahl, P. J., Kreivi, J.-P., and Akusjarvi, G. (1989). Role of the branch site/3' splice site region in adenovirus-2 E1A pre-mRNA alternative splicing: evidence for 5' and 3' splice site cooperation. *Nucleic Acids Res.* **17,** 925–938.

Utans, U., and Kramer, A. (1990). Splicing factor SF4 is dispensable for the assembly of a functional splicing complex and participates in the subsequent steps of the splicing reaction. *EMBO J.* **9,** 4119–4126.

Valcárcel, J., Singh, R., Zamore, P. D., and Green, M. R. (1993). The protein Sex-lethal antagonizes the splicing factor U2AF to regulate alternative splicing of *transformer* pre-mRNA. *Nature (London)* **362,** 171–175.

Vankan, P., McGuigan, C., and Mattaj, I. W. (1992). Roles of U4 and U6 snRNAs in the assembly of splicing complexes. *EMBO J.* **11**, 335–343.

Vijayraghavan, U., Company, M., and Abelson, J. (1989). Isolation and characterization of pre-mRNA splicing mutants of *Saccharomyces cerevisiae*. *Genes Dev.* **3**, 1206–1216.

Voelker, R. A., Gibson, W., Graves, J. P., Sterling, J. F., and Eisenberg, M. T. (1991). The *Drosophila suppressor of sable* gene encodes a polypeptide with regions similar to those of RNA-binding proteins. *Mol. Cell. Biol.* **11**, 894–905.

Wang, J., and Pederson, T. (1990). A 62,000 molecular weight spliceosome protein cross-links to the intron polypyrimidine tract. *Nucleic Acids Res.* **18**, 5995–6001.

Wang, J., Cao, L.-G., Wang, Y.-L., and Pederson, T. (1991). Localization of pre-messenger RNA at discrete nuclear sites. *Proc. Natl. Acad. Sci. U.S.A.* **88**, 7391–7395.

Wansink, D. G., Schul, W., van der Kraan, I., van Steensel, B., van Driel, R., and de Jong, L. (1993). Fluorescent labelling of nascent RNA reveals transcription by RNA polymerase II in domains scattered throughout the nucleus. *J. Cell Biol.* **122**, 283–293.

Warner, J. R. (1987). Applying genetics to the splicing problem. *Genes Dev.* **1**, 1–3.

Wassarman, D. A., and Steitz, J. A. (1991). Alive with DEAD proteins. *Nature (London)* **349**, 463–464.

Wassarman, D. A., and Steitz, J. A. (1992). Interactions of small nuclear RNA's with precursor messenger RNA during *in vitro* splicing. *Science* **257**, 1918–1925.

Wassarman, K. M., and Steitz, J. A. (1993). Association with terminal exons in pre-mRNAs: a new role for the U1 snRNP? *Genes Dev.* **7**, 647–659.

Watakabe, A., Inoue, K., Sakamoto, H., and Shimura, Y. (1989). A secondary structure at the 3' splice site affects the *in vitro* splicing reaction of mouse immunoglobulin μ chain pre-mRNAs. *Nucleic Acids Res.* **17**, 8159–8169.

Watakabe, A., Tanaka, K., and Shimura, Y. (1993). The role of exon sequences in splice site selection. *Genes Dev.* **7**, 407–418.

Whittaker, E., and Beggs, J. D. (1991). The yeast PRP8 protein interacts directly with pre-mRNA. *Nucleic Acids Res.* **19**, 5483–5489.

Whittaker, E., Lossky, M., and Beggs, J. D. (1990). Affinity purification of spliceosomes reveals that the precursor RNA processing protein PRP8, a protein in the U5 small nuclear ribonucleoprotein particle, is a component of yeast spliceosomes. *Proc. Natl. Acad. Sci. U.S.A.* **87**, 2216–2219.

Winkelmann, G., Bach, M., and Lührmann, R. (1989). Evidence from complementation assays *in vitro* that U5 snRNP is required for both steps of mRNA splicing. *EMBO J.* **8**, 3105–3112.

Wise, J. A. (1993). Guides to the heart of the spliceosome. *Science* **262**, 1978–1979.

Wolff, T., and Bindereif, A. (1992). Reconstituted mammalian U4/U6 snRNP complements splicing: a mutational analysis. *EMBO J.* **11**, 345–359.

Wu, J. Y., and Maniatis, T. (1993). Specific interactions between proteins implicated in splice site selection and regulated alternative splicing. *Cell* **75**, 1061–1070.

Wu, J., and Manley, J. (1989). Mammalian pre-mRNA branch site selection by U2 snRNP involves base pairing. *Genes Dev.* **3**, 1553–1561.

Wu, J., and Manley, J. (1991). Base pairing between U2 and U6 snRNAs is necessary for splicing of a mammalian pre-mRNA. *Nature (London)* **352**, 818–821.

Wu, Z., Murphy, C., Callan, H. G., and Gall, J. G. (1991). Small nuclear ribonucleoproteins and heterogeneous nuclear ribonucleoproteins in the amphibian germinal vesicle: loops, spheres, and snurposomes. *J. Cell Biol.* **113**, 465–483.

Wyatt, J. R., Sontheimer, E. J., and Steitz, J. A. (1992). Site-specific cross-linking of mammalian U5 snRNP to the 5' splice site before the first step of pre-mRNA splicing. *Genes Dev.* **6**, 2542–2553.

Xing, Y., Johnson, C. V., Dobner, P. R., and Lawrence, J. B. (1993). Higher level organization of individual gene transcription and RNA splicing. *Science* **259**, 1326–1330.

Xu, R., Teng, J., and Cooper, T. A. (1993). The cardiac troponin T alternative exon contains a novel purine-rich positive splicing element. *Mol. Cell. Biol.* **13**, 3660–3674.

Yeakley, J. M., Hedjran, F., Morfin, J.-P., Merillat, N., Rosenfeld, M. G., and Emeson, R. B. (1993). Control of calcitonin/calcitonin gene-related peptide pre-mRNA processing by constitutive intron and exon elements. *Mol. Cell. Biol.* **13**, 5999–6011.

Zachar, Z., Davison, D., Garza, D., and Bingham, P. M. (1985). A detailed developmental and structural study of the transcriptional effects of insertion of the copia transposon into the *white* locus of *Drosophila melanogaster. Genetics* **111**, 495–515.

Zachar, Z., Chou, T.-B., and Bingham, P. M. (1987a). Evidence that a regulatory gene autoregulates splicing of its transcript. *EMBO J.* **6**, 4105–4111.

Zachar, Z., Garza, D., Chou, T., Goland, J., and Bingham, P. M. (1987b). Molecular cloning and genetic analysis of the *suppressor of white apricot* locus from *Drosophila melanogaster. Mol. Cell. Biol.* **7**, 2498–2505.

Zachar, Z., Kramer, J., Mims, I. P., and Bingham, P. M. (1993). Evidence for channeled diffusion of pre-mRNAs during nuclear RNA transport in metazoans. *J. Cell Biol.* **121**, 729–742.

Zahler, A. M., Lane, W. S., Stolk, J. A., and Roth, M. B. (1992). SR proteins: a conserved family of pre-mRNA splicing factors. *Genes Dev.* **6**, 837–847.

Zahler, A. M., Neugebauer, K. M., Lane, W. S., and Roth, M. B. (1993). Distinct functions of SR proteins in alternative pre-mRNA splicing. *Science* **260**, 219–222.

Zamore, P. D., and Green, M. R. (1989). Identification, purification and biochemical characterization of U2 small nuclear ribonucleoprotein auxiliary factor. *Proc. Natl. Acad. Sci. U.S.A.* **86**, 9243–9247.

Zamore, P. D., and Green, M. R. (1991). Biochemical characterization of U2 snRNP auxiliary factor: an essential pre-mRNA splicing factor with a novel intranuclear distribution. *EMBO J.* **10**, 207–214.

Zamore, P. D., Patton, J. G., and Green, M. R. (1992). Cloning and domain structure of the mammalian splicing factor U2AF. *Nature (London)* **355**, 609–614.

Zeitlin, S., and Efstratiadis, A. (1984). *In vitro* splicing products of the rabbit β-globin pre-mRNA. *Cell* **39**, 589–602.

Zhang, M., Zamore, P. D., Carmo-Fonesca, M., Lamond, A. I., and Green, M. R. (1992). Cloning and intracellular localization of the U2 small nuclear ribonucleoprotein auxiliary factor small subunit. *Proc. Natl. Acad. Sci. U.S.A.* **89**, 8769–8773.

Zhuang, Y., and Weiner, A. M. (1986). A compensatory base change in U1 snRNA suppresses a 5' splice site mutation. *Cell* **46**, 827–835.

Zhuang, Y., and Weiner, A. M. (1989). A compensatory base change in human U2 snRNA can suppress a branch site mutation. *Genes Dev.* **3**, 1545–1552.

Zhuang, Y., Lueng, H., and Weiner, A. M. (1987). The natural 5' splice site of simian virus 40 large T antigen can be improved by increasing the base complementarity to U1 RNA. *Mol. Cell. Biol.* **7**, 3018–3020.

Zillmann, M., Rose, S. D., and Berget, S. M. (1987). U1 small nuclear ribonucleoproteins are required early during spliceosome assembly. *Mol. Cell. Biol.* **7**, 2877–2883.

Zuo, P., and Manley, J. L. (1993). Functional domains of the human splicing factor ASF/SF2. *EMBO J.* **12**, 4727–4737.

INDEX

A

abnormal chemosensory jump (acj) mutant, *Drosophila melanogaster*, 147–148
achaete-scute (AS-C) complex, *Drosophila melanogaster*, 5
achaete (T5) gene, *Drosophila melanogaster*
  in neurogenesis, 16
  in sex determination, 8
Actin-binding proteins, genes for, *Drosophila melanogaster*, 107
Actin gene family, *Drosophila melanogaster*, 106–107
α-Actinin, *cis*-acting mutations, effects on alternative splicing, 258–259
Activin gene, *Xenopus laevis*, 41–43
Adenovirus E1A transcripts, production by alternative splice sites, 229–231
aj42 mutant, *Drosophila melanogaster*, 171
Andante (And) mutant, *Drosophila melanogaster*, 169–170
Anterior–posterior axis, formation
  in, *Drosophila melanogaster*
    gap to pair-rule gene patterns, 11–12
    pair-rule interactions, 12–14
    periodicity gradients, 10–11
  in, *Xenopus laevis*
    Hox gene role, 56–57
    patterns
      induction, 54–55
      lability, 53–54
      posterior dominance, 55–56
      prepattern, 53–54
Aromatic sulfates, as sulfur source, *Neurospora crassa*, 192–194
ars gene, in sulfur circuit, *Neurospora crassa*, 191–192

Aryl sulfatase, in sulfur circuit, *Neurospora crassa*, 192–194
ASF-SF2, in alternative splicing, 220–225
ATPases, in RNA splicing, 226–228
aurora mutant, *Drosophila melanogaster*, 102–103

B

bicoid (bcd) gene, *Drosophila melanogaster*, 11–12
Biological rhythms, *Drosophila melanogaster*
  aj42 mutant, 171
  Andante (And) mutant, 169–170
  Clock (Clk) mutant, 169
  disconnected (disco) mutant, 170–171
  ebony (e) mutant, 169–170
  period (per) gene
    behavioral effects, 161–162
    molecular biology, 162–166
    sequence similarities with *arnt, Ahr,* and *sim,* 167–169
    threonine–glycine repeats, 162–166
    transcript localization, 164–166
  PER protein
    circadian cycling, 166–167
    expression, 164–166
    transcript localization, 164–166

C

cacophony (cac) mutant, *Drosophila melanogaster*, 152–154
Cadherins, neural plate, *Xenopus laevis*, 66–67
cdc2 gene family, in, *Drosophila melanogaster*, 81–83

ISBN 0-12-017631-9

90040

9 780120 176311